Kinase Drug Discovery

RSC Drug Discovery Series

Editor-in-Chief
Professor David Thurston, *London School of Pharmacy, UK*

Series Editors:
Dr David Fox, *Pfizer Global Research and Development, Sandwich, UK*
Professor Salvatore Guccione, *University of Catania, Italy*
Professor Ana Martinez, *Instituto de Quimica Medica-CSIC, Spain*
Dr David Rotella, *Montclair State University, USA*

Advisor to the Board:
Professor Robin Ganellin, *University College London, UK*

Titles in the Series:
1: Metabolism, Pharmacokinetics and Toxicity of Functional Groups: Impact of Chemical Building Blocks on ADMET
2: Emerging Drugs and Targets for Alzheimer's Disease; Volume 1: Beta-Amyloid, Tau Protein and Glucose Metabolism
3: Emerging Drugs and Targets for Alzheimer's Disease; Volume 2: Neuronal Plasticity, Neuronal Protection and Other Miscellaneous Strategies
4: Accounts in Drug Discovery: Case Studies in Medicinal Chemistry
5: New Frontiers in Chemical Biology: Enabling Drug Discovery
6: Animal Models for Neurodegenerative Disease
7: Neurodegeneration: Metallostasis and Proteostasis
8: G Protein-Coupled Receptors: From Structure to Function
9: Pharmaceutical Process Development: Current Chemical and Engineering Challenges
10: Extracellular and Intracellular Signaling
11: New Synthetic Technologies in Medicinal Chemistry
12: New Horizons in Predictive Toxicology: Current Status and Application
13: Drug Design Strategies: Quantitative Approaches
14: Neglected Diseases and Drug Discovery
15: Biomedical Imaging: The Chemistry of Labels, Probes and Contrast Agents
16: Pharmaceutical Salts and Cocrystals
17: Polyamine Drug Discovery
18: Proteinases as Drug Targets
19: Kinase Drug Discovery

How to obtain future titles on publication:
A standing order plan is available for this series. A standing order will bring delivery of each new volume immediately on publication.

For further information please contact:
Book Sales Department, Royal Society of Chemistry, Thomas Graham House, Science Park, Milton Road, Cambridge, CB4 0WF, UK
Telephone: +44 (0)1223 420066, Fax: +44 (0)1223 420247, Email: books@rsc.org
Visit our website at http://www.rsc.org/Shop/Books/

Kinase Drug Discovery

Edited by

Richard A. Ward and Frederick Goldberg
AstraZeneca, Macclesfield, UK

RSC Publishing

RSC Drug Discovery Series No. 19

ISBN: 978-1-84973-174-4
ISSN: 2041-3203

A catalogue record for this book is available from the British Library

Published by The Royal Society of Chemistry,
Thomas Graham House, Science Park, Milton Road,
Cambridge CB4 0WF, UK

Registered Charity Number 207890

For further information see our web site at www.rsc.org

Preface

Kinases, also known as phosphotransferases, are a family of enzymes that catalyse the transfer of a phosphate from donor molecules such as ATP (adenosine triphosphate) to their specific substrates. Over 500 protein kinases have now been identified in the human kinome, which have been the main focus of Kinase Drug Discovery in recent years. There are currently around 13 FDA-approved small molecule kinase inhibitors (listed at the end of this Preface) along with a handful of biological therapeutics, with over 500 additional small molecules in active development targeting kinases. These agents have in turn delivered very significant benefits to patients that can be measured as significant life extension or significant improvement in quality of life in diseases like cancer and inflammation. The kinase family has therefore been a rich source of new targets and opportunities for the pharmaceutical industry but has also presented significant challenges. Due to the sequence and structural similarity of kinases in the kinase domain, specifically the ATP binding site, targeting kinases in drug discovery was initially assumed by many people to be extremely difficult, if not impossible. The affinity of kinases to ATP was also a concern, which was often manifested in large potency drop-offs from enzyme to cell assays. However, selective kinase inhibitors were first discovered by exploiting amino acids around the ATP pocket which are not in contact with ATP. The similarity of kinases in the immediate vicinity of the ATP site but increasing differences outside of this region has resulted in an explosion of interest by medicinal chemistry over the last decade. The mining of activity data for compounds across increasing numbers of kinases along with the increasing success in solving X-ray crystal structures have allowed kinases to be approached by drug discovery scientists as a target class. A variety of lead generation approaches have been specifically tailored to exploit this family such as the re-profiling of leads against different kinase targets,

RSC Drug Discovery Series No. 19
Kinase Drug Discovery
Edited by Richard A. Ward and Frederick Goldberg
© Royal Society of Chemistry 2012
Published by the Royal Society of Chemistry, www.rsc.org

scaffold or template-hopping along with structural hybridizations and SAR transfer. Methods of screening compounds and identifying leads have also evolved with new assay formats and approaches such as fragment-based lead generation (FBLG). In addition to the significant focus around targeting the active form of the kinase through direct competition with ATP, inactive conformations of kinases have also been targeted. Kinases have been observed to have a variety of inactive conformations through movement of the conserved DFG-loop and *c*-helix which has increased the diversity of chemical opportunities. Targeting such conformations can increase selectivity and potentially overcome the large enzyme to cell potency drop-offs from direct competition with ATP. There may also be kinetic advantages to targeting inactive conformations, where slow off rates (long residence times) have been linked to enhanced *in vivo* efficacy. Novel allosteric pockets have also been identified which can be targeted by a small molecule in the kinase (and adjacent) domains along with compounds binding to the kinase-ATP complex.

Throughout this book we aim to summarise the challenges and the opportunities which the kinase family has presented to drug discovery scientists over recent years. One view is that the protein kinase target class has now largely been exploited. However, feedback from the clinic is likely to result in significant additional learning and opportunities over the coming years. Rather than focussing on case studies of specific targets we have aimed to cover broader themes. In particular, our aim for this book is to focus on the 'hot topics' in kinase drug discovery which we believe will be key areas in the coming years.

Rather than seeing the conserved nature of the ATP site as a hindrance, David Drewry and co-authors (Chapter 1) examine how medicinal chemists are exploiting our increased understanding of the kinome by looking at cross-reactivity and tuning out broader kinase activity. Also described is a summary of available assays and methods to understand the selectivity of kinase inhibitors. In Chapter 2 Iain Simpson goes on to review the cutting edge approaches to identifying hits against kinases, then Martin Swarbrick covers the issues that medicinal chemists face when optimising those leads into candidate drugs. In a related topic Walter Ward (Chapter 4) then covers the different mechanisms available to targeting kinase inhibition and how to exploit them to drive SAR. One particular hot topic of kinase drug discovery is how we should respond to clinical feedback. One way we can apply this feedback in drug discovery is covered by Jack Bikker in Chapter 5, reviewing kinase mutations and resistance. Whereas the primary focus of the first 5 chapters is on the dominant theme of protein kinases for oncology indications, Chapters 6-9 focus on other diverse areas of interest to drug discovery. Jeroen Verheijen and co-authors cover the progress made and future opportunities within non-protein kinases in Chapter 6, focussing on sugar, nucleoside and lipid kinases. Andrew Ratcliffe then covers non-oncology applications for kinase inhibitors in Chapter 7. Although the majority of kinase inhibitors in development are targeted at oncology indications, important opportunities remain in non-oncology disease indications where kinase inhibitors have the potential to address other significant unmet medical needs. Kinase activators are outside the scope of Andrew Ratcliffe's review, but are picked up by Kevin Guertin who presents a case study of allosteric activators of glucokinase for the treatment of type 2 diabetes in Chapter 8. In this chapter he reviews the history of glucokinase activators and discusses their novel mechanism of action. Another significant unmet medical need where kinase drug discovery could have a future impact is in human parasitic diseases. To that end Andrew Wilks goes beyond the human kinome in Chapter 9, where he discusses the therapeutic opportunities that the non-human kinomes present and the early but exciting drug discovery efforts that have been made thus far. To conclude the book Carlos Garcia-Echeverria has kindly contributed his thoughts on the future of kinase drug discovery, on such topics as new methods to overcome resistance, novel mechanisms and pseudokinases.

Although each of these chapters can be read as an individual standalone review of the topic being covered, there are themes which are picked up in multiple chapters which allow the readers to review key areas from multiple authors.

We would like to thank all of the authors and co-authors for their hard work in contributing the chapters for this book. In particular we would like to acknowledge their efforts to stay committed to the book during a very challenging time for drug discovery scientists in industry and academia. The authors of this book have brought a combined experience of well over 200 years in drug discovery research, across dozens of research organisations and covering multiple disciplines and specialities.

Richard A Ward, PhD, is a Computational Chemist working within the
Oncology iMed at AstraZeneca, Alderley Park, UK
Frederick W Goldberg, PhD, is a Medicinal Chemist working within the
Oncology iMed at AstraZeneca, Alderley Park, UK

Imatinib(Novartis), Bcr-Abl

Gefitinib(AstraZeneca), EGFR

Sorafenib(Onyx/Bayer), Multiple

Dasatinib(BMS), Multiple

Sunitinib(Pfizer), Multiple

Erlotinib(OSI/Genentech/Roche), EGFR

Nilotinib(Novartis), Bcr-Abl

Lapatinib(GSK), Erb1/2

Vandetanib(AstraZeneca), VEGFR/RET

Pazopanib(GSK)
VEGFR2/PDGFR/c-Kit

Temsirolimus(Wyeth), mTORC1

Everolimus(Novartis), mTORC1

Sirolimus(Wyeth), mTORC1

FDA Approved Small Molecule Kinase Inhibitors

Contents

RSC Drug Discovery Series No. 19
Kinase Drug Discovery
Edited by Richard A. Ward and Frederick Goldberg
© Royal Society of Chemistry 2012
Published by the Royal Society of Chemistry, www.rsc.org

Chapter 6 Non-Protein Kinases as Therapeutic Targets **161**
Jeroen C. Verheijen, David J. Richard and Arie Zask

Chapter 10   The Future of Kinase Drug Discovery 286
Carlos García-Echeverría

The Kinome and its Impact on Medicinal Chemistry

DAVID H. DREWRY[a], PAUL BAMBOROUGH[b],
KLAUS SCHNEIDER†[b] AND GARY K. SMITH[c]

[a] Chemical Biology; Computational and Structural Chemistry,
GlaxoSmithKline, 20 T.W. Alexander Drive, Research Triangle Park, NC
27709, USA; [b] Computational & Structural Chemistry, GlaxoSmithKline,
Medicines Research Centre, Gunnels Wood Road, Stevenage, Hertfordshire
SG1 2NY, UK; [c] Screening and Compound Profiling; Platform Technology
and Science, GlaxoSmithKline, 5 Moore Dr, Research Triangle Park, NC
27709, USA

1.1 Introduction

The "Kinome" describes the protein kinase component of the human genome,
exhaustively compiled in 2002 by Manning *et al.*[1] A search of available human
sequence sources using a hidden Markov model identified 478 ePK (eukaryotic
protein kinase) genes, 491 ePK domains, and 40 "atypical" protein kinases. The
tree-like classification developed from this has become the defining image in the
field of protein kinase pharmaceutical research and has been used as the framework
for the display of numerous properties. The ePKs are subdivided into eight main
groups (Table 1.1), extending the previous classification of Hanks and Hunter.[2]

For the last few years protein kinases have been among the most actively
studied pharmaceutical targets. In 2005 it was estimated that around one in
three discovery efforts targeted protein kinases.[3] Federov *et al.* surveyed the

RSC Drug Discovery Series No. 19
Kinase Drug Discovery
Edited by Richard A. Ward and Frederick Goldberg
© Royal Society of Chemistry 2012
Published by the Royal Society of Chemistry, www.rsc.org
†Current Address: Bioanalytical Chemistry, Central Analytics, Merck KGaA, Frankfurter Straße
250, 64293 Darmstadt, Germany

Table 1.1   Kinase group classifications based on phylogeny.

Kinase Group Name abbreviation	Details
AGC	related to protein kinases A, G and C
CaMK	including the Calmodulin-regulated kinases
CMGC	containing the Cyclin-dependent, Mitogen-activated, Glycogen synthase, and CDK-like kinases
TK	tyrosine kinases
STE	including homologues of the yeast Sterile kinases
CK1	including Casein Kinase 1
TKL	Tyrosine Kinase-Like
RCG	Receptor Guanylate Cyclase

landscape of kinase inhibitor publications and patents,[4] and noted that the number of patents declined from 2006–2009, perhaps suggesting that industrial research is moving on to other areas, although this could also show that it is becoming increasingly difficult to differentiate novel compounds from prior art. However, it was also found that not all kinases have received equal attention and that the majority of protein kinases have not been studied at all. A similar trend is seen when the count of compounds from journals and patent literature, rather than the number of publications, is plotted on the kinome tree as shown in Figure 1.1 [Illustration reproduced courtesy of Cell Signaling Technology, Inc. (http://www.cellsignal.com)].[5] Certain tyrosine kinases, CMGC kinases, and AGC kinases have been intensively studied. In contrast, the majority of kinases have no published activity data whatsoever. A significant number of targets have no published inhibition data apart from the small number of compounds arising from high throughput profiling efforts.[6–12] It is also apparent that these "untouched" kinases are unevenly distributed and that there are entire branches of the kinome tree for which small molecule inhibitors are unknown. Targets for which inhibitors exist with the selectivity necessary to make them useful tool compounds are even fewer. This distribution probably mirrors the body of literature describing the functions of these kinases, of which $\sim 50\%$ are said to be largely uncharacterised.[4] Many kinases still have unknown function, and as a result there has been little incentive to seek to develop inhibitors.

This pattern is partly repeated when the analysis is restricted to the ten kinase inhibitors (Figure 1.2) that have been approved and marketed (Table 1.2). At least so far as it is possible to judge from the published stories of their discovery, nine out of the ten drugs were originally conceived as inhibitors of various tyrosine or tyrosine-like kinases. With evolution in screening technology it is possible to profile these clinical compounds against an ever expanding list of kinases. They are generally less selective than initial reports suggested, but range from the highly selective, such as Lapatinib, which

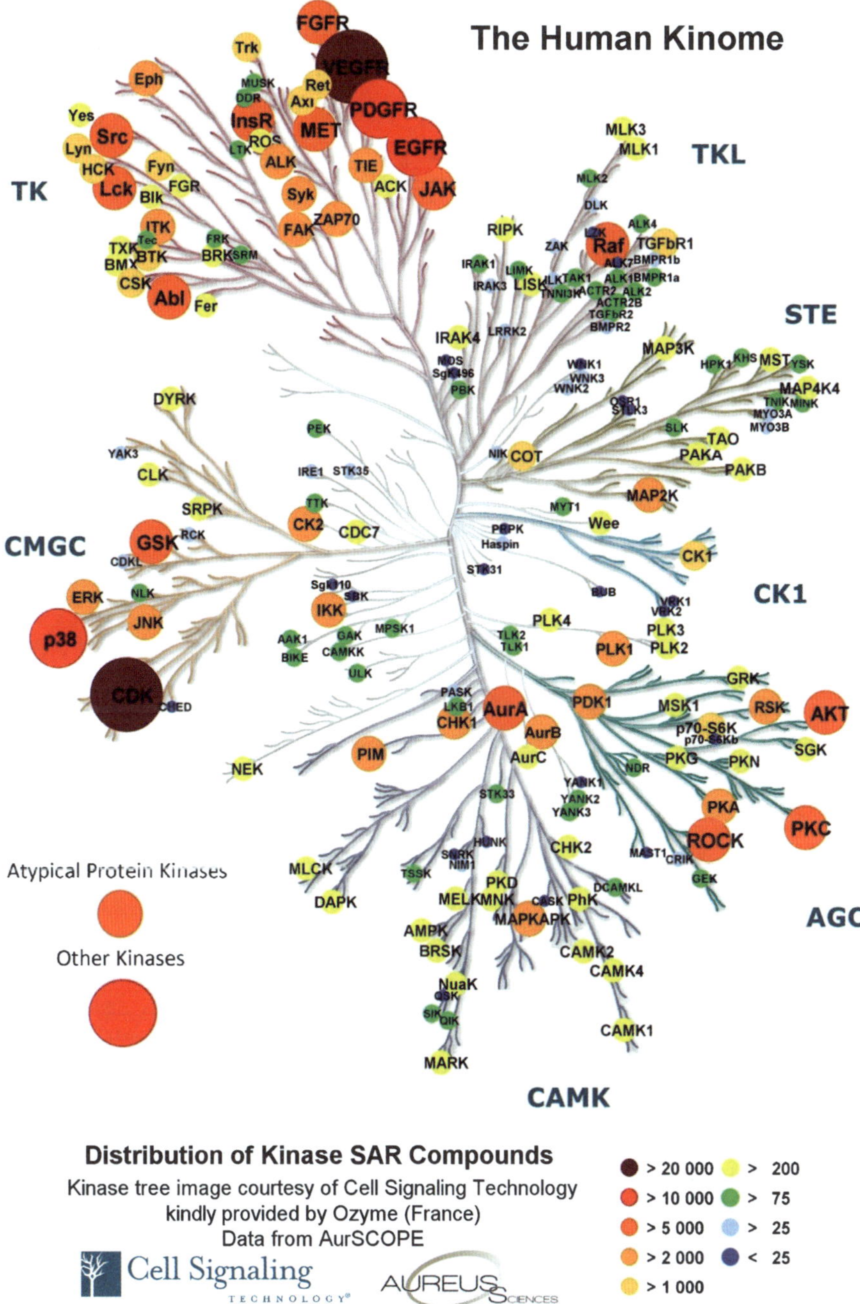

Figure 1.1   Kinome tree plot illustrating number of compounds with inhibition data reported in patents and journal articles for each kinase.

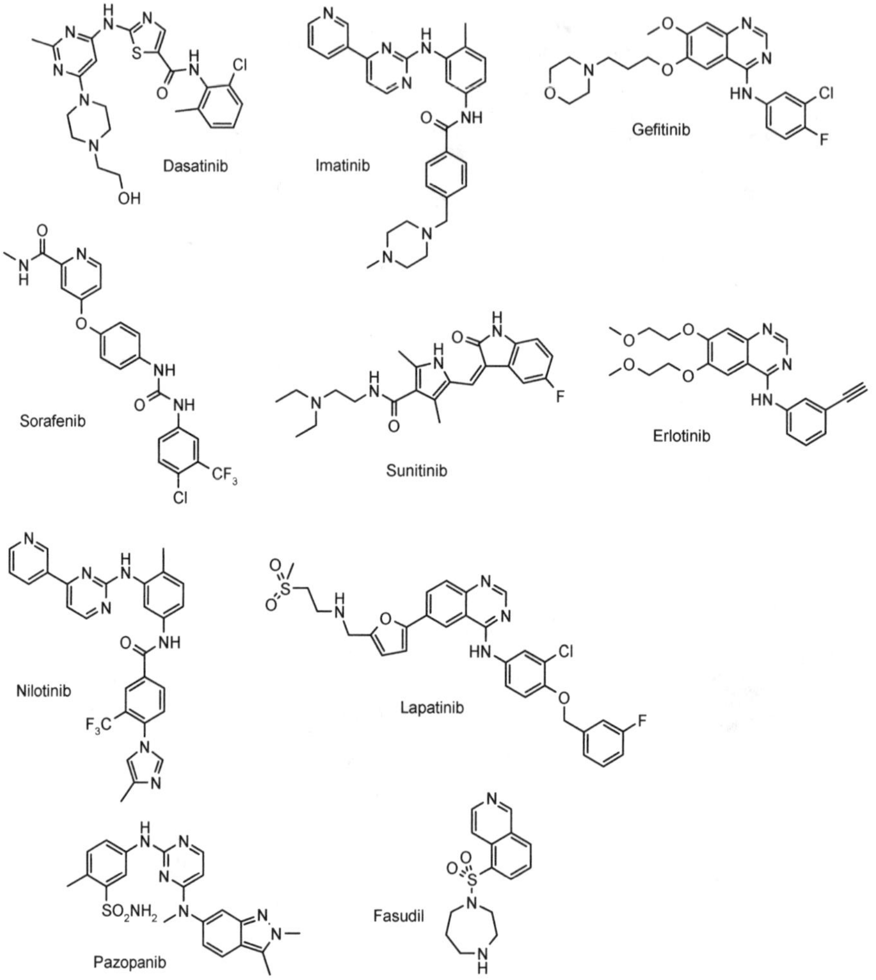

Dasatinib

Imatinib

Gefitinib

Sorafenib

Sunitinib

Erlotinib

Nilotinib

Lapatinib

Pazopanib

Fasudil

Figure 1.2 Structures of approved small molecule kinase inhibitors.

targeted EGFR and ErbB2, to the fairly promiscuous, such as Sunitinib, which bound to multiple kinases.[7]

These compounds share another feature in common besides the intent to inhibit tyrosine kinases. Nine of the ten kinase drugs are approved for oncology indications. Whilst the clinical attrition rate of kinase compounds is significantly lower than that for other antitumor agents,[13] there are few examples of success for other indications to date. Fasudil, an inhibitor of ROCK (an AGC family kinase) is the exception, approved for acute cardiovascular disease in Japan. To some extent this may indicate that there has been a greater focus on cancer indications in kinase R&D organizations, perhaps because these are now precedented targets. This cannot be the whole story, since considerable effort has been expended in other areas, notably for inflammatory diseases.[14]

Table 1.2 Approved small molecule kinase inhibitors.

Name	Manufacturer	Approved for	Intended target(s)	MW
Dasatinib / Sprycel [a]	BMS	CML, ALL	Abl/Src family	488
Gleevec / Imatinib [b]	Novartis	CML, ALL, MDS/ MPD, ASM, HES, CEL, DFSP, GIST	Abl, KIT, PDGFR	493
Iressa / Gefitinib [c]	AstraZeneca	NSCLC	EGFR	446
Sorafenib / Nexavar [d]	Bayer / Onyx	Hepato/renal cell carcinoma	Raf	464
Sutent / Sunitinib [e]	Pfizer	GIST, renal cell carcinoma	VEGFR, PDGFR, FLT3, KIT	398
Tarceva / Erlotinib [f]	Genentech / Roche	NSCLC, pancreatic cancer	EGFR	393
Tasigna / Nilotinib [g]	Novartis	CML	Abl	529
Tykerb / Lapatinib [h]	GSK	Breast cancer	EGFR, ErbB2	580
Votrient / Pazopanib [i]	GSK	Renal cell carcinoma	VEGFR	437
Fasudil (Japan) [j]	Asahi	Cerebral vasospasm	ROCK	291

[a] L. J. Lombardo, F. Y. Lee, P. Chen, D. Norris, J. C. Barrish, K. Behnia, S. Castaneda, L. A. Cornelius, J. Das, A. M. Doweyko, C. Fairchild, J. T. Hunt, I. Inigo, K. Johnston, A. Kamath, D. Kan, H. Klei, P. Marathe, S. Pang, R. Peterson, S. Pitt, G. L. Schieven, R. J. Schmidt, J. Toarski, M. L. Wen, J. Wityak and R. M. Borzilleri, *J. Med. Chem.*, 2004, 47, 6658. [b] R. Capdeville, E. Buchdunger, J. Zimmermann and A. Matter, *Nat. Rev. Drug Discovery*, 2002, 1, 493. [c] A. E. Wakeling, S. P. Guy, J. R. Woodburn, S. E. Ashton, B. J. Curry, A. J. Barker and K. H. Gibson, *Cancer Res.*, 2002, 62, 5749. [d] S. Wilhelm, C. Carter, M. Lynch, T. Lowinger, J. Dumas, R. A. Smith, B. Schwartz, R. Simantov and S. Kelley, *Nat. Rev. Drug Discovery*, 2006, 5, 835. [e] L. Sun, C. Liang, S. Shirazian, Y. Zhou, T. Miller, J. Cui, J. Y. Fukuda, J. Y. Chu, A. Nematalla, X. Wang, H. Chen, A. Sislta, T. C. Luu, F. Tang, J. Wei and C. Tang, *J. Med. Chem.*, 46, 1116. [f] J. D. Moyer, E. G. Barbacci, K. K. Iwata, L. Arnold, B. Boman, A. Cunningham, C. DiOrio, J. Doty, M. J. Morin, M. P. Moyer, M. Neveu, V. A. Pollack, L. R. Pustilnik, M. M. Reynolds, D. Sloan, A. Theleman and P. Miller, *Cancer Res.*, 1997, 57, 4838. [g] E. Weisberg, P. W. Manley, W. Breitenstein, J. Bruggen, S. W. Cowan-Jacob, A. Ray, B. Huntly, D. Fabbro, G. Fendrich, E. Hall-Meyers, A. L. Kung, J. Mestan, G. Q. Daley, L. Callahan, L. Catley, C. Cavazza, M. Azam, D. Neuberg, R. D. Wright, D. G. Gilliland and J. D. Griffin, *Cancer Cell*, 2005, 7, 129. [h] D. W. Rusnak, K. Lackey, K. Affleck, E. R. Wood, K. J. Alligood, N. Rhodes, B. R. Keith, D. M. Murray, W. B. Knight, R. J. Mullin and T. M. Gilmer, *Mol. Cancer Ther.*, 2001, 1, 85. [i] P. A. Harris, A. Boloor, M. Cheung, R. Kumar, R. M. Crosby, R. G. Davis-Ward, A. H. Epperly, K. W. Hinkle, R. N. Hunter III, J. H. Johnson, V. B. Knick, C. P. Laudeman, D. K. Luttrell, R. A. Mook, R. T. Nolte, S. K. Rudolph, J. R. Szewczyk, A. T. Truesdale, J. M. Veal, L. Wang and J. A. Stafford, *J. Med. Chem.*, 2008, 51, 4632. [j] T. Asano, I. Ikegaki, S. Satoh, M. Seto and Y. Sasaki, *Cardiovasc. Drug Rev.*, 1998, 16, 76.

On the positive side, if drugs for chronic conditions can be developed, they should not suffer from the problems of emerging resistance that have plagued kinase oncology drugs.[15] However, there are significant challenges that must be overcome before kinase inhibitors can be useful for chronic diseases. The first limiting factor is the understandable unwillingness to accept risks of side-effects in chronic disease that may be acceptable in severe acute illness. All medications may suffer from on-target side-effects, but kinase drugs seem more likely than most to suffer from unexpected off-target effects because of their closely related binding sites and well-established cross-activity. A secondary problem is that a chronic disease treatment may place a greater importance on oral, twice-daily dosing in a tablet form than acute disease. To some extent these two factors may be related. Many of the cancer compounds listed in Table 1.2 have relatively high molecular weights compared to the average for oral drugs of 337.[16] Presumably this has been brought about in part by the need to build in extra features in order to achieve greater kinase selectivity. Compounds targeting the inactive DFG-out and C-helix-out conformations are often especially large. Larger molecules generally have poorer pharmacokinetic properties, so if indeed larger molecules are needed to attain the requisite selectivity, it may be harder to find orally active compounds with appropriate selectivity for chronic diseases.

This may be made harder still depending on the choice of primary kinase target. It has been shown that tyrosine kinase inhibitors are more likely to show cross-inhibition of moderately closely related kinases than are inhibitors of CMGC kinases, for example.[11] In the case of chronic diseases demanding a low-risk profile and greater selectivity, the historical emphasis on tyrosine kinases as targets might have contributed to the low success rate. Yet, with a few exceptions, target validation for most kinases on other kinome branches has been slower to emerge.

Table 1.3 lists some promising compounds in development for inflammation, and structures are shown in Figure 1.3. The furthest advanced of these is Tasocitinib, a JAK inhibitor currently in Phase III trials for rheumatoid arthritis (RA). Following behind are several compounds in Phase II trials for various diseases, notably SYK. Both JAK and SYK are tyrosine kinases. One

Table 1.3 Some kinase inhibitors in development for inflammation indications.

Name	Manufacturer	Phase	Intended target(s)	MW
CP-690,550 / Tasocitinib	Pfizer	Phase III for RA	JAK	312
R935788 / Fostamatinib	Rigel	Phase II for RA	SYK	514*
PH-797804	Pfizer	Phase II for RA, neuropathic pain	p38α	477
GW856553 / Losmapimod	GSK	Phase II for RA, COPD, CV	p38α	383
VX-702	Vertex	Phase II, discontinued	p38α	404
Pamapimod	Pfizer	Phase II, discontinued	p38α	406

*phosphate pro-drug

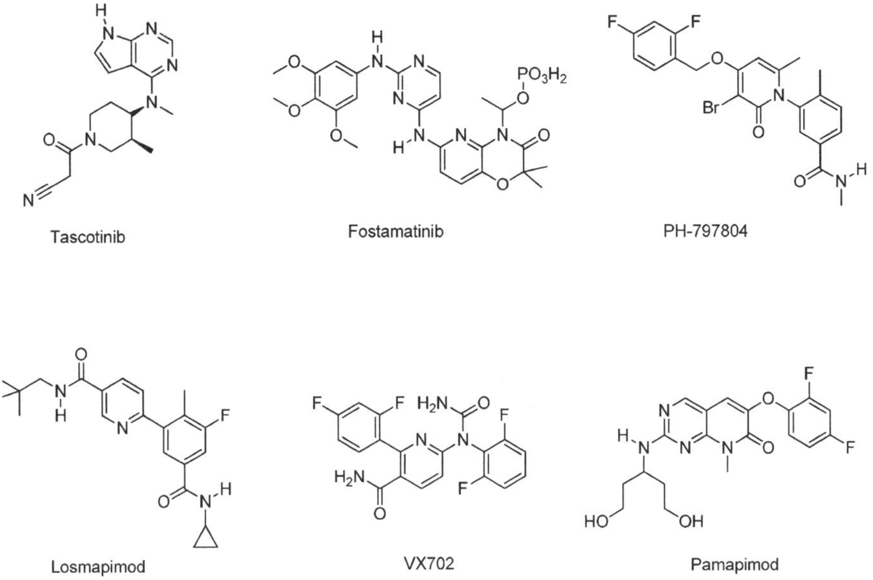

Tascotinib

Fostamatinib

PH-797804

Losmapimod

VX702

Pamapimod

Figure 1.3   Structures of some kinase inhibitors studied for inflammation indications.

of the most intensively studied targets that is not a tyrosine kinase is p38 MAP kinase.[4]

It is instructive to consider the reasons for the failure of clinical p38 inhibitors, a target which has been extensively pursued for RA for many years without success.[17, 18] Early failures can be attributed to the use of a limited number of closely related chemical series and to incompletely determined selectivity profiles. For example, early pyridinyl imidazoles such as SB-203580 were thought to be p38-selective but are now known to inhibit JNKs and other targets.[10] When truly selective molecules such as VX-702 finally entered Phase II clinical trials for RA, side-effects were relatively minor.[19] The surprising result was the lack of efficacy of bioavailable inhibitors for reasons that remain unclear. It is striking that early signs of efficacy on inflammatory biomarkers diminished over time, suggesting that an unknown compensation mechanism engages in the RA disease process, and calling into question the established RA disease models. These results may have finally invalidated p38 as a target for RA, but its suitability as a target for other diseases is perhaps enhanced by the lack of severe side-effects. Interestingly, a p38 inhibitor has achieved positive proof of concept in a study of dental pain.[20]

This example reinforces the critical importance of reliable target validation for a given disease. It seems likely that other kinases yet to emerge could impact chronic diseases but not suffer from such unforeseen target/disease related problems. If these are not tyrosine kinases, it may be easier to obtain selective inhibitors to avoid off-target effects while remaining in drug-like chemical space. While research continues into tyrosine kinases for chronic diseases, other kinases with fewer close homologues are emerging as targets, for example mTOR, IKKβ, MK2, and lipid kinases, and these may prove more tractable, provided their intrinsic functions allow for a safe therapeutic window.

As mentioned above, nine of the ten approved kinase inhibitor drugs were originally conceived as inhibitors of various TK or TKL kinases. If success in kinase drug discovery is measured by drugs on the market, then success has been limited to oncology, and to this branch of the kinome. In order to move towards a wider exploitation of kinases as drug targets, we need small molecule inhibitors with suitable potency, selectivity and cell activity that can be used to elucidate the biology of unexplored kinases. To identify these chemical probes we need to have assays to understand inhibition profiles, chemical starting points for optimization, measurements of selectivity, and strategies to modulate selectivity. We will now touch on each of these areas.

1.2   Kinome Scale Assays

Understanding the affinity of new inhibitors across multiple kinases will be important to exploring the therapeutic value and liability of the unexplored kinase targets. The next section of this review will focus on the commercial assays available to assess broad kinase inhibitor potency. These assays are currently expanding the breadth of the kinase world and in the future will enable production of inhibitors for most of the kinases on the kinome.

The publication of the kinome tree dramatically increased the challenge and potential for kinase therapeutics. This challenge was reflected in both the recognition that selectivity within the protein kinase family was defined to a universe of about 518 individuals and in the recognition that an assay for each individual kinase was needed to assess the selectivity of compounds in development. It was immediately understood that the presence of a highly conserved ATP binding site throughout the kinome would make achieving selectivity challenging.

In contrast, recognition of the conserved ATP site also led to the realization that "lead hopping" from an inhibitor of one kinase to an inhibitor of another kinase, closely or more distantly related, was feasible. It may be possible to find ligands for each member of the tree *via* this lead hopping and cross screening through the kinome. Indeed GSK has used this strategy to develop a number of kinase leads and candidates. For example, Figure 1.4 shows the source of GSK kinase leads or Starts of Chemistry from 2002-1Q2009.

The data show that from 2002 to 1Q2009 the largest source of leads or SoC over this period consistently came from the kinase cross-screen (X-screen). The X-screen process is depicted schematically in Figure 1.5. Over half (53%) of all leads identified derived from this approach. For our purposes, a hit is an active compound that comes out of a screen, and a lead is a more advanced compound that has significant promise such that one has confidence that an optimization effort will lead to a clinical candidate. The other two target class

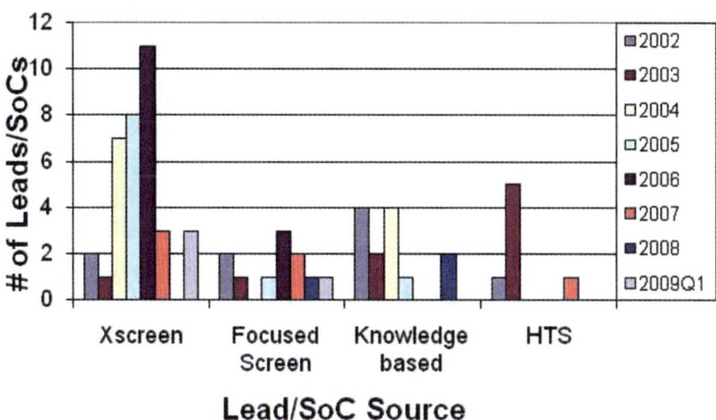

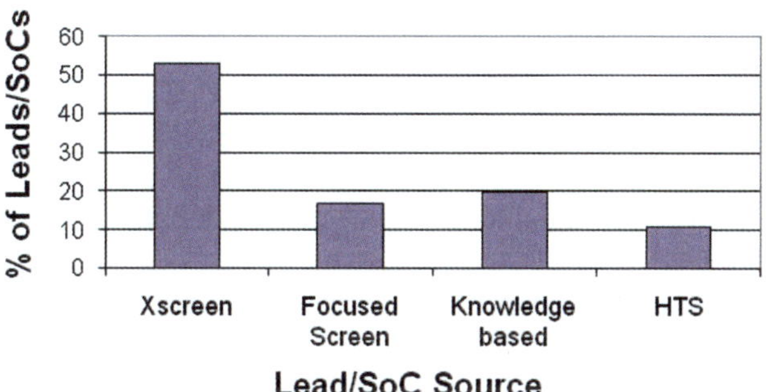

Figure 1.4   Figure 1.4a shows the leads or Starts of Chemistry (SoCs) that derived from GSK screening strategies from 2002 to 2009, and Figure 1.4b shows the percentages over the entire 2002 – 1Q2009 period. In the figures, "X-screen" (cross screen) represents the leads derived from the kinase process in which all new molecules from all ongoing kinase programs were assayed on a weekly or monthly basis against all active kinase targets. This process, shown in Figure 1.5, allowed us to rapidly determine if a compound that had been made for one program would be an inhibitor for a kinase target of another program. It also allowed us to determine the selectivity of all inhibitors for all of these kinases in parallel. "Focused screen" represents leads derived from screens of an in house 10 to 13 thousand compound Kinase Chemistry Sets (KCS) designed with known kinase inhibition or kinase ATP site "hinge region" binding potential. "Knowledge based" leads/SoC derive from *de novo* design based on crystallography, modelling or lead hopping from in house or published inhibitors. These approaches were generally used only when other approaches proved unsuccessful. Finally, HTS leads/SoC derive from high throughput hits that had not been previously found in one of the other methodologies.

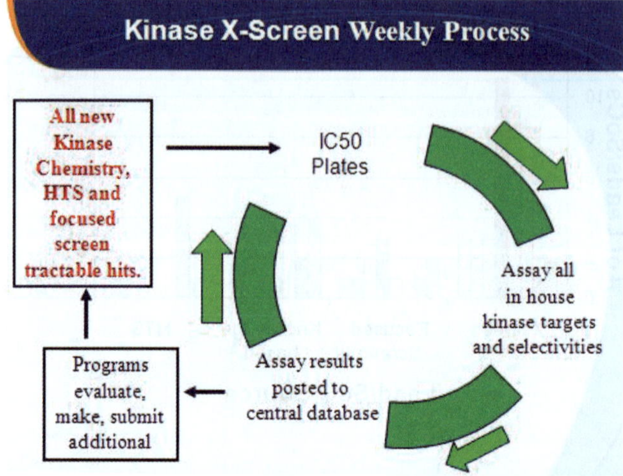

Figure 1.5   The Kinase Cross-Screen (X-Screen) process to identify starting points for kinase programs.

based screening methodologies produced 17% and 20% of the leads/SoC, respectively. In total, about 90% of kinase leads/SoC from 2002 to 1Q2009 derived from target class based focused screening – a combination of X-screening, focused screening, and knowledge based design. HTS accounted for about 11% of the leads/SoC.

Stavenger and colleagues described an excellent example of the power of the cross screen to identify a starting point that can be turned into a useful tool molecule.[21] In this case the molecule shown on the left in Figure 1.6 was made for our MSK1 program. The Rho kinase 1 (Rock1) program routinely screened all new kinase chemistry and found this compound to be a 19 nM IC50 hit. However, the molecule was not selective for Rock1. It exhibited mid to low nM inhibition of other AGC kinases including RSK1, p70/S6K, MSK1 and sub-micromolar inhibition of a number of other kinases. Application of several rounds of medicinal chemistry to the hit provided the elaborated molecule shown on the right of Figure 1.6 with enhanced selectivity and

MSK inhibitor identified as ROCK
hit via kinase X-screen

ROCK tool molecule with enhanced selectivity
and PK properties suitable for in vivo experiments

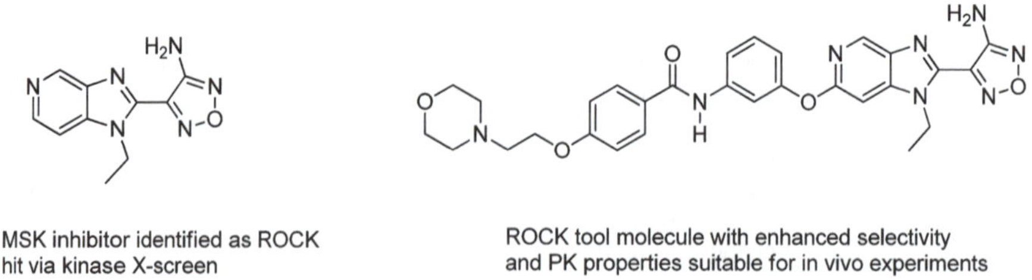

Figure 1.6   ROCK inhibitor identified from Kinase X-Screen.

useable pharmacokinetic parameters that the team utilized for *in vivo* target validation experiments.

We believe that the target class based focused screening was so successful because of the highly conserved ATP binding site in all protein kinases (and many lipid kinases). The Kinase Compound Set (KCS) was designed to target this ATP site, and the X-Screen exposed the target to the newest GSK kinase chemistry thought processes (largely targeting this site). These two approaches provided 70% of chemical starting points. Finally, knowledge based screening of compounds designed for this site adds another 20% of historical chemical starting points.

These on-going kinase medicinal efforts at GSK and other companies increased the demand for additional kinase selectivity assays to thoroughly assess development and tool potential for these compounds coming from corporate and academic screens. To address this within GSK, a small, representative kinome panel ($\sim$15% of the kinome at any time) was developed and screened weekly along with targets. In general we found this quite predictive of broader kinome selectivity. However, we did not find it feasible to develop a panel of kinase assays approaching the full scale of the kinome for selectivity or hit identification, and because of known and unknown liabilities associated with inhibiting the other $\sim$85% of the kinome, therapeutic project teams sought the means to assay the broader kinome. This limitation at GSK and other pharmaceutical companies was one impetus for the development of commercial kinome panels.

The table in Appendix 1.1 shows the status of kinome assay panels (as of October 2010) that have been developed (and are still growing) for nine of the major contributors to this field. In the table, the first column lists the kinases from the Sugen kinome paper,[1] plus some lipid kinases, mutant kinases and specific constructs or activation states. The next five columns include gene names and aliases along with family information from the Sugen publication. The remaining columns list the kinase assays commercially available from the suppliers; these assays are aligned to the Sugen names. Within the individual supplier columns, the nomenclature of the supplier was maintained where feasible in order to facilitate using this information when communicating with suppliers.

The sizes of these panels vary from about 200 kinases at Caliper to over 400 at Ambit. The assay counts include the lipid kinases, mutant kinases and kinases that are available in more than one assay condition (the number of wild type kinases in any of the panels is less than the total). These assay panels enable assessment of broad selectivity and liability information needed for lead and candidate characterization. They also enable broad screening for chemical starting points for kinases across the kinome in parallel. Indeed, we reported doing this in 2008 against the 203 kinases available in the Ambit panel at that time, which will be discussed elsewhere in this review.

The kinome panels at the nine suppliers in Appendix 1.1 are not all the same, and the supplier diversity provides value. They differ in composition, kinase constructs, assay technology, assay conditions, kinase and ATP concentration, and in kinase substrate selection. This diversity enables the kinome scientist to select the kinase and, sometimes, the assay and conditions to answer specific questions.

By combining assays from the nine providers in Appendix 1.1, inhibitor binding for up to 407 wild type protein or lipid kinases can be assessed. The data in Appendix 1.1 shows that the kinase families can be well covered by a combination of the commercial assay panels. Where one panel may be deficient, other panels provide coverage.

All of the providers offer screening at a single concentration of compound (giving %Inhibition at that concentration) as well as IC50 or K_d determination, and they all also provide some mechanism of action work for all kinases in the panel. In addition, all providers offer screens against a number of clinically important mutant protein or lipid kinases. Most providers also enable the scientist to view target profiling and selectivity data graphically, with a visualization tool. Some also have nonhuman and non-mammalian kinases available (not included in this review or Appendix 1.1).

Assay formats among the kinase panels differ greatly. The Ambit panel is the largest, and its technology is the most unique. The Ambit panel includes ~375 wt (wild type) human assays (including some lipid kinases), 54 mutants and 10 additional specific activation states. The latter enable the scientist to assess affinity to some kinases under different states of activation.[22] The Ambit format is a binding assay and not a substrate turnover assay. The assay platform is presented in two similar formats. In one format, human kinases are expressed on the exterior of T7 phage where they are able to bind bead-immobilized nonspecific ATP site kinase inhibitors.[7] This process immobilizes the phage to the bead. In the presence of a test inhibitor of interest in solution, the phage is displaced from the bead. By varying its concentration, the dissociation constant (K_d) of inhibitor in solution can be determined for the kinase by measuring the loss of phage from the bead. More recently, HEK-293 cell expression has been used for some kinases to enable better control of the kinase activation state, and PCR of an attached DNA label has been used to detect the kinase on the bead.[10] Since these are binding assays and do not require substrate turnover, the assay of inactive constructs and states may be more feasible than under the constraints of a catalytic assay. As such, Ambit also offers some kinases in both un-activated and activated states to enable exploration effects of these options on compound affinity. The Ambit panel is now offered by DiscoveRx and can be accessed here: http://www.discoverx.com/technology/technology-kinomescan.php.

All other panels of isolated kinases use catalytic assays and measure the conversion of a substrate (either ATP or peptide/lipid) to product. Because these are catalytic assays, they are all run in the presence of ATP plus a substrate. The ability to assay inhibitors at multiple ATP concentrations provides the scientist with the ability to obtain some information on whether the inhibitor is competitive with ATP or not and to rapidly eliminate the undesired mechanism. An additional advantage of catalytic assays is the assurance that the protein in the assay is properly folded and in an active state. This can also be a limitation since, as discussed, interesting inactive states may be less accessible in these assays. However, in this regard some of the catalytic panels also provide some kinases in both an un-activated and activated state or with more than one protein substrate.

Millipore, Proqinase, Reaction Biology (RBC) and Signalchem use a ^{33}P-ATP filter binding method for most kinase assays. In these methods, the gamma phosphate of gamma ^{33}P-ATP is transferred by the kinase to product, and after filter capture, ^{33}P-peptide or lipid is counted on the filter. An advantage of these radiophosphate-ATP type assays is the direct measurement of phosphate incorporation into the product, a method with low artifact incidence. These types of assay also provide the ability to vary ATP for assessing binding mechanism; however, they do pose some limitations on the upper concentrations of ATP that can realistically be used in the assay. This concentration limit is typically the ATP K_m or a few-fold higher.

The Millipore human kinase panel consists of $\sim$236 wt kinases (including some lipid kinases; fluorescence energy transfer technology is used for lipid kinases). They also provide assays for 40 mutant kinases and for 11 kinases with special substrates or in activated states. As with most catalytic assay suppliers, whole panel screens are typically done at one ATP concentration (10 μM ATP for Millipore), but assays at the ATP K_m can be requested. For more information on the Millipore panel, see http://www.millipore.com/drugdiscovery/dd3/kinaseprofiler.

The Proqinase assay panel currently consists exclusively of protein kinases: $\sim$285 wt protein kinases, 27 mutants and 5 kinases with special substrates or in activated states. Their standard panel is run at 1 μM ATP. For more information on the Proqinase panel, see http://proqinase.com/content/view/27.

The RBC kinase panel consists of $\sim$325 wt kinases (including some lipid kinases), 35 mutants and 6 kinases with special substrates or in activated states. A review of the technology as well as most of the other technologies discussed here was recently published by authors from RBC.[23] This paper provides a nice overview of many of the standard catalytic assays used by RBC and other kinome assay providers. For more information on the RBC panel, see http://www.reactionbiology.com/pages/kinase.htm.

The Signalchem human kinase panel consists of $\sim$273 wt kinases (including some lipid kinases), 3 mutant kinases and 17 kinases with special binding partners or constructs. For more information, see http://www.signalchem.com/discovery.php?type=1&catfiler=12.

Nanosyn, Caliper Life Sciences, and most Carna catalytic assays rely on a microfluidics capillary electrophoresis (CE) technology commercialized by Caliper Life Sciences. These are non-radioactive assays that measure the change in electrophoretic mobility of the substrate (usually a fluorescent-labelled peptide or lipid) upon phosphorylation. Both substrate and product are measured in these assays enabling increased assay precision. Since unlabelled ATP is used in the assays, wide ATP concentrations can be requested. For IC50 and mechanism of action follow up work, this is a distinct advantage for the CE technology over radiophosphate-ATP type assays since ATP concentration is only limited by solubility.

The Nanosyn panel consists of ~231 wt kinases (including some lipid kinases), 22 mutants and 3 kinases with special substrates or in activated states. For more information on Nanosyn profiling and use of CE, see http:// www.nanosyn.com/?id=39&page=products-screening_profiling.

Caliper Life Sciences also provides a CE based kinase panel. The Caliper Life Sciences panel consists of ~177 wt kinases (including some lipid kinases) and 12 mutant kinases. This company provides kinases and substrates on assay plates and is not strictly a screening service, but it could be a good way forward for labs that have access to a Caliper CE instrument. The company does also provide a smaller panel as a screening service. For more information about both of these, see http://www.caliperls.com/products/reagents/kinase-profiling/ profilerpro-kinase-selectivity-assay-kits.htm.

The Carna human kinase panel consists of ~275 wt kinases (including some lipid kinases), 29 mutants and 2 kinases with alternate binding partners. The standard ATP concentration for the Carna panel is the ATP apparent K_m, and half of the panel can also be run at up to 1 mM ATP. As indicated above, the Carna assay technology is primarily CE, but they also offer some assays that are not amenable to CE, and these are run primarily in ELISA format. Similar to Caliper Life Sciences, Carna also offers large subset of their kinase panel as kits for Caliper CE instruments. Details and specific kinase can be found at https://www.carnabio.com/english/product/assay-msa.html.

The Life Technologies/Invitrogen (LT) human kinase panel consists of ~281 wt kinases (including some lipid kinases), 23 mutants and 4 kinases with special binding partners or assay conditions. The LT panel uses three assay formats. The primary format is a fluorescent resonance energy transfer (FRET) format based on differential proteolytic cleavage of substrate and phosphorylated peptide product. Phosphorylation suppresses cleavage by the protease to maintain FRET; most of the LT kinases are assayed in this format. A second technology used measures ADP production upon ATP phosphate transfer to substrate. The third technology is a non-catalytic, binding assay. In this methodology, a tagged kinase is fluorescently labelled *via* an anti-tag antibody; then an ATP site inhibitor with a fluorescent tag is bound to the enzyme to enable FRET. Displacement of the fluorescent ATP site inhibitor by test compounds quenches the FRET. These three assay formats are all radioactive-ATP free. As with CE, this enables the use of very wide ATP concentrations for follow up mechanism of action work. These types of assays are susceptible to artifacts. In the case of the FRET assays, fluorescence quenching and extrinsic fluorescence from test compounds must be monitored and avoided. In the case of the ADP assay, ADP production from irrelevant ATPase activity in the kinase preparation can be an issue. As such, for this technology, it is critical that ADP production is substrate (protein, peptide or lipid) dependent and that the ADP production is shown to be coupled to product synthesis. For more information on the LT panel and formats, see http://www.invitrogen.com/site/us/en/home/Products-and-Services/Services/ Screening-and-Profiling-Services/SelectScreen-Profiling-Service/experience-selectscreen/expect-more-from-us.html.

In addition to these *in vitro* screening panels of engineered kinase constructs for screening inhibitor binding, a very different approach to kinome scale inhibitor screening has recently been reported by two groups.[24,25] Two companies, Cellzome and Kinaxo, have taken the lead in commercializing the technology. Their methods use chemical proteomics methods and mass spectrometry based peptide sequencing to screen for inhibitor affinity against over half of the kinome in whole cells, cell extracts and some tissues. The methods are related to the Ambit screening technology in that they make use of the conserved ATP binding site of kinases to capture many kinases on affinity beads. They differ from the Ambit technology in that they use the beads to capture the endogenous native kinases in cell or tissue preparations rather than an engineered construct. Not surprisingly, the particular kinases detected and assessable in the assay depend upon the cell or tissue type screened.

In this technology, the cell or tissue preparation is incubated with the beads to allow specific binding of the kinases, and then the beads are washed extensively under non-denaturing conditions to remove nonspecific proteins. The specifically bound kinases (and kinase binding partners) are eluted under denaturing conditions, subjected to proteolysis to liberate constituent peptides, and the latter are subsequently analyzed and sequenced by mass spectrometry methods.

Similar to the Ambit technology, determination of the affinity of kinase inhibitors to the kinases of the "panel" are carried out *via* competition experiments with the soluble inhibitors. Key to the value and success of this work is assessment of binding specificity and determination of inhibitor binding affinities *via* competition experiments with soluble ligands. In these experiments, the cells or extracts are exposed to the affinity beads in the presence and absence of kinase inhibitors of interest at varying concentrations. As the concentration of the soluble inhibitor increases, the amount of kinase detected on the bead decreases according to K_d for the ligand. The value of the K_d can be calculated from dose response curves. In this way, the method is able to assess the selectivity of a kinase inhibitor *versus* all of the endogenous kinases detected in the kinome experiment in parallel. It can be argued that this type of K_d and selectivity determination has value over those determined by classical *in vitro* experiments, such as those in Appendix 1.1, since these experiments assess affinity of the full length proteins in their natural surroundings under closer to natural conditions (activation state, binding proteins, membranes, and other post translational modifications).

These chemical proteomics experiments also provide additional interesting information. They provide some assessment of kinase protein binding partners. These partners are retained on the beads during the wash steps by binding indirectly to the beads through the bound kinase. Then in the presence of the soluble ligand they will indirectly be "competed" from the beads along with the bound kinase and with the same apparent "affinity" as the kinase to which they are bound. This can identify or confirm kinase binding partners, and has the potential to find compounds that either enhance or destroy the interaction of a kinase with a particular partner.

This is an evolving technology, and ongoing improvements are increasing quality. For example two modifications reported by Sharma and colleagues include incorporation of SILAC (Stable Isotope Labelling by Amino acids in Cell culture) technology and establishment of equilibrium binding to improve data quality and kinase detection ability.[25] Further information on the use of the technology by Kinaxo and Cellzome can be found here: http://www.kinaxo.de/tec_kinaff.html and here: http://www.cellzome.com/technology.html, respectively. The business models of the two companies are different. The Kinaxo model includes a fee for service screening, while the Cellzome model is a drug discovery and development deal and they do not include a fee for a service screening option on their website.

The diversity of kinome assays discussed and readily available to the scientist provides an exciting and growing toolbox to find and define the next generation of kinase inhibitors for important diseases. The remainder of this review will discuss chemistry's response to these and other tools provided by the discovery and publication of the kinome.

1.3 Starting Points for Kinase Probe Discovery: The benefits of chemical connectivity

Most reported kinase inhibitors are competitive with ATP. Kinase inhibitors are often designed to mimic ATP at least in a general sense, and they recapitulate some of the interactions ATP makes with the kinase, and exploit conserved features in the binding site. A consequence of this is that kinase inhibitors can show a great deal of cross kinase inhibition. Although this "chemical connectivity" between kinases can be considered a curse in terms of identifying exquisitely selective inhibitors, it is a blessing in terms of identifying a chemical starting point for a new kinase of interest. The publication of kinase inhibition data is heavily skewed towards certain targets, but there are in fact published small molecule inhibitors for a wide range of kinases covering all the branches of the kinome. At least seven kinase inhibitor profiling experiments have been carried out that illustrate the chemical connectivity of kinases and the potential for discovering inhibitors for a range of kinases by screening a relatively small set of kinase inhibitors (Table 1.4).

A number of important lessons can be gleaned from these reports. Importantly, screening sets of diverse kinase inhibitors against a panel of kinases almost always yields hits for every kinase. In the study published by Bamborough *et al.*, at least one hit was found for each of the 203 kinases

Table 1.4 Scope of published large-scale kinase profiling experiments revealing the chemical connectivity of kinases

Study: # of inhibitors x # of kinases	Reference
28 inhibitors × 24 kinases	S. P. Davies, H. Reddy, M. Caivano and P. Cohen, *Biochem. J.*, 2000, **351**, 95
20 inhibitors × 119 kinases	M. A. Fabian, W. H. Biggs III, D. K. Treiber, C. E. Atteridge, M. D. Azimioara, M. G. Benedetti, T. A. Carter, P. Ciceri, P. T. Edeen, M. Floyd, J. M. Ford, M. Galvin, J. L. Gerlach, R. M. Grotzfeld, S. Herrgard, D. E. Insko, M. A. Insko, A. G. Lai, J-M. Lelias, S. A. Mehta, Z. V. Milanov, A. M. Velasco, L. M. Wodicka, H. K. Patel, P. P. Zarrinkar and D. J. Lockhart, *Nat. Biotechnol.*, 2005, **23**, 329
65 inhibitors × 70 kinases	J. Bain, L. Plater, M. Elliott, N. Shpiro, C. J. Hastie, H. McLauchlan, I. Klevernic, J. S. C. Arthur, D. R. Alessi and P. Cohen, *Biochem. J.*, 2007, **408**, 297
156 inhibitors × 60 kinases	O. Fedorov, B. Marsden, V. Pogacic, P. Rellos, S. Muller, A. N. Bullock, J. Schwaller, M. Sundstrom and S. Knapp, *Proc. Nat. Acad. Sci. U. S. A.*, 2007, **104**, 20523
38 inhibitors × 317	M. W. Karaman, S. Herrgard, D. K. Treiber, P. Gallant, C. E. Atteridge, B. T. Campbell, K. W. Chan, P. Ciceri, M. I. David, P. T. Edeen, R. Faraoni, M. Floyd, J. P. Hunt, D. J. Lockhart, Z. V. Milanov, M. J. Morrison, G. Pallares, H. K. Patel, S. Pritchard, L. M. Wodicka and P. P. Zarrinkar, *Nat. Biotechnol.*, 2008, **26**, 127
577 inhibitors × 203 kinases	P. Bamborough, D. Drewry, G. Harper, G. K. Smith and K. Schneider, *J. Med. Chem.*, 2008, **51**, 7898
21 851 inhibitors × 317–402 kinases	S. L. Posy, M. A. Hermsmeier, W. Vaccaro, K-H. Ott, G. Todderud, J. S. Lippy, G. L. Trainor, D. A. Loughney and S. R. Johnson, *J. Med. Chem.*, 2011, **54**, 54

screened.[11] In the recent large scale profiling experiment described by Posy *et al.*, the same observation was made: each of the 402 kinases tested was potently bound by at least one compound.[12] PIM1 is an unusual kinase in that it has a proline in the hinge region of the active site, thus masking an NH that usually makes a hydrogen bond with a hydrogen bond acceptor group on kinase inhibitors. In spite of this atypical binding site, inhibitors can be identified by screening inhibitors of kinases with more standard binding sites. A strategy of screening a set of compounds related to known kinase inhibitors will likely uncover chemical starting points for less well-studied kinases.

These profiling experiments indicate that hits can usually be identified for each kinase, but also that different kinases have very different hit rates. In the largest scale profiling experiment, hit rates for the panel of kinases studied varied from 0.03 to 18%.[12] This suggests that the chemical tractability of kinases, defined as finding different chemotypes that bind, varies considerably. It is important to note that the hit rate for a particular kinase can vary from study to study. Hit rates may vary when different compound sets are screened or when different assay formats are used. As more profiling experiments are completed it may be possible to understand the reasons behind varied hit rates and prioritize targets for which we are more likely to find reasonably selective starting points.

A useful observation is that there is a statistical tendency for kinases with higher sequence similarity to bind similar compounds (Figure 1.7). Bamborough *et al.* found that when two kinases show over 40–50% sequence identity in the kinase domain it is likely that they will show reasonable SAR similarity, and compounds are likely to behave similarly against them.[11] In their large scale study, Posy *et al.* also found that kinases with higher sequence identity had greater activity homology.[12] Specifically, they noted that half of the kinase pairs with 80–90% sequence identity had > 60% inhibitor activity homology. Only 20% of the kinase pairs with much lower homology, between 50–60%, had > 60% inhibitor activity homology. It is interesting that this 40–50% homology rule does not apply equally to all branches of the kinome. For example, pairs of kinases in the AGC and CMGC branches that show 40–50% sequence identity have lower SAR similarity than pairs with the same sequence identity in other branches.[11] The tyrosine kinase branch is also a special case. SAR similarity between kinases with a given sequence similarity is higher in the tyrosine kinase branch than in other branches of the kinome. This suggests that obtaining selective inhibitors within the tyrosine kinase branch may be harder than other branches. Useful selectivity can be achieved, though, as exemplified by Lapatinib, a dual EGFR and ErbB2 inhibitor.[7,26]

Although sequence similarity often correlates with SAR similarity, all of the published studies reveal exceptions. In some cases surprising selectivity can be observed within kinase sub-families. For example, in their profiling study of serine-threonine kinases, Fedorov and colleagues found inhibitors that bound more potently to CLK1 than CLK3, and examples of imidazopyridazines that were more potent inhibitors of PIM1 than the closely related PIM2.[9] There are many examples of compounds designed for tyrosine kinases that inhibit serine-threonine kinases and *vice versa*. Posy and co-workers found that only half of the MST3 actives they identified bound YSK1, even though these two kinases share 88% sequence identity, and differ by only one residue in the active site.[12] In other cases, pairs of kinases with low sequence identity and high activity homology can be found. This suggests that simply looking at kinases close in sequence will not always be sufficient to identify troublesome off-target kinases. Broader data sets, detailed medicinal chemistry SAR studies, and X-ray crystallography of bound inhibitors should help us understand the

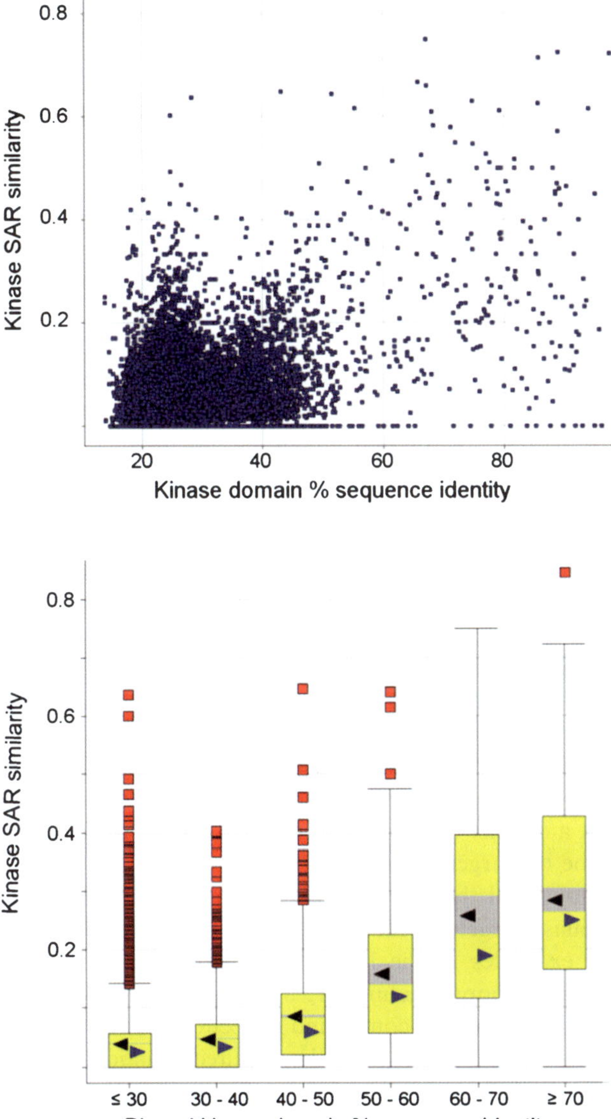

Figure 1.7 Relationship between kinase SAR similarity and kinase domain % sequence identity.

reasons behind unexpected selectivity and unexpected similarity of inhibition profiles. Despite the inevitable exceptions, this 40–50% sequence identity rule of thumb provides a good framework for consideration of selectivity issues, in that it reasonably suggests that while monitoring selectivity one should pay close attention to kinases of similar sequence.

In terms of compounds, large scale profiling experiments have revealed both very selective compounds and very promiscuous compounds. Promiscuity is not necessarily compound class dependent, as specificity can vary a great deal among compounds with the same scaffold. For example, Lapatinib is a rather selective kinase inhibitor, and Erlotinib and AZD-1152HQPA, two other inhibitors with a quinazoline core, are significantly less so.[27] From a medicinal chemistry standpoint this observation that selectivity is not solely related to the hinge binding core of the molecules makes sense, because the scaffold makes interactions with the target(s) and the groups attached to any given scaffold probe the active site and make interactions. A given scaffold does not only inhibit within one kinase family, either. In the study undertaken by Posy *et al.*, it became clear that many chemotypes have examples that inhibit kinases in each family.[12] They did not find many chemotypes that bind selectively to only a few families. This suggests that the building blocks attached to the core ring system can be used to manipulate potency (and thus selectivity) at various kinases. Incorporating a greater diversity of building blocks improves the chances of finding selective compounds with a given template.

Finally, kinase inhibitor profiling experiments suggest that compounds that inhibit the same target but are structurally dissimilar are likely to have a different off-target kinase inhibition profile. If one is seeking to understand if a particular kinase is causing a desired phenotype, this observation has important implications. When using kinase inhibitors as probes of cellular function, it is advantageous to use at least two different chemotypes. The more tool compounds that give the desired phenotype, the easier it will be to connect the kinase activity (or pattern of kinase activity) to the effect. Along similar lines, if looking for a back-up compound to a compound in the clinic, it would be wise to pick a distinct chemotype as this will reduce the risk of the back-up having the same off-target activity.

We have established that a collection of kinase inhibitors can be screened against new kinase targets to identify starting points for optimization. There are numerous examples in the literature of this sort of "target hopping". Substituted 2,4-diaminopyrimidines are a common kinase inhibitor motif and well represented in corporate compound collections. They are a useful kinase inhibitor template because they have the required functionality for binding to the hinge region of a kinase (N1 of the pyrimidine and the NH of the two position form the critical hydrogen bonds), and they are readily synthesized with chemistry that allows incorporation of a range of substituents in the 2-, 4-, and 5- positions. The ability to incorporate a range of functionalities allows for exploration of the kinase active sites to find ways to enhance selectivity and modulate compound properties. Figure 1.8 shows some example 2,4-diamino-pyrimidine kinase inhibitors.

These inhibitors all share the same 2,4-diaminopyrimidine core, yet were made to target different kinases. One can see that the scientists exploring the SAR have incorporated vastly different groups in the 2-, 4-, and 5- positions in order to achieve either potency at the desired target or selectivity over other

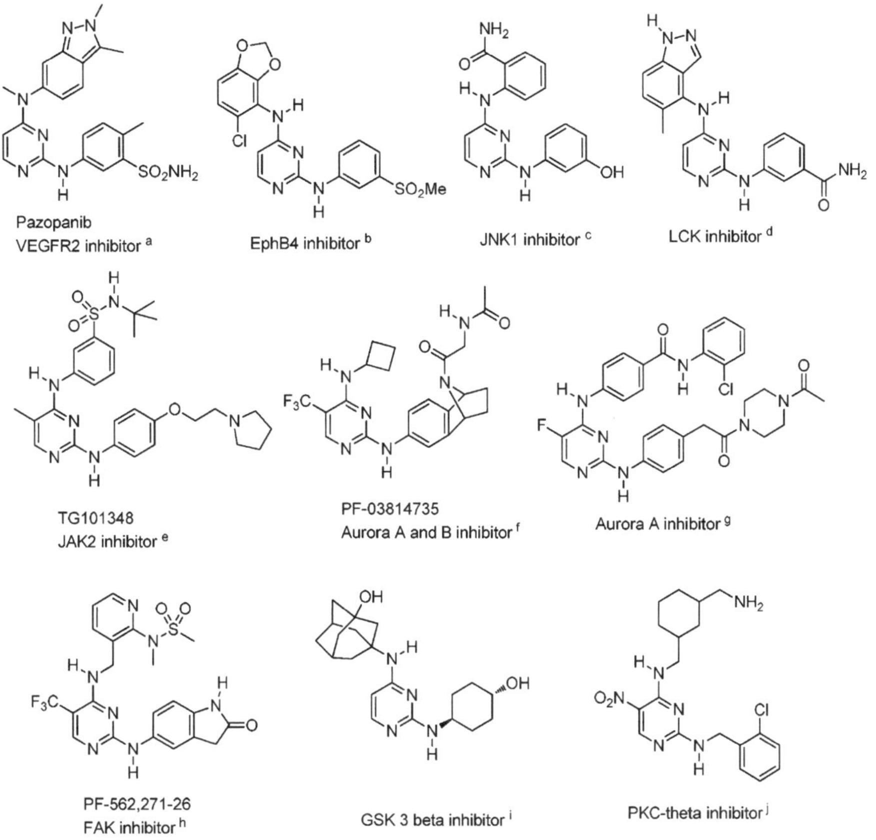

Pazopanib
VEGFR2 inhibitor [a]

EphB4 inhibitor [b]

JNK1 inhibitor [c]

LCK inhibitor [d]

TG101348
JAK2 inhibitor [e]

PF-03814735
Aurora A and B inhibitor [f]

Aurora A inhibitor [g]

PF-562,271-26
FAK inhibitor [h]

GSK 3 beta inhibitor [i]

PKC-theta inhibitor [j]

Figure 1.8 A range of 2,4-diamino pyrimidine kinase inhibitors.

targets. The 4-position contains differently substituted anilines, an example of a benzyl amine, and aliphatic amines with different shapes. The 2-position also has substituted anilines and a benzyl amine, and a hydroxylated aliphatic amine. There are a variety of 5-position groups (H, F, CH_3, CF_3, NO_2) which may make direct interactions with the kinase, and likely modify the electronics of the pyrimidine and affect the conformational preference of the 4- position group. Full selectivity data for these compounds are not published so the overall selectivity patterns achieved are unknown, but it is clear that these sorts of large variations at the different positions leads to a range of kinase profiles.

1.4 Measures of Selectivity

As discussed, the close homology within the ATP binding pocket is a defining characteristic for protein kinases and constitutes the promise and the challenge for kinase drug discovery. For targeting the ATP-binding pocket and its surroundings in a particular kinase target with the aim to identify a safe and efficacious inhibitor, the high likelihood for off-target kinase activity demands the continuous monitoring of kinase activity across related kinases during lead optimization and, at appropriate intervals in the process of lead discovery and optimization, across a large part of the kinome. The latter is important in order to avoid unexpected activity that could call in doubt conclusions drawn from target validation through to efficacy and toxicity studies and is necessary for the identification of truly selective inhibitors. It is no surprise that recognition of the importance of wider selectivity profiling has grown with the availability of larger panels of kinase assays. And as these have developed to cover a substantial portion of the kinome, various groups have utilized these assays to define selectivity profiles of kinase tool compounds, clinical candidates, kinase inhibitors on the market and even representative sets of screening collections as listed above. The information obtained from profiling kinase inhibitors over an increasingly large part of the kinome represents a paradigm shift for kinase drug discovery. On one side, it raises the important question of how best to utilize and to mine this information for driving kinase lead discovery and optimization programs. On the other side, the capability to gather quantitative data on the interaction of closely related, but – in many details – still diverse protein structures against the immense structural and physico-chemical diversity offered by small molecules is bound to provide new insights and new options that will be beneficial to the field.

The first challenge for a medicinal chemist on using wider-kinome profiling data is in how to analyse the complex data set. The visual representation of kinase inhibition data onto the kinome tree as first demonstrated by researchers from Ambit Biosciences has been widely adopted.[7] It allows for a quick inspection of data for a single compound and can be useful to compare data across a small number of compounds. This is largely a qualitative analysis and does not allow for spotting subtle differences between compounds or to rank compounds with complex selectivity profiles. A number of quantitative approaches to define and rank selectivity of compounds have been proposed. Karaman and co-workers suggested the use of a selectivity score (S) for data from the Ambit kinase panel.[10] The score is derived by counting the number of kinases with binding affinity below 3 μM divided by the number of kinases tested. In effect, this score represents the fraction of kinases bound by the inhibitor with a value between 0 for no binding to any kinase and 1 for a completely unselective compound. This score can be useful for ranking of inhibitors, in particular for data from large kinase panels, although the lack of differentiation between very strong and somewhat weaker binding affinities in calculation of the score has to be recognized. Another scoring algorithm proposed by Graczyk uses the Gini coefficient which is a measure of the inequality of a distribution and well established for comparing income distributions.[28] Non-selective inhibitors are characterized by Gini values close to zero (*e.g.* Staurosporine, Gini 0.150). Highly selective compounds exhibit Gini values close to 1 (*e.g.* PD184352 Gini 0.905). Both these scores have recently been compared to another measure of kinase inhibitor selectivity derived by a thermodynamics-based partition index.[29] This latter score is based on the concept of thermodynamic equilibrium of the compound with the full panel of kinases. The fraction of bound inhibitor at each target defines the 'partition' index P. A value of P = 0 indicates no occupancy and a value of 1 indicates 100% occupancy. Thus, this measure provides a selectivity score for a kinase of interest, in contrast to the overall selectivity measure of the Ambit score and Gini coefficient. Cheng and colleagues demonstrate the utility of the thermodynamics-based partition index for ranking selectivity of a compound across small panels of kinases for which the other scores are not appropriate due to the small numbers. They clearly demonstrate that the P score is a useful measure in analysis of small focused panel data that are common in hit-to-lead and lead optimization efforts. A comparison of the P score with S score and Gini coefficient based on the published data for a larger panel indicates that the P score provides a complementary measure to S score and Gini coefficient and may have advantages for defining the selectivity of a compound targeted at inhibition of multiple kinases.

Further challenges in dealing with large kinome profiling data are (1) at which point to profile compounds and (2) how to deal with data that demonstrate strong inhibition of one or several kinases by a compound previously expected to be relatively selective. With the cost of profiling continuing to come down, profiling at an early stage, *i.e.* by testing of a lead series under consideration for lead optimization, is advisable. Including profiling data as part of the decision-making process in lead series selection may avoid surprises during lead optimization. Additionally, broad cross-screening data at an early stage may allow prioritization of a few lead series with complementary inhibition profiles, improving the chances of identifying a clinical candidate. Increasingly, kinome-wide profiling data will be available from the screening of kinase inhibitor sets. New discovery programs will have data at hand at the stage of target selection and can use this for selection of tool compounds for target validation, for planning early chemistry strategy and also to set up initial co-crystallization experiments to gain structural data at a much earlier stage than previously possible.

1.5   Examples of Selectivity Improvement

Although a relatively small percentage of the kinome has been targeted with small molecules, a relatively larger percentage has been studied by X-ray crystallography. Scientists have been able to use crystallographic information (or homology models), kinase screening data, and empirical medicinal chemistry to improve selectivity for kinases of interest. Kinase inhibitor starting points typically have flat, heteroaromatic cores that bind in the hydrophobic purine site and make one, two or three hydrogen bonds to the hinge region of the kinase, and project groups of varied shaped sizes and functionalities towards various sub-pockets of the kinase.[30–34] Selectivity is gained by exploiting differences in the residues lining the pockets occupied by the inhibitors or growing the inhibitor into new pockets (Figure 1.9). Several representative examples are described.

McInnes and co-workers used literature data, crystal structures, homology models and docking experiments to hypothesize that an acidic residue in CDK4 (E144) was a key contact that was exploited to make compounds with enhanced selectivity for CDK4 over CDK2, which has a glutamine in this position (Q131).[35] They used this information to improve selectivity in an aniline-pyrimidine series by adding an appropriately placed positive charge on the side chain reaching towards the solvent front.

Scientists at Pfizer converted a compound with an IC50 of 26 nM on EGFR and an IC50 of 2100 nM on the closely related kinase erbB2 into a compound with the opposite selectivity profile (erbB2 IC50 = 55 nM; EGFR IC50 = 7000 nM).[36] The compounds were designed based on the observation of a key difference in one amino acid near the entrance to a large lipophilic pocket (Ser783 in erbB2 and Cys755 in EGFR). They probed this region in the active site with their chemistry efforts and were able to modulate the selectivity profile. The most selective molecule has a pyridine nitrogen which they postulated could make a water mediated hydrogen bond with Ser783.

Heerding and co-workers at GSK started their search for an AKT inhibitor with an aminofurazan hinge binder on an imidazopyridine scaffold that was more potent on ROCK1 and MSK1 than on AKT2 (ROCK1 IC50 = 8 nM, MSK1 IC50 = 21 nM, AKT2 IC50 = 1000 nM).[37] Changes were found at the 7-position of the imidazopyridine core that improved AKT2 potency, but they had little impact on selectivity. In contrast, work at the C4 position of the imidazopyridine led to compounds with enhanced selectivity. Modelling

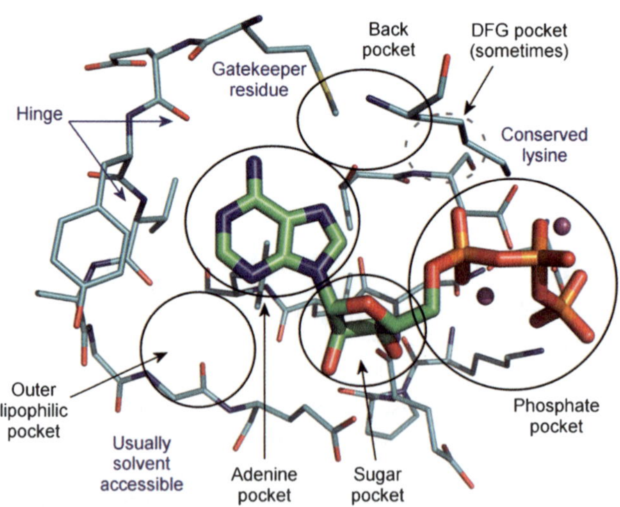

Figure 1.9   Schematic of crystal structure of ATP bound to a kinase illustrating key interactions and pockets used to modulate potency, selectivity and compound properties.

suggested that an alkyne group could project a substituent into the back pocket of AKT2. These compounds were indeed potent at AKT2, and the changes were not well tolerated by MSK1 and ROCK1, presumably due to the different residues that line this pocket in these three enzymes.

Semones and colleagues at GSK developed selective inhibitors of the receptor tyrosine kinase Tie2 starting from a well known p38 inhibitor.[38] They synthesized compounds that differed in the substituent at the 4-position of the imidazole, a position that accesses a specificity pocket, often referred to as the "back pocket" and identified a naphthyl substituent with significantly enhanced Tie2 potency and a drop off in P38 potency. They docked the compound into a homology model of Tie 2, and results suggested that the back pocket of Tie 2 is deeper than the back pocket of p38, explaining the advantage gained by the naphthyl group. Further analyses led the team to introduce substituents on the 6-position of the naphthyl ring to drive even greater selectivity enhancements. These design efforts enabled the team to identify a compound that showed efficacy in an *in vivo* model of angiogenesis.

BI 2536 is a Polo-like kinase 1 (PLK1) inhibitor currently in clinical trials for oncology indications. X-Ray crystallography of BI 2536, analysis of the inhibition profiles of BI 2536 and close analogues on a panel of kinases, and comparison of active site residues led to a detailed description of selectivity-determining residues in PLK1.[39] The crystal structure reveals that the ortho methoxy group on the aniline substituent binds in a pocket created by Leu 132

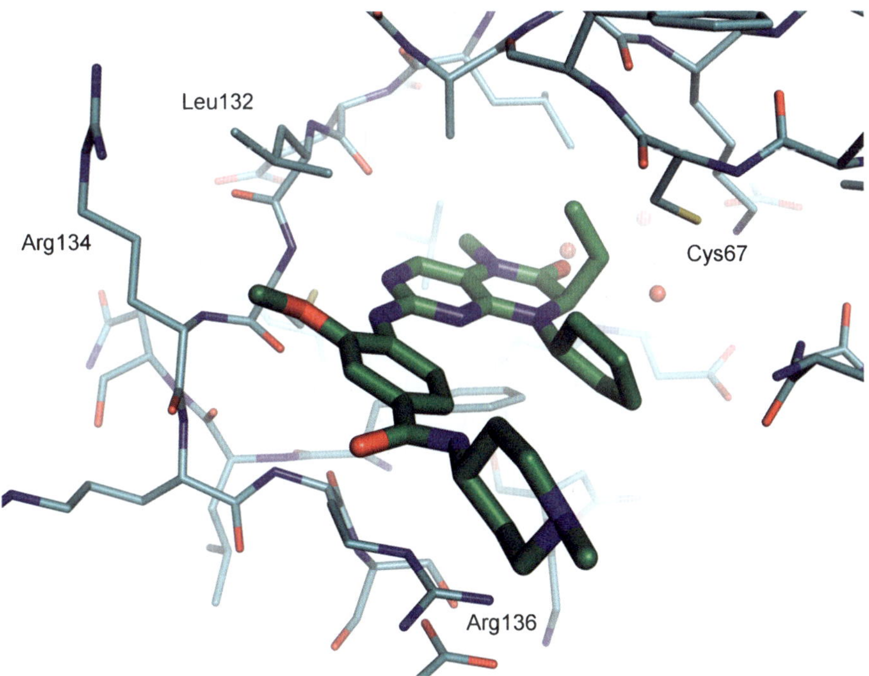

Figure 1.10   Crystal structure of BI 2536 bound to PLK1.

of the hinge region (Figure 1.10). Many kinases have an amino acid with a larger side chain here, and thus the addition of this substituent dramatically cleans up the inhibition profile. Removing this methoxy group leads to a compound with a more promiscuous inhibition profile. Interactions away from the hinge region also contribute to the potency and selectivity of BI 2536. The ethyl group binds in a small pocket that is formed by a number of residues including the side chain of Cys67. Cys67 is rotated relative to its position in a crystal structure with another inhibitor in order to allow the ethyl side chain to fit. Most of the other kinases looked at in this study have a valine at this position, which may make ideal positioning of the ethyl and cyclopentyl groups of the inhibitor more difficult. This study demonstrates that the overall potency and selectivity profile of a compound arises from the whole suite of interactions that are made, and crystallography can help explain observed selectivity and guide scientists in the optimization of these different regions to build in or enhance selectivity.

1.6 Conclusions

Kinases catalyze a very simple reaction, and are a critical component in signalling pathways. Often, these signalling pathways are mis-regulated in disease, and kinase inhibition can be beneficial. Kinases have proven to be tractable and important targets as evidenced by the growing list of compounds that have been approved and are helping patients. Because of the chemical connectivity in this protein family, achieving the desired selectivity profile can be an imposing challenge, and yet there are numerous examples of scientists improving selectivity through the integration of medicinal chemistry, screening, computational chemistry, and X-ray crystallography. Numerous groups have now demonstrated that the chemical connectivity also means that hits and leads can emerge for "new" kinases by screening of focused sets. The availability of broad assay panels (in a variety of screening formats) and a range of chemotypes enables the selection of the best chemical starting points for a particular kinase. Interestingly, only a small percentage of the kinome has been explored with small molecules although it is likely that inhibitors of other kinases in the "un-targeted kinome" will have therapeutic benefit. With an ever-growing range of kinase inhibitor chemotypes, broad assay panels, and highly fruitful crystallography efforts, kinase scientists are well positioned to identify molecules that can elucidate the function of previously un-targeted kinases and ultimately help patients.

References

1. G. Manning, D. B. Whyte, R. Martinez, T. Hunter and S. Sudarsanam, *Science*, 2002, **298**, 1912.
2. S. K. Hanks and T. Hunter, *FASEB J.*, 1995, **9**, 576.
3. H. Weinmann and R. Metternich, *ChemBioChem*, 2005, **6**, 455.

4. O. Fedorov, S. Muller and S. Knapp, *Nat. Chem. Biol.*, 2010, **6**, 166.

5. Aureus website reference. www.aureus.com

6. S. P. Davies, H. Reddy, M. Caivano and P. Cohen, *Biochem. J.*, 2000, **351**, 95.

7. M. A. Fabian, W. H. Biggs III, D. K. Treiber, C. E. Atteridge, M. D. Azimioara, M. G. Benedetti, T. A. Carter, P. Ciceri, P. T. Edeen, M. Floyd, J. M. Ford, M. Galvin, J. L. Gerlach, R. M. Grotzfeld, S. Herrgard, D. E. Insko, M. A. Insko, A. G. Lai, J-M. Lelias, S. A. Mehta, Z. V. Milanov, A. M. Velasco, L. M. Wodicka, H. K. Patel, P. P. Zarrinkar and D. J. Lockhart, *Nat. Biotechnol.*, 2005, **23**, 329.

8. J. Bain, L. Plater, M. Elliott, N. Shapiro, C. J. Hastie, H. McLauchlan, I. Klevernic, J. S. C. Arthur, D. R. Alessi and P. Cohen, *Biochem. J.*, 2007, **408**, 297.

9. O. Fedorov, B. Marsden, V. Pogacic, P. Rellos, S. Muller, A. N. Bullock, J. Schwaller, M. Sundstrom and S. Knapp, *Proc. Nat. Acad. Sci. U. S. A.*, 2007, **104**, 20523.

10. M. W. Karaman, S. Herrgard, D. K. Treiber, P. Gallant, C. E. Atteridge, B. T. Campbell, K. W. Chan, P. Ciceri, M. I. Davis, P. T. Edeen, R. Faraoni, M. Floyd, J. P. Hunt, D. J. Lockhart, Z. V. Milanov, M. J. Morrison, G. Pallares, H. K. Patel, S. Pritchard, L. M. Wodicka and P. P. Zarrinkar, *Nat. Biotechnol.*, 2008, **26**, 127.

11. P. Bamborough, D. Drewry, G. Harper, G. K. Smith and K. Schneider, *J. Med. Chem.*, 2008, **51**, 7898.

12. S. L. Posy, M. A. Hermsmeier, W. Vaccaro, K-H. Ott, G. Todderud, J. S. Lippy, G. L. Trainor, D. A. Loughney and S. R. Johnson, *J. Med. Chem.*, 2011, **54**, 54.

13. I. Walker and H. Newell, *Nat. Rev. Drug Discovery*, 2009, **8**, 15.

14. S. Muller and S. Knapp, *Expert Opin. Drug Discovery*, 2010, **5**, 867.

15. See for example:
    a. B. Liegl, I. Kepten, C. Le, M. Zhu, G. D. Demetri, M. C. Heinrich, C. M. D. Fletcher, C. L. Corless and J. A. Fletcher, *J. Pathol.*, 2008, **216**, 64.
    b. D. Bixby and M. Talpaz, *Leukemia*, 2011, **25**, 7.
    c. D. Milojkovic and J. Apperley, *Clin. Cancer Res.*, 2009, **15**, 7519.
    d. J. A. Engelma and P. A. Janne, *Clin. Cancer Res.*, 2008, **14**, 2895.

16. A. L. Gill, M. Verdonk, R. G. Boyle and R. Taylor, *Curr. Top. Med. Chem.*, 2007, **7**, 1408.

17. D. M. Goldstein, A. Kuglstatter, Y. Lou and M. J. Soth, *J. Med. Chem.*, 2010, **53**, 2345.

18. M. Genovese, *Arthritis Rheum.*, 2009, **60**, 317.

19. N. Demanja, R. S. Kauffman and G. T. Spencer-Green, *Arthritis Rheum*, 2009, **60**, 1232.

20. S. E. Tong, S. E. Daniels, T. Nontano, S. Chang and P. Desjardins, *Clin. Pharmacol. Ther.*, 2004, **75**, 3.

21. R. A. Stavenger, H. Cui, S. E. Dowdell, R. G. Franz, D. E. Gaitanopoulos, K. B. Goodman, M. A. Hilfiker, R. L. Ivy, J. D. Leber, J. P. Mariano, Jr., H-J. Oh, A. Q. Viet, W. Xu, G. Ye, D. Zhang, Y. Zhao, L. J. Jolivette, M. S. Head, S. F. Semus, P. A. Elkins, R. B. Kirkpatrick, E. Dul, S. S. Khandekar, T. Yi, D. K. Jung, L. L. Wright, G. K. Smith, D. J. Behm, C. P. Doe, R. Bentley, Z. X. Chen, E. Hu and D. Lee, *J. Med. Chem.*, 2007, **50**, 2.

22. L. M. Wodicka, P. Ciceri, M. I. Davis, J. P. Hunt, M. Floyd, S. Salerno, X. H. Hua, J. M. Ford, R. C. Armstrong, P. P. Zarrinkar and D. K. Treiber, *Chem. Biol.*, 2010, **17**, 1241.

23. H. Ma, S. Deacon and K. Horiuchi, *Expert Opin. Drug Discov*ery, 2008, **3**, 607.

24. M. Bantscheff, D. Eberhard, Y. Abraham, S. Bastuck, M. Boesche, S. Hobson, T. Mathieson, J. Perrin, M. Raida, C. Rau, V. Reader, G. Sweetman, A. Bauer, T. Bouwmeester, C. Hopf, U. Kruse, G. Neubauer, N. Ramsden, J. Rick, B. Kuster and G. Drewes, *Nat. Biotechnol.*, 2007, **25**, 1035.

25. K. Sharma, C. Weber, M. Bairlein, Z. Greff, G. Keri, J. Cox, J.V. Olsen and H. Daub, *Nat. Methods*, 2009, **6**, 741.

26. D. W. Rusnak, K. Lackey, K. Affleck, E. R. Wood, K. J. Alligood, N. Rhodes, B. R. Keith, D. M. Murray, W. B. Knight, R. J. Mullin and T. M. Gilmer, *Mol. Cancer Ther.*, 2001, **1**, 85

27. D. M. Goldstein, N. S. Gray and P. P. Zarrinkar, *Nature Rev. Drug Discov*ery, 2008, **7**, 391.

28. P. P. Graczyk, *J. Med. Chem.*, 2007, **50**, 5773.

29. A. C. Cheng, J. Eksterowicz, S. Geuns-Meyer and Y. Sun, *J. Med. Chem.*, 2010, **53**, 4502.

30. V. Birault, C. J. Harris, J. Le, M. Lipkin, R. Nerella and A. Stevens, *Curr. Med. Chem.*, 2006, **13**, 1735.

31. F. Zuccotto, E. Ardini, E. Casale and M. Angiolini, *J. Med. Chem.*, 2010, **53**, 2681.

32. M. Sawa, *Mini-Rev. Med. Chem.*, 2008, **8**, 1291.

33. A. Ghose, T. Herbertz, D. A. Pippin, J. M. Salvino and J. P. Mallamo, *J Med. Chem.*, 2008, **51**, 5149.

34. J. Liao, *J. Med. Chem.*, 2007, **50**, 409.

35. C. McInnes, S. Wang, S. Anderson, J. O'Boyle, W. Jackson, G. Kontopidis, C. Meades, M. Mezna, M. Thomas, G. Wood, D. P. Lane and P. M. Fischer, *Chem. Biol.*, 2004, **11**, 525.

36. S. Bhattacharya, E. D. Cox, J. C. Kath, A. M. Mathiowetz, J. Morris, J. D. Moyer, L. Pustilnik, K. Radifi, D. T. Richter, C. Su and M. D. Wessel, *Biochem. Biophys. Res. Commun.*, 2003, **307**, 267.

37. D. Heerding, N. Rhodes, J. D. Leber, T. J. Clark, R. M. Keenan, L. V. Lafrance, M. Li, I. G. Safonov, D. T. Takata, J. W. Venslavsky, D. S. Yamashita, A. E. Choudhry, R. A. Copeland, Z. Lai, M. D. Schaber, P. J. Tummino, S. L. Strum, E. R. Wood, D. R. Duckett, D. Eberwein, V. B.

Knick, T. J. Lansing, R. T. McConnell, A. Zhang, E. A. Minthorn, N. O. Concha, G. L. Warren and R. Kumar, *J. Med. Chem.*, 2008, **51**, 5663.

38. M. Semones, Y. Feng, N. Johnson, J. L. Adams, J. Winkler and M. Hansbury, *Bioorg. Med. Chem. Lett.*, 2007, **17**, 4756.

39. M. Kothe, D. Kohls, S. Low, R. Coli, G. R. Rennie, F. Feru, C. Kuhn and Y-H. Ding, *Chem. Biol. Drug. Des.*, 2007, **70**, 540.

Appendix 1.1: Selection of suppliers of commercially available kinase assay for compound screening

Kinase Name	Gene Name	Alias	Group	Family	Sub-family	Millipore Assay	RBC Assay	Ambit Assay	Nanosyn Assay	ProQinase Assay	InVitrogen Assay	Carna Assay	Caliper Assay Kit	Signalchem
A6	PTK9		Atypical	A6										
A6r	PTK9L		Atypical	A6										
AAK1	KIAA1048		Other	NAK				AAK1						
ABL	ABL1		TK	Abl		Abl	ABL1		ABL1	ABL1 wt	ABL1	ABL(ABL1)	ABL	ABL1 (h)
ABL (E255K)							ABL1 (E255K)			ABL1 E255K	ABL1 E255K	ABL(ABL1) [E255K]		
Abl (H396P)						Abl (H396P)	ABL1 (H396P)		ABL (H396P)				Abl (H396P)	
Abl (M351T)						Abl (M351T)	ABL1 (M351T)		ABL (M351T)					
Abl (Q252H)						Abl (Q252H)	ABL1 (Q252H)		ABL (Q252H)				Abl (Q252H)	
Abl (T315I)						Abl (T315I)	ABL1 (T315I)		ABL (T315I)	ABL1 T315I	ABL1 T315I	ABL(ABL1) [T315I]	Abl (T3135I)	
Abl (Y253F)						Abl (Y253F)	ABL1 (Y253F)			ABL1 Y253F	ABL1 Y253F		ABL (Y253F)	
ABL1 (G250E)							ABL1 (G250E)			ABL1 G250E	ABL1 G250E		ABL (G250E)	
ABL1(E255K)-phosph								ABL1(E255K)-phosph					ABL (E255K)	
ABL1(F317I)-nonphosph								ABL1(F317I)-non-phosph						
ABL1(F317I)-phosph								ABL1(F317I)-phosph						
ABL1(F317L)-nonphosph								ABL1(F317L)-nonphosph						
ABL1(F317L)-phosph								ABL1(F317L)-phosph						
ABL1(H396P)-nonphosph								ABL1(H396P)-nonphosph						
ABL1(H396P)-phosph								ABL1(H396P)-phosph						
ABL1(M351T)-phosph								ABL1(M351T)-phosph						
ABL1(Q252H)-nonphosph								ABL1(Q252H)-nonphosph						
ABL1(Q252H)-phosph								ABL1(Q252H)-phosph						
ABL1(T315I)-nonphosph								ABL1(T315I)-non-phosph						
ABL1(T315I)-phosph								ABL1(T315I)-phosph						
ABL1(Y253F)-phosph								ABL1(Y253F)-phosph						
ABL1-nonphosph								ABL1-nonphosph						

Appendix. (Continued)

Kinase Name	Gene Name	Alias	Group	Family	Sub-family	Millipore Assay	RBC Assay	Ambit Assay	Nanosyn Assay	ProQinase Assay	InVitrogen Assay	Carna Assay	Caliper Assay Kit	Signalchem
ABL1-phosph								ABL1-phosph						
ACK	ACK1		TK	Ack		ACK1	ACK1	TNK2	ACK (TNK2)	ACK1	TNK2 (ACK)	ACK(TNK2)		ACK
ACTR2	ACVR2		TKL	STKR	Type2			ACVR2A						
ACTR2B	ACVR2B		TKL	STKR	Type2		ACVR2B	ACVR2B			ACVR2B			
ADCK1	LOC57143		Atypical	ABC1	ABC1-B									
ADCK2	MGC20727		Atypical	ABC1	ABC1-C									
ADCK3	CABC1		Atypical	ABC1	ABC1-A			ADCK3						
ADCK4	FLJ12229		Atypical	ABC1	ABC1-A			ADCK4						
ADCK5	TIG43574		Atypical	ABC1	ABC1-B									
ADRBK1 (GRK2)											ADRBK1 (GRK2)			GRK2
ADRBK2 (GRK3)											ADRBK2 (GRK3)			
AKT1	AKT1		AGC	AKT		PKBá	AKT1	AKT1	AKT1	AKT1	AKT1 (PKB alpha)	AKT1	AKT1	AKT1
AKT1ÄPH									AKT1ÄPH					
AKT2	AKT2		AGC	AKT		PKBâ	AKT2	AKT2	AKT2	AKT2	AKT2 (PKB beta)	AKT2	AKT2	AKT2 (His)
AKT2ÄPH									AKT2ÄPH					
AKT2 (GST)														AKT2 (GST)
AKT3	AKT3		AGC	AKT		PKBä	AKT3	AKT3	AKT3	AKT3	AKT3 (PKB gamma)	AKT3	Akt3	AKT3
AKT3ÄPH									AKT3ÄPH					
ALK	ALK		TK	Alk	RTK	ALK	ALK	ALK	ALK	ALK	ALK	ALK	ALK	
ALK [F1174L]												ALK [F1174L]		
ALK [R1275Q]												ALK [R1275Q]		
ALK1	ACVRL1		TKL	STKR	Type1		ALK1/ACVRL1	ACVRL1		ACV-RL1				ALK1
ALK2	ACVR1		TKL	STKR	Type1		ALK2/ACVR1	ACVR1		ACV-R1	ACVR1 (ALK2)			ALK2
ALK4	ACVR1B		TKL	STKR	Type1	ALK4	ALK4/ACVR1B	ACVR1B		ACV-R1B	ACVR1B (ALK4)			ALK4
ALK5							ALK5/TGFBR1							
ALK7	tar7690		TKL	STKR	Type1		ALK7/ACVR1C Casein(mg/ml) +Mn ~2% at 200nM Rxn 300nM							
ALK-EML4												EML4-ALK		
ALK-NPM1												NPM1-ALK		
AlphaK1	MIDORI		Atypical	Alpha										
AlphaK2	HAK		Atypical	Alpha										
AlphaK3	FLJ22670		Atypical	Alpha										
AMPKa1														
AMPK(A1/B1/G1)	PRKAA1		CAMK	CAMKL	AMPK	AMPK (á1, âl, ä1)	AMPK(A1/B1/G1)		AMPK A1/B1/G1		AMPK A1/B1/G1	AMPKá1/â1/ä1(PRKAA1/B1/G1)	AMPKa1	AMPK (A1/B1/G1)

Appendix. *(Continued)*

Kinase Name	Gene Name	Alias	Group	Family	Sub-family	Millipore Assay	RBC Assay	Ambit Assay	Nanosyn Assay	ProQinase Assay	InVitrogen Assay	Carna Assay	Caliper Assay Kit	Signalchem
AMPK(A1/B1/G2)							AMPK(A1/B1/G2)							AMPK (A1/B1/G2)
AMPK(A1/B1/G3)							AMPK(A1/B1/G3)							AMPK (A1/B1/G3)
AMPK(A1/B2/G1)							AMPK(A1/B2/G1)							AMPK (A1/B2/G1)
AMPK(A2/B1/G1)						AMPK (á2, â1, â1)	AMPK(A2/B1/G1)		AMPK A2/B1/G1		AMPK A2/B1/G1	AMPKá2/â1/â1(PRKAA2/B1/G1)	AMPK-a2/ß1/g1	AMPK (A2/B1/G1)
AMPK(A2/B2/G1)							AMPK(A2/B2/G1)							AMPK (A2/B2/G1)
AMPKa2	PRKAA2		CAMK	CAMKL	AMPK			AMPK-alpha2						
AMPK (A2/B2/G2)														AMPK (A2/B2/G2)
AMPK (A2/B2/G3)														AMPK (A2/B2/G3)
ANKK1	LOC255239	SgK288	TKL	RIPK				ANKK1						
ANKRD3	ANKRD3		TKL	RIPK				RIPK4						
ANPa	NPR1		RGC	RGC										
ANPb	NPR2		RGC	RGC										
ARAF	ARAF1		TKL	RAF			ARAF							
ARG	ABL2		TK	Abl		Arg	ARG	ABL2	ARG	ABL2	ABL2 (Arg)	ARG(ABL2)	Arg	ABL2
ATM	ATM		Atypical	PIKK	ATM									
ATR	ATR		Atypical	PIKK	ATR									
AurA	STK15	STK6	Other	AUR		Aurora-A	Aurora A	AURKA	AURORA-A	Aurora-A	AURKA (Aurora A)	AurA(AURKA)	AurA	AURORA A (h)
AurA(AURKA)/TPX2												AurA(AURK-A)/TPX2		
AurB	STK12		Other	AUR		Aurora-B	Aurora B	AURKB	AURORA-B	Aurora-B	AURKB (Aurora B)		AurB	AURORA B
AurB(AURKB)/INCENP												AurB(AURKB)/INCENP		
AurC	STK13	Aur 3	Other	AUR			Aurora C	AURKC	AURORA-C	Aurora-C	AURKC (Aurora C)	AurC(AURKC)	AurC	AURORA C
AXL	AXL		TK	Axl	RTK	Axl	AXL	AXL	AXL	AXL	AXL	AXL	AXL	AXL
BARK1	ADRBK1	GRK2	AGC	GRK	BARK		GRK2			GRK2				
BARK2	ADRBK2	GRK3	AGC	GRK	BARK		GRK3			GRK3				
BCKDK	BCKDK		Atypical	PDHK										
BCR	BCR		Atypical	BCR										
BIKE	BIKE		Other	NAK				BIKE						
BLK	BLK		TK	Src		Blk (mouse)	BLK	BLK	BLK	BLK	BLK	BLK	BLK	BLK
BMPR1A	BMPR1A	Alk3	TKL	STKR	Type1		ALK3	BMPR1A		BMPR1A	BMPR1A (ALK3)	BMPR1A		

Appendix. (*Continued*)

Kinase Name	Gene Name	Alias	Group	Family	Sub-family	Millipore Assay	RBC Assay	Ambit Assay	Nanosyn Assay	ProQinase Assay	InVitrogen Assay	Carna Assay	Caliper Assay Kit	Signalchem
BMPR1B	BMPR1B	Alk6	TKL	STKR	Type1			BMPR1B						
BMPR2	BMPR2		TKL	STKR	Type2			BMPR2						
BMX	BMX		TK	Tec		Bmx	BMX/ETK	BMX	BMX	BMX	BMX	BMX	BMX	BMX
BRAF	BRAF		TKL	RAF			BRAF	BRAF	BRAF	B-RAF wt	BRAF	BRAF		
BRAF (V599E)							BRAF (V599E)		BRAF (V599E)		BRAF V599E			BRAF
BRAF(V600E)								BRAF(V600E)		B-RAF V600E		BRAF [V600E]		BRAF (V599E)
BRD2	BRD2		Atypical	BRD										
BRD3	BRD3		Atypical	BRD										
BRD4	BRD4		Atypical	BRD										
BRDT	BRDT		Atypical	BRD										
BRK	PTK6		TK	Src		BRK	BRK	BRK	PTK6 (BRK)	BRK	PTK6 (Brk)	BRK(PTK6)		BRK
BRSK1	tar7716		CAMK	CAMKL	BRSK	BrSK1	BRSK1	BRSK1	BRSK1	BRSK1	BRSK1 (SAD1)	BRSK1	BRSK1	
BRSK2	STK29		CAMK	CAMKL	BRSK	BrSK2	BRSK2	BRSK2	BRSK2			BRSK2	BRSK2	
BTK	BTK		TK	Tec		BTK	BTK	BTK	BTK	BTK	BTK	BTK	BTK	BTK
BTK (E41K)							BTK (E41K)							
BTK (R28H)						BTK (R28H)								
BUB1	BUB1		Other	BUB										
BUBR1	BUB1B		Other	BUB										
CaMK1a	CAMK1		CAMK	CAMK1		CamKI	CAMK1a	CAMK1			CAMK1 (CaMK1)	CaMK1á(CAM-K1)	CaMK1a	CAMK1
CaMK1b	TIG32230		CAMK	CAMK1			CAMK1b							CAMK1â
CaMK1d	LOC57118		CAMK	CAMK1		CaMKId	CAMK1d	CAMK1D	CaMK1ä	CAMK1D	CAMK1D (CaMKI delta)	CaMK1ä(CAM-K1D)	CaMK1d	CAMK1ä
CaMK1g	DJ272-L16.1		CAMK	CAMK1			CAMK1g	CAMK1G						CAMK1ä (1-330)
CAMK1ä (Full)														CAMK1ä (Full)
CaMK2a	CAMK2A		CAMK	CAMK2			CAMK2a	CAMK2A	CaMK2á	CAMK2A	CAMK2A (CaMKII alpha)	CaMK2á(CAM-K2A)	CaMK2a	CAMK2á
CaMK2b	CAMK2B		CAMK	CAMK2		CaMKII â	CAMK2b	CAMK2B	CaMK2â	CAMK2B	CAMK2B (CaMKII beta)	CaMK2â(CAM-K2B)	CaMK2b	CAMK2â
CaMK2d	CAMK2D		CAMK	CAMK2		CaMKII ä	CAMK2d	CAMK2D	CaMK2ä	CAMK2D	CAMK2D (CaMKII delta)	CaMK2ä(CAM-K2D)	CaMK2d	CAMK2ä
CaMK2g	CAMK2G		CAMK	CAMK2		CaMKII ä	CAMK2g	CAMK2G				CaMK2ä(CAM-K2G)	CaMK2g	CAMK2ä
CaMK4	CAMK4		CAMK	CAMK1		CaMKIV	CAMK4	CAMK4	CaMK4	CAMK4	CAMK4 (CaMKIV)	CaMK4	CaMK4	CAMK4
CaMKK1	DKFZ-p761M0423		Other	CAMKK	Meta		CAMKK1	CAMKK1		CAMKK1	CAMKK1 (CAMKKA)			CAMKK1
CaMKK2	CAMKK2		Other	CAMKK	Meta		CAMKK2	CAMKK2		CAMKK2	CAMKK2 (CaMKK beta)			CAMKK2

Appendix. (*Continued*)

Kinase Name	Gene Name	Alias	Group	Family	Sub-family	Millipore Assay	RBC Assay	Ambit Assay	Nanosyn Assay	ProQinase Assay	InVitrogen Assay	Carna Assay	Caliper Assay Kit	Signalchem
caMLCK	LOC91807		CAMK	MLCK			MYLK3	MLCK		MYLK3	MLCK (MLCK2/ MYLK3)			MYLK3
CASK	CASK		CAMK	CASK				CASK						
CCK4	PTK7		TK	CCK4	RTK									
CCRK	CCRK		CMGC	CDK										
CDC2L2								CDC2L2						
CDC7	CDC7L1		Other	CDC7								CDC7/ASK		
CDK1/CycE										CDK1/CycE				
CDK1/cyclin A	CDC2	CDK1	CMGC	CDK	CDC2		CDK1/cyclin A			CDK1/CycA				CDK1/CyclinA1
CDK1/CyclinA2														CDK1/CyclinA2
CDK1/cyclinB	CDC2	CDK1	CMGC	CDK	CDC2	CDK1/cyclinB	CDK1/cyclin B		CDK1/cyclinB	CDK1/CycB1	CDK1/cyclin B	CDC2/CycB1	CDC2/Cycline B1	CDK1/CyclinB1
CDK10	CDK10		CMGC	CDK	CDK10									
CDK11	KIAA1028	PITSLRE	CMGC	CDK	CDK8			CDK11						
CDK2	CDK2		CMGC	CDK	CDC2			CDK2					CDK2	CDK2/CyclinA1
CDK2/cyclinA						CDK2/cyclinA	CDK2/cyclin A		CDK2/cyclinA	CDK2/CycA	CDK2/cyclin A	CDK2/CycA2		CDK2/CyclinA2
CDK2/CyclinA2														CDK2/CyclinE1
CDK2/cyclinE						CDK2/cyclinE	CDK2/cyclin E		CDK2/cyclinE	CDK2/CycE		CDK2/CycE1		
CDK3	CDK3		CMGC	CDK	CDC2			CDK3					CDK3	
CDK3/cyclinE						CDK3/cyclinE	CDK3/cyclin E		CDK3/cyclinE	CDK3/CycE		CDK3/CycE1		CDK3/CyclinE1
CDK4/cyclin D3							CDK4/cyclin D3	CDK4-cyclinD3		CDK4/CycD3		CDK4/CycD3		CDK4/CyclinD3
CDK4/cyclinD	CDK4		CMGC	CDK	CDK4		CDK4/cyclin D1	CDK4-cyclinD1	CDK4/cyclinD	CDK4/CycD1				CDK4/CyclinD1
CDK5	CDK5		CMGC	CDK	CDK5			CDK5						
cdk5/p25						cdk5/p25	CDK5/p25			CDK5/p25NCK	CDK5/p25	CDK5/p25	CDK5/p25	CDK5/p25
CDK5/p29														CDK5/p29
CDK5/p35						CDK5/p35	CDK5/p35		CDK5/p35	CDK5/p35NCK	CDK5/p35			CDK5/p35
CDK6	CDK6		CMGC	CDK	CDK4				CDK6					
CDK6/cyclin D1							CDK6/cyclin D1			CDK6/CycD1				CDK6/CyclinD1
CDK6/cyclinD3						CDK6/cyclinD3	CDK6/cyclin D3					CDK6/CycD3		CDK6/CyclinD3
CDK7	CDK7		CMGC	CDK	CDK7			CDK7						
CDK7/cyclinH/MAT1						CDK7/cyclinH/ MAT1	CDK7/cyclin H		CDK7?Cyclin H?MNAT1	CDK7/CycH/ MAT1	CDK7/cyclin H/ MNAT1	CDK7/CycH/ MAT1		CDK7/CyclinH/ MNAT1
CDK8	CDK8		CMGC	CDK	CDK8			CDK8						
CDK8/cyclin C							CDK8/cyclin C			CDK8/CycC	CDK8/cyclin C			CDK8/CyclinC
CDK9	CDK9		CMGC	CDK	CDK9			CDK9						
CDK9/cyclin K							CDK9/cyclin K			CDK9/CycK	CDK9/cyclin K			CDK9/CyclinK

Appendix. (*Continued*)

Kinase Name	Gene Name	Alias	Group	Family	Sub-family	Millipore Assay	RBC Assay	Ambit Assay	Nanosyn Assay	ProQinase Assay	InVitrogen Assay	Carna Assay	Caliper Assay Kit	Signalchem
CDK9/CycT						CDK9/cyclinT1	CDK9/cyclin T1		CDK9/Cyclin T1	CDK9/CycT	CDK9/cyclin T1	CDK9/CycT1		
CDKL1	CDKL1		CMGC	CDKL					CDKL1					
CDKL2	CDKL2		CMGC	CDKL					CDKL2					
CDKL3	CDKL3		CMGC	CDKL					CDKL3					
CDKL4	TIG35919		CMGC	CDKL										
ChaK1	TRPM7_v1		Atypical	Alpha	ChaK									
ChaK2	TRPM6		Atypical	Alpha	ChaK				TRPM6					
CHED	CDC2L5		CMGC	CDK	CRK7				CDC2L5					
CHK1	CHEK1		CAMK	CAMKL	CHK1	CHK1	CHK1	CHEK1	CHEK1	CHK1	CHEK1 (CHK1)	CHK1(CHEK1)	CHK1	CHK1 (GST)
CHK1 (His)														
CHK2	CHEK2		CAMK	RAD53		CHK2	CHK2	CHEK2	CHEK2	CHK2	CHEK2 (CHK2)	CHK2(CHEK2)	CHK2	CHK2
CHK2 (I157T)						CHK2 (I157T)	CHK2 (I157T)							
CHK2 (R145W)						CHK2 (R145W)								
CK1a	CSNK1A1		CK1	CK1		CK1	CK1a1	CSNK1A1	CK1á	CK1-alpha1	CSNK1A1 (CK1 alpha 1)	CK1á(CSNK1-A1)	CSNK1A1	
CK1a2	tar7724		CK1	CK1				CSNK1A1L						
CK1d	CSNK1D		CK1	CK1		CK1ä	CK1d	CSNK1D		CK1-delta	CSNK1D (CK1 delta)	CK1ä(CSNK1-D)	CK1d	
CK1e	CSNK1E		CK1	CK1			CK1epsilon	CSNK1E		CK1-epsilon	CSNK1E (CK1 epsilon)	CK1å(CSNK1-E)	CK1e	
CK1gl	CSNK1G1		CK1	CK1		CK1ä1	CK1gl	CSNK1G1	CK1gl	CK1-gamma1	CSNK1G1 (CK1 gamma 1)	CK1ä1(CSNK1-G1)	CK1-ä1 (CSNK1G1)	
CK1g2	CSNK1G2		CK1	CK1		CK1ä2	CK1g2	CSNK1G2	CK1g2	CK1-gamma2	CSNK1G2 (CK1 gamma 2)	CK1ä2(CSNK1-G2)	Casein kinase 1g2 (CK1g2)	
CK1g3	CSNK1G3		CK1	CK1		CK1ä3	CK1g3	CSNK1G3	CK1g3	CK1-gamma3	CSNK1G3 (CK1 gamma 3)	CK1ä3(CSNK1-G3)	CK1g3 (CSNK1G3)	
CK2a1	CSNK2A1		Other	CK2		CK2	CK2a	CSNK2A1	CK2	CK2-alpha1	CSNK2A1 (CK2 alpha 1)	CK2á1/b(CSNK2A1/B)		CK2á1
CK2a2	CSNK2A2		Other	CK2		CK2á2	CK2a2	CSNK2A2		CK2-alpha2	CSNK2A2 (CK2 alpha 2)	CK2á2/b(CSNK2A2/B)		CK2á2
CLIK1	STK35		Other	NKF4				STK35						
CLIK1L	tar7705		Other	NKF4										
CLK1	CLK1		CMGC	CLK			CLK1	CLK1	CLK1	CLK1	CLK1	CLK1		CLK1
CLK2	CLK2		CMGC	CLK		CLK2	CLK2	CLK2	CLK2	CLK2	CLK2	CLK2	CLK2	CLK2
CLK3	CLK3		CMGC	CLK		CLK3	CLK3	CLK3	CLK3	CLK3	CLK3	CLK3		CLK3
CLK4	CLK4		CMGC	CLK			CLK4	CLK4		CLK4	CLK4			
COT	MAP3K8		STE	STE-Unique			COT1/MAP3K8			COT	MAP3K8 (COT)	COT(MAP3K8)		COT
CRIK	CIT	Citron Kinase	AGC	DMPK				CIT				CRIK(CIT)		
CRK7	CrkRS		CMGC	CDK	CRK7									

Appendix. *(Continued)*

Kinase Name	Gene Name	Alias	Group	Family	Sub-family	Millipore Assay	RBC Assay	Ambit Assay	Nanosyn Assay	ProQinase Assay	InVitrogen Assay	Carna Assay	Caliper Assay Kit	Signalchem
CSK	CSK		TK	Csk		CSK	CSK	CSK	CSK	CSK	CSK	CSK		CSK (untag)
CSK (GST)														CSK (GST)
Src (T341M)						Src (T341M)	c-Src (T341M)							
CTK	MATK	CHK	TK	Csk			CTK/MATK	CTK	CTK (MATK)	MATK	MATK (HYL)	CTK(MATK)		
CYGD	GUCY2D		RGC	RGC										
CYGF	GUCY2F		RGC	RGC										
DAPK1	DAPK1		CAMK	DAPK		DAPK1	DAPK1	DAPK1	DAPK1	DAPK1	DAPK1	DAPK1	DAPK1	DAPK1
DAPK2	DAPK2		CAMK	DAPK		DAPK2	DAPK2	DAPK2		DAPK2				DAPK2
DAPK3	DAPK3		CAMK	DAPK		ZIPK	ZIPK/DAPK3	DAPK3	ZIPK (DAPK3)	DAPK3	DAPK3 (ZIPK)		ZIPK (DAPK3)	DAPK3
DCAMKL1	DCAMKL1		CAMK	DCAM-KL				DCAMKL1					DCaMKL1	DCAMKL1
DCAMKL2	tar7698		CAMK	DCAM-KL		DCAMKL2	DCAMKL2	DCAMKL2	DCAMKL2	DCAMKL2	DCAMKL2 (DCK2)	DCAMKL2	DCaMKL2	DCAMKL2
DCAMKL3	KIAA1765		CAMK	DCAM-KL				DCAMKL3						
DDR1	DDR1		TK	DDR	RTK		DDR1	DDR1	DDR1		DDR1	DDR1		DDR1
DDR2	DDR2		TK	DDR	RTK	DDR2	DDR2	DDR2	DDR2	DDR2	DDR2	DDR2	DDR2	DDR2
DLK	MAP3K12		TKL	MLK	LZK		DLK/ MAP3K12 MBP ~4% at 200nM. Rxn 200 nM	DLK				DLK(MAP3-K12)		
DMPK1	DMPK		AGC	DMPK	GEK	DMPK	DMPK	DMPK		DMPK	DMPK			DMPK
DMPK2	tar7733		AGC	DMPK	GEK			DMPK2	DMPK2					
DNAPK	PRKDC		Atypical	PIKK	DNAPK	DNAPK	DNA-PK			DNA-PK	DNA-PK			
Domain2_GCN2	GCN2		STE	STE-Unique										
Domain2_JAK1	JAK1		TK	JakB										
Domain2_JAK2	JAK2		TK	JakB										
Domain2_JAK3	JAK3		TK	JakB										
Domain2_MSK1	MSK1		CAMK	RSKb	MSKb									
Domain2_MSK2	RPS6KA4		CAMK	RSKb	MSKb									
Domain2_Obscn	tar7694		CAMK	Trio										
Domain2_RSK1	RPS6KA2		CAMK	RSKb	RSKb									
Domain2_RSK2	RPS6KA3		CAMK	RSKb	RSKb									
Domain2_RSK3	RPS6KA1		CAMK	RSKb	RSKb									
Domain2_RSK4	RPS6KA6		CAMK	RSKb	RSKb									
Domain2_SPEG	tar7777		CAMK	Trio										
Domain2_TYK2	TYK2		TK	JakB										
DRAK1	STK17A		CAMK	DAPK		DRAK1	DRAK1/STK17A	DRAK1		STK17A	STK17A (DRAK1)			
DRAK2	STK17B		CAMK	DAPK				DRAK2						

Appendix. (*Continued*)

Kinase Name	Gene Name	Alias	Group	Family	Sub-family	Millipore Assay	RBC Assay	Ambit Assay	Nanosyn Assay	ProQinase Assay	InVitrogen Assay	Carna Assay	Caliper Assay Kit	Signalchem
DYRK1A	tar7715		CMGC	DYRK	Dyrk1		DYRK1/ DYRK1A	DYRK1A	DYRK1A	DYRK1A	DYRK1A	DYRK1A	DYRK1a	DYRK1á
DYRK1B	DYRK1B	Mirk	CMGC	DYRK	Dyrk1		DYRK1B	DYRK1B	DYRK1B	DYRK1B	DYRK1B	DYRK1B	DYRK1B	
DYRK2	DYRK2	Yak1	CMGC	DYRK	Dyrk2	DYRK2	DYRK2	DYRK2	DYRK2			DYRK2		
DYRK3	DYRK3	Yak3	CMGC	DYRK	Dyrk2		DYRK3	DYRK3	DYRK3	DYRK3	DYRK3	DYRK3	DYRK3	DYRK3
DYRK4	DYRK4		CMGC	DYRK	Dyrk2		DYRK4			DYRK4	DYRK4		DYRK4	
eEF2K	HSU93850		Atypical	Alpha	eEF2K	eEF-2K	EEF2K		EEF2K	EEF2K	EEF2K	EEF2K		EEF2K
EGFR	EGFR		TK	EGFR	RTK	EGFR	EGFR	EGFR	EGFR	EGF-R wt	EGFR (ErbB1)	EGFR	EGFR	EGFR
EGFR (L858R)						EGFR (L858R)	EGFR (L858R)	EGFR(L858R)	EGFR (L858R)	EGF-R L858R	EGFR (ErbB1) L858R	EGFR [L858R]		
EGFR (L861Q)						EGFR (L861Q)	EGFR (L861Q)	EGFR(L861Q)	EGFR (L861Q)	EGF-R L861Q	EGFR (ErbB1) L861Q	EGFR [L861Q]		
EGFR (T790M)						EGFR (T790M)	EGFR (T790M)	EGFR(T790M)	EGFR (T790M)	EGF-R T790M	EGFR (ErbB1) T790M	EGFR [T790M]	EGFR (T790M)	
EGFR (T790M, L858R)						EGFR (T790M, L858R)	EGFR (T790M, L858R)		EGFR (T790M L858R)	EGF-R T790M/ L858R	EGFR (ErbB1) T790M L858R	EGFR [T790M/ L858R]	EGFR (T790M, L858R)	
EGFR [d746-750/T790M]												EGFR [d746-750/T790M]		
EGFR [d746-750]												EGFR [d746-750]		
EGFR(E746-A750del)								EGFR(E746-A750del)						
EGFR(G719C)								EGFR(G719C)						
EGFR(G719S)								EGFR(G719S)						
EGFR(L747-E749del, A750P)								EGFR(L747-E749del, A750P)						
EGFR(L747-S752del, P753S)								EGFR(L747-S752del, P753S)						
EGFR(L747-T751del,Sins)								EGFR(L747-T751del,Sins)						
EGFR(L858R,T790M)								EGFR(L858R,-T790M)						
EGFR(S752-I759del)								EGFR(S752-I759del)						
EphA1	EPHA1		TK	Eph	RTK	EphA1	EPHA1	EPHA1	EPHA1	EPHA1	EPHA1	EPHA1	EPHA1	EPHA1 (h)
EphA10	FLJ33655		TK	Eph	RTK									
EphA2	EPHA2		TK	Eph	RTK	EphA2	EPHA2	EPHA2	EPHA2	EPHA2	EPHA2	EPHA2	EPHA2	EPHA2
EphA3	EPHA3		TK	Eph	RTK	EphA3	EPHA3	EPHA3	EPHA3	EPHA3	EPHA3	EPHA3		EPHA3
EphA4	EPHA4		TK	Eph	RTK	EphA4	EPHA4	EPHA4	EPHA4	EPHA4	EPHA4	EPHA4	EPHA4	EPHA4
EphA5	EPHA5		TK	Eph	RTK	EphA5	EPHA5	EPHA5	EPHA5	EPHA5	EPHA5	EPHA5	EPHA5	
EphA6	EPHA6		TK	Eph	RTK		EPHA6	EPHA6				EPHA6		EPHA6

Appendix. *(Continued)*

Kinase Name	Gene Name	Alias	Group	Family	Sub-family	Millipore Assay	RBC Assay	Ambit Assay	Nanosyn Assay	ProQinase Assay	InVitrogen Assay	Carna Assay	Caliper Assay Kit	Signalchem
EphA7	EPHA7		TK	Eph	RTK	EphA7	EPHA7	EPHA7	EPHA7	EPHA7	EPHA7	EPHA7		
EphA8	EPHA8		TK	Eph	RTK	EphA8	EPHA8	EPHA8	EPHA8	EPHA8	EPHA8	EPHA8	EPHA8	
EphB1	EPHB1		TK	Eph	RTK	EphB1	EPHB1	EPHB1	EPHB1	EPHB1	EPHB1	EPHB1	EPHB1	EPHB1
EphB2	EPHB2		TK	Eph	RTK	EphB2	EPHB2	EPHB2	EPHB2	EPHB2	EPHB2	EPHB2	EPHB2	EPHB2
EphB3	EPHB3		TK	Eph	RTK	EphB3	EPHB3	EPHB3	EPHB3	EPHB3	EPHB3	EPHB3	EPHB3	EPHB3
EPHB4	EPHB4		TK	Eph	RTK	EphB4	EPHB4	EPHB4	EPHB4	EPHB4	EPHB4	EPHB4	EPHB4	EPHB4
EphB6	EPHB6		TK	Eph	RTK			EPHB6						
Erk1	MAPK3	p44, MAPK1	CMGC	MAPK	ERK	MAPK1	ERK1/MAPK3	ERK1	MAPK3 (ERK1)	ERK1	MAPK3 (ERK1)	Erk1(MAPK3)	Erk1	ERK1
Erk2	MAPK1	p42, MAPK2	CMGC	MAPK	ERK	MAPK2	ERK2/MAPK1	ERK2	MAPK1 (ERK2)	ERK2	MAPK1 (ERK2)	Erk2(MAPK1)	Erk2	ERK2
Erk3	MAPK6		CMGC	MAPK	ERK			ERK3						
Erk4	MAPK4		CMGC	MAPK	ERK			ERK4						
Erk5	MAPK7		CMGC	MAPK	ERK			ERK5				Erk5(MAPK7)		
Erk7	tar7702		CMGC	MAPK	Erk7									
ERK8		MAPK15						ERK8						
FAK	PTK2		TK	Fak		FAK	FAK/PTK2	FAK	FAK (PTK2)	FAK	PTK2 (FAK)	FAK(PTK2)		FAK
FASTK	FASTK		Atypical	FAST										
FER	FER		TK	Fer		Fer	FER	FER	FER	FER	FER	FER	FER	FER
FES	FES		TK	Fer		Fes	FES/FPS	FES	FES	FES	FES (FPS)	FES	FES	FES
FGFR1	FGFR1		TK	FGFR	RTK	FGFR1	FGFR1	FGFR1	FGFR1	FGF-R1 wt	FGFR1	FGFR1	FGFR1	FGFR1 (FLT2)
FGFR1 (V561M)						FGFR1 (V561M)	FGFR1 (V561M)			FGF-R1 V561M			FGFR1 (V561M)	FGFR1 (V561M)
FGFR2	FGFR2		TK	FGFR	RTK	FGFR2	FGFR2	FGFR2	FGFR2	FGF-R2	FGFR2	FGFR2	FGFR2	FGFR2
FGFR2 (N549H)						FGFR2 (N549H)	FGFR2 (N549H)		FGFR2 (N549H)				FGFR2 (N549H)	
FGFR3	FGFR3		TK	FGFR	RTK	FGFR3	FGFR3	FGFR3	FGFR3	FGF-R3 wt	FGFR3	FGFR3	FGFR3	FGFR3
FGFR3 (K650E)							FGFR3 (K650E)		FGFR3 (K650E)	FGF-R3 K650E	FGFR3 K650E	FGFR3 [K650E]	FGFR3 (K650E)	
FGFR3 [K650M]												FGFR3 [K650M]		
FGFR3(G697C)								FGFR3(G697C)						
FGFR4	FGFR4		TK	FGFR	RTK	FGFR4	FGFR4	FGFR4	FGFR4	FGF-R4	FGFR4	FGFR4	FGFR4	FGFR4
FGFR4 [N535K]												FGFR4 [N535K]		
FGFR4 [V550E]												FGFR4 [V550E]		
FGFR4 [V550L]												FGFR4 [V550L]		
FGR	FGR		TK	Src		Fgr	FGR	FGR	FGR	FGR	FGR	FGR	FGR	FGR
FLT1	FLT1	VEGFR1	TK	VEGFR	RTK	Flt1	FLT1/VEGFR1	FLT1	FLT-1	VEGF-R1	FLT1 (VEGFR1)	FLT1	FLT1	FLT1
FLT3	FLT3		TK	PDGFR	RTK	Flt3	FLT3	FLT3	FLT-3	FLT3 wt	FLT3	FLT3	FLT3	FLT3

Appendix. (*Continued*)

Kinase Name	Gene Name	Alias	Group	Family	Sub-family	Millipore Assay	RBC Assay	Ambit Assay	Nanosyn Assay	ProQinase Assay	InVitrogen Assay	Carna Assay	Caliper Assay Kit	Signalchem
Flt3 (D835Y)						Flt3 (D835Y)	FLT3 (D835Y)	FLT3(D835Y)	FLT-3-D835Y	FLT3 D835Y	FLT3 D835Y		Flt3 (D835Y)	
FLT3(D835H)								FLT3(D835H)						
FLT3(ITD)								FLT3(ITD)		FLT3 ITD				
FLT3(K663Q)								FLT3(K663Q)						
FLT3(N841I)								FLT3(N841I)						
FLT3(R834Q)								FLT3(R834Q)						
FLT4	FLT4	VEGFR3	TK	VEGFR	RTK	Flt4	FLT4/VEGFR3	FLT4	FLT-4	VEGF-R3	FLT4 (VEGFR3)	FLT4	FLT4	
FMS	CSF1R		TK	PDGFR	RTK	Fms	FMS	CSF1R	FMS (CSFR)	CSF1-R	CSF1R (FMS)	FMS(CSF1R)	FMS (CSF1R)	FMS
Fms (Y969C)						Fms (Y969C)								
FRAP	mTOR		Atypical	PIKK	FRAP	mTOR	mTOR/FRAP1	MTOR		mTOR	FRAP1 (mTOR)			
FRAP (mTOR/FKBP12)						mTOR/FKBP12								
FRK	FRK		TK	Src		PTK5	FRK/PTK5	FRK	PTK5	FRK	FRK (PTK5)	FRK	FRK	FRK
Fused	KIAA1278	STK36	Other	ULK				STK36						
FYN	FYN		TK	Src		Fyn	FYN	FYN	FYN	FYN	FYN	FYN	FYN	FYN A
G11	STK19		Atypical	G11										
GAK	GAK		Other	NAK				GAK						
GCK	MAP4K2		STE	STE20	KHS	GCK	GCK/MAP4K2	MAP4K2	GCK (MAP4K2)	MAP4K2	MAP4K2 (GCK)		GCK (MAP4K2)	GCK
GCN2	GCN2		Other	PEK	GCN2			GCN2(Kin.Dom.2, S808G)						
GPRK4	GPRK2L		AGC	GRK	GRK		GRK4	GRK4		GRK4	GRK4			
GPRK5	GPRK5		AGC	GRK	GRK	GRK5	GRK5			GRK5	GRK5			GRK5
GPRK6	GPRK6		AGC	GRK	GRK	GRK6	GRK6			GRK6	GRK6			GRK6
GPRK7	tar7722		AGC	GRK	GRK	GRK7	GRK7	GRK7		GRK7	GRK7			GRK7
GSK3A	GSK3A		CMGC	GSK		GSK3á	GSK3a	GSK3A	GSK3á	GSK3-alpha	GSK3A (GSK3 alpha)	GSK3á(GSK3-A)	GSK3a	GSK3á
GSK3B	GSK3B		CMGC	GSK		GSK3â	GSK3b	GSK3B	GSK3â	GSK3-beta	GSK3B (GSK3 beta)	GSK3á(GSK3-B)	GSK3ß	GSK3á
H11	E2IG1		Atypical	H11										
Haspin	GSG2		Other	Haspin		haspin	Haspin		Haspin (GSG2)	GSG2	GSG2 (Haspin)	Haspin(GSG2)		
HCK	HCK		TK	Src		Hck	HCK	HCK	HCK	HCK	HCK	HCK	HCK	HCK
Hck, activated						Hck, activated								
HER2/ErbB2	ERBB2		TK	EGFR	RTK		ERBB2/HER2	ERBB2	ERBB2	ERBB2	ERBB2 (HER2)	HER2(ERBB2)		HER2
HER3/ErbB3	ERBB3		TK	EGFR	RTK			ERBB3						
HER4/ErbB4	ERBB4		TK	EGFR	RTK	ErbB4	ERBB4/HER4	ERBB4	ERBB4 (HER4)	ERBB4	ERBB4 (HER4)	HER4(ERBB4)	HER4 (ERBB4)	HER4
HH498	LOC51086	TNNI3K, CARK	TKL	MLK	HH498	HH498			TNNI3K					
HIPK1	Nbak2		CMGC	DYRK	HIPK	HIPK1	HIPK1	HIPK1	HIPK1	HIPK1	HIPK1 (Myak)	HIPK1	HIPK1	HIPK1

Appendix. (Continued)

Kinase Name	Gene Name	Alias	Group	Family	Sub-family	Millipore Assay	RBC Assay	Ambit Assay	Nanosyn Assay	ProQinase Assay	InVitrogen Assay	Carna Assay	Caliper Assay Kit	Signalchem
HIPK2	HIPK2		CMGC	DYRK	HIPK	HIPK2	HIPK2	HIPK2	HIPK2	HIPK2	HIPK2	HIPK2	HIPK2	
HIPK3	HIPK3		CMGC	DYRK	HIPK	HIPK3	HIPK3	HIPK3	HIPK3	HIPK3	HIPK3 (YAK1)	HIPK3		HIPK3
HIPK4	tar7725		CMGC	DYRK	HIPK		HIPK4	HIPK4	HIPK4	HIPK4	HIPK4	HIPK4		HIPK4
HPK1	MAP4K1		STE	STE20	KHS			HPK1						
HRI	HRI		Other	PEK			EIF2AK1/HRI	EIF2AK1		HRI				
HSER	GUCY2C		RGC	RGC										
HUNK	HUNK		CAMK	CAMKL	HUNK			HUNK						
ICK	KIAA0936		CMGC	RCK				ICK						ICK
IGF1R	IGF1R		TK	InsR	RTK	IGF-1R	IGF1R	IGF1R	IGF1R	IGF1-R	IGF1R	IGF1R	IGF1R	IGF1R
IGF-1R, activated						IGF-1R, activated								
IKKa	CHUK		Other	IKK		IKKá	IKKa/CHUK	IKK-alpha	IKKa	IKK-alpha	CHUK (IKK alpha)	IKKá(CHUK)		
IKKb	IKK2		Other	IKK		IKKâ	IKKb/IKBKB	IKK-beta	IKKb	IKK-beta	IKBKB (IKK beta)	IKKâ(IKBKB)	IKKb	
IKKe	IKKE	IKK3	Other	IKK			IKKe/IKBKE	IKK-epsilon	IKKe	IKK-epsilon	IKBKE (IKK epsilon)	IKKá(IKBKE)	IKBKE (IKKe)	
ILK	ILK		TKL	MLK	ILK									
INSR	INSR		TK	InsR	RTK	IR	IR	INSR	INSR	INS-R	INSR	INSR	INSR	InsR
IR, activated						IR, activated								
IRAK1	IRAK1		TKL	IRAK		IRAK1	IRAK1	IRAK1		IRAK1	IRAK1	IRAK1		
IRAK2	IRAK2		TKL	IRAK										IRAK2
IRAK3	IRAK-M		TKL	IRAK				IRAK3						
IRAK4	LOC51135		TKL	IRAK		IRAK4	IRAK4	IRAK4	IRAK4	IRAK4	IRAK4	IRAK4	IRAK4	IRAK4
IRE1	ERN1		Other	IRE				ERN1						
IRE2	TIG44020		Other	IRE										
IRR	INSRR		TK	InsR	RTK	IRR	IRR/INSRR	INSRR	IRR	INSR-R	INSRR (IRR)	IRR(INSRR)		IRR
ITK	ITK		TK	Tec		Itk	ITK	ITK	ITK	ITK	ITK	ITK	ITK	ITK
JAK1	JAK1		TK	JakA			JAK1	JAK1(JH1domain-catalytic)	JAK1	JAK1*	JAK1	JAK1		
JAK1(JH2domain-pseudokinase)								JAK1(JH2domain-pseudokinase)						
JAK2	JAK2		TK	JakA		JAK2	JAK2	JAK2(JH1domain-catalytic)	JAK2	JAK2*	JAK2	JAK2	JAK2	
JAK2 (V647F)							JAK2 (V647F)							
JAK2 JH1 JH2											JAK2 JH1 JH2			
JAK2 JH1 JH2 V617F											JAK2 JH1 JH2 V617F			
JAK3	JAK3		TK	JakA		JAK3	JAK3	JAK3(JH1domain-catalytic)	JAK3	JAK3	JAK3	JAK3		JAK3
JNK1	MAPK8		CMGC	MAPK	JNK	JNK1á1	JNK1	JNK1		JNK1	MAPK8 (JNK1)	JNK1(MAPK8)		JNK1
JNK2	MAPK9		CMGC	MAPK	JNK	JNK2á2	JNK2	JNK2	JNK2	JNK2	MAPK9 (JNK2)	JNK2(MAPK9)		JNK2
JNK3	MAPK10		CMGC	MAPK	JNK	JNK3	JNK3	JNK3		JNK3	MAPK10 (JNK3)	JNK3(MAP-K10)		JNK3

Appendix. *(Continued)*

Kinase Name	Gene Name	Alias	Group	Family	Sub-family	Millipore Assay	RBC Assay	Ambit Assay	Nanosyn Assay	ProQinase Assay	InVitrogen Assay	Carna Assay	Caliper Assay Kit	Signalchem
KDR	KDR	VEGFR2, FLK1	TK	VEGFR	RTK	KDR	KDR/VEGFR2	VEGFR2	KDR	VEGF-R2	KDR (VEGFR2)	KDR	KDR	KDR
KHS1	MAP4K5		STE	STE20	KHS		KHS/MAP4K5	MAP4K5		MAP4K5	MAP4K5 (KHS1)			KHS1
KHS2	MAP4K3		STE	STE20	KHS			MAP4K3						
KIS	KIS		Other	Other-Unique										
KIT	KIT		TK	PDGFR	RTK	c-Kit	c-Kit	KIT	KIT	KIT wt	KIT	KIT	KIT	c-KIT
KIT (T670I)							c-Kit (T670I)		KIT (T670I)	KIT T670I	KIT T670I	KIT [T670I]	KIT[T670I]	
KIT (V560G)						c-Kit (V560G)	c-Kit (V560G)		KIT (V560G)			KIT [V560G]		
KIT (V654A)						c-Kit (V654A)	c-Kit (V654A)		KIT (V654A)		KIT V654A	KIT [V654A]		c-KIT(V654A)
KIT(A829P)								KIT(A829P)						
KIT(D816H)							c-Kit (D816H)	KIT(D816H)						
KIT(D816V)						c-Kit (D816V)	c-Kit (D816V)	KIT(D816V)	KIT (D816V)			KIT [D816V]		
KIT(L576P)								KIT(L576P)						
KIT(V559D)								KIT(V559D)						
KIT(V559D,T670I)								KIT(V559D,T670-I)						
KIT(V559D,V654A)								KIT(V559D,V654-A)						
KSR1	KSR		TKL	RAF										
KSR2	tar7726		TKL	RAF										
LATS1	LATS1		AGC	NDR				LATS1						
LATS2	LATS2		AGC	NDR				LATS2				LATS2		
LCK	LCK		TK	Src		Lck	LCK	LCK	LCK	LCK	LCK	LCK	LCK	LCK
Lck, activated						Lck, activated								
LIMK1	LIMK1		TKL	LISK	LIMK	LIMK1	LIMK1	LIMK1		LIMK1	LIMK1	LIMK1		LIMK1
LIMK2	LIMK2		TKL	LISK	LIMK		LIMK2	LIMK2		LIMK2	LIMK2			
LKB1	STK11		CAMK	CAMKL	LKB	LKB1	LKB1	LKB1				LKB1(STK11)/MO25á/STRADá		LKB1/MO25a/STRADa
LMR1	AATK		TK	Lmr										
LMR2	KIAA1079		TK	Lmr										
LMR3	tar7720		TK	Lmr										
LOK	LOK	STK10	STE	STE20	SLK	LOK	LOK/STK10	LOK	LOK (STK10)			LOK(STK10)	LOK	LOK
LRRK1	FLJ23119		TKL	LRRK										
LRRK2	tar7728		TKL	LRRK			LRRK2	LRRK2	LRRK2	LRRK2 wt	LRRK2			LRRK2
LRRK2 (G2019S)							LRRK2 (G2019S)	LRRK2(G2019S)		LRRK2 G2019S	LRRK2 G2019S			

Appendix. (*Continued*)

Kinase Name	Gene Name	Alias	Group	Family	Sub-family	Millipore Assay	RBC Assay	Ambit Assay	Nanosyn Assay	ProQinase Assay	InVitrogen Assay	Carna Assay	Caliper Assay Kit	Signalchem
LRRK2 (I2020T)							LRRK2 (I2020T)			LRRK2 I2020T				
LRRK2 (R1441C)							LRRK2 (R1441C)			LRRK2 R1441C				
LTK	LTK		TK	Alk	RTK		TYK1/LTK	LTK	LTK (TYK1)	LTK	LTK (TYK1)	LTK	LTK	
LYN	LYN		TK	Src		Lyn	LYN	LYN	LYN	LYN				
LYNA									LYNA		LYN A	LYNa	LYNa	LYN A
LYN B							LYN B		LYNB		LYN B	LYNb	LYNb	LYN B
LZK	MAP3K13		TKL	MLK	LZK			LZK						
MAK	MAK		CMGC	RCK				MAK						
MAP2K1	MAP2K1	MEK1, MKK1	STE	STE7		MEK1	MEK1	MEK1	MEK1	MEK1	MAP2K1 (MEK1)	MAP2K1		MEK1
MAP2K1 - S218D S222D											MAP2K1 (MEK1) S218D S222D			
MAP2K2	MAP2K2	MKK2	STE	STE7			MEK2	MEK2		MEK2	MAP2K2 (MEK2)	MAP2K2		MEK2
MAP2K3	MAP2K3	MKK3	STE	STE7			MEK3	MEK3			MAP2K3 (MEK3)	MAP2K3		
MAP2K4	MAP2K4	MKK4	STE	STE7		MKK4		MEK4				MAP2K4		
MAP2K5	MAP2K5		STE	STE7				MEK5				MAP2K5		
MAP2K6	MAP2K6	MKK6	STE	STE7		MKK6	MKK6	MEK6			MAP2K6 (MKK6)	MAP2K6		
MAP2K6 - S207D/T211D										MKK6 S207D/ T211D				
MAP2K6 - S207E T211E											MAP2K6 (MKK6) S207E T211E			
MAP2K7	MAP2K7	MKK7	STE	STE7		MKK7â		MKK7				MAP2K7		
MAP3K1	MAP3K1	MEKK1	STE	STE11				MAP3K1				MAP3K1		
MAP3K2	MAP3K2	MEKK2	STE	STE11			MEKK2	MAP3K2		MEKK2	MAP3K2 (MEKK2)	MAP3K2		MEKK2
MAP3K3	MAP3K3	MEKK3	STE	STE11			MEKK3	MAP3K3		MEKK3	MAP3K3 (MEKK3)	MAP3K3		MEKK3
MAP3K4	MAP3K4		STE	STE11				MAP3K4				MAP3K4		
MAP3K5	MAP3K5	Ask1, MEKK5	STE	STE11		Ask1	ASK1/MAP3K5	ASK1		ASK1	MAP3K5 (ASK1)	MAP3K5		ASK1
MAP3K6	MAP3K6	ASK2	STE	STE11				ASK2						
MAP3K7	LOC286417		STE	STE11				MAP3K15		MAP3K7/ MAP3K7IP1	MAP3K7/ MAP3K7IP1 (TAK1-TAB1)			
MAP3K8	FLJ23074		STE	STE11				YSK4						
MAPKAPK2	MAPKAP-K2		CAMK	MAPK-APK	MAPK-APK	MAPKAP-K2	MAPKAPK2	MAPKAPK2	MAPKAP-K2	MAPKAPK2	MAPKAPK2	MAPKAPK2	MAPKAPK2	MAPKAPK2 (GST)
MAPKAPK2 (His)														MAPKAPK2 (His)
MAPKAPK3	MAPKAP-K3		CAMK	MAPK-APK	MAPK-APK	MAPKAP-K3	MAPKAPK3		MAPKAP-K3	MAPKAPK3	MAPKAPK3	MAPKAPK3	MAPKAPK3	MAPKAPK3
MAPKAPK5	MAPKAP-K5	PRAK	CAMK	MAPK-APK	MAPK-APK	PRAK	MAPKAPK5/ PRAK	MAPKAPK5	PRAK	MAPKAPK5	MAPKAPK5 (PRAK)	MAPKAPK5	MAPKAPK5	MAPKAPK5

Appendix. *(Continued)*

Kinase Name	Gene Name	Alias	Group	Family	Sub-family	Millipore Assay	RBC Assay	Ambit Assay	Nanosyn Assay	ProQinase Assay	InVitrogen Assay	Carna Assay	Caliper Assay Kit	Signalchem
MARK1	MARK		CAMK	CAMKL	MARK	MARK1	MARK1	MARK1	MARK1	MARK1	MARK1 (MARK)	MARK1	MARK1	MARK1
MARK2	EMK1		CAMK	CAMKL	MARK	PAR-1Bá	MARK2/PAR-1Ba	MARK2	MARK2 (PAR1)	MARK2	MARK2	MARK2	MARK2	MARK2
MARK3 (F/S and S/G)	MARK3	CTAK	CAMK	CAMKL	MARK		MARK3		MARK3 C-TAK1 (natural polymorph of MARK3)	MARK3	MARK3	MARK3	c-TAK1	MARK3
MARK4	MARKL1		CAMK	CAMKL	MARK		MARK4		MARK4	MARK4	MARK4	MARK4	MARK4	MARK4
MAST1	SAST		AGC	MAST										
MAST2	MAST205		AGC	MAST										
MAST3	KIAA0561		AGC	MAST				MARK3						
MAST4	KIAA0303		AGC	MAST				MARK4						
MASTL	FLJ14813		AGC	MAST				MAST1						
MELK	KIAA0175		CAMK	CAMKL	MELK	MELK	MELK	MELK	MELK	MELK	MELK	MELK	MELK	MELK
MER	MERTK		TK	Axl	RTK	Mer	c-MER	MERTK	MER	MERTK	MERTK (cMER)	MER(MERTK)	Mer	c-MER
MET	MET		TK	Met	RTK	Met	c-MET	MET	MET	MET wt	MET (cMet)	MET	MET	MET (h)
Met (D1246H)						Met (D1246H)								
Met (D1246N)						Met (D1246N)								
MET (M1250T)							c-MET (M1250T)	MET(M1250T)		MET M1250T	MET M1250T		Met (M1250T)	
Met (M1268T)						Met (M1268T)			MET (M1268T)					
Met (Y1248C)						Met (Y1248C)								
Met (Y1248D)						Met (Y1248D)								
Met (Y1248H)						Met (Y1248H)								
MET F1200I										MET F1200I				
MET Y1230A										MET Y1230A				
MET Y1230C										MET Y1230C				
MET Y1230D										MET Y1230D				
MET Y1230H										MET Y1230H				
MET(Y1235D)									MET(Y1235D)			MET [Y1235D]		
MISR2	AMHR2		TKL	STKR	Type2									
MLK1	MAP3K9		TKL	MLK	MLK	MLK1	MLK1/MAP3K9	MLK1		MAP3K9	MAP3K9 (MLK1)	MLK1(MAP3-K9)		
MLK2	MAP3K10		TKL	MLK	MLK		MLK2/MAP3K10	MLK2		MAP3K10	MAP3K10 (MLK2)	MLK2(MAP3-K10)		
MLK3	MAP3K11		TKL	MLK	MLK		MLK3/MAP3K11	MLK3		MAP3K11	MAP3K11 (MLK3)	MLK3(MAP3-K11)		MLK3
MLK4	KIAA1804		TKL	MLK	MLK									
MLKL	tar7704		TKL	TKL-Unique										

Appendix. (Continued)

Kinase Name	Gene Name	Alias	Group	Family	Sub-family	Millipore Assay	RBC Assay	Ambit Assay	Nanosyn Assay	ProQinase Assay	InVitrogen Assay	Carna Assay	Caliper Assay Kit	Signalchem
MNK1	MKNK1		CAMK	MAPK-APK	MNK		MNK1	MKNK1	MNK1	MKNK1	MKNK1 (MNK1)	MNK1(MKN-K1)	MINK (MINK1)	MNK1
MNK2	GPRK7		CAMK	MAPK-APK	MNK	Mnk2	MNK2	MKNK2	MNK2	MKNK2	MKNK2 (MNK2)	MNK2(MKN-K2)	MNK1 (MKNK1)	MNK2
MOK	RAGE		CMGC	RCK										
MOS	MOS		Other	MOS								MOS		
MPSK1	STK16		Other	NAK			STK16	STK16		TSF1	STK16 (PKL12)			
MRCKa	PK428		AGC	DMPK	GEK	MRCKá	MRCKa/CDC42BPA	MRCKA	MRCKa	CDC42BPA	CDC42 BPA (MRCKA)	MRCKá(CD-C42BPA)		MRCKá
MRCKb	CDC42BP-B		AGC	DMPK	GEK	MRCKâ	MRCKb/CDC42BPB	MRCKB	MRCKb	CDC42BPB	CDC42 BPB (MRCKB)	MRCKá(CD-C42BPB)		MRCKá
MSK1	MSK1	RPS6KA5	AGC	RSK	MSK	MSK1	MSK1/RPS6KA5		MSK1	RPS6KA5	RPS6KA5 (MSK1)	MSK1(RPS6K-A5)	MSK1	MSK1
MSK2	RPS6KA4		AGC	RSK	MSK	MSK2	MSK2/RPS6KA4		MSK2	RPS6KA4	RPS6KA4 (MSK2)	MSK2(RPS6K-A4)	MSK2	
MSSK1	STK23		CMGC	SRPK		MSSK1	MSSK1/STK23	SRPK3	MSSK1 (STK23)	STK23	STK23 (MSSK1)	MSSK1(STK23)		MSSK1
MST1	MST1		STE	STE20	MST	MST1	MST1/STK4	MST1	MST1	MST1	STK4 (MST1)	MST1(STK4)	MST1 (STK4)	MST1
MST2	MST2		STE	STE20	MST	MST2	MST2/STK3	MST2	MST2	MST2	STK3 (MST2)	MST2(STK3)	MST2	STK3
MST3	MST3		STE	STE20	YSK	MST3	MST3/STK24	MST3	MST3	MST3	STK24 (MST3)	MST3(STK24)	MST3 (STK24)	MST3
MST4	MASK		STE	STE20	YSK		MST4	MST4	MST4	MST4	MST4	MST4		MST4
MUSK	MUSK		TK	Musk	RTK	MuSK	MUSK	MUSK	MUSK	MUSK	MUSK	MUSK		MUSK
MYO3A	MYO3A		STE	STE20	NinaC			MYO3A						
MYO3B	MYO3B		STE	STE20	NinaC		MYO3B	MYO3B						MYO3â
MYT1	PKMYT1		Other	WEE				PKMYT1						MYT1
NDR1	NDR		AGC	NDR			STK38/NDR1	NDR1				NDR1(STK38)		NDR
NDR2	KIAA0965	STK38L	AGC	NDR				NDR2				NDR2(STK38-L)		
NEK1	NEK1		Other	NEK			NEK1	NEK1	NEK1	NEK1	NEK1	NEK1	NEK1	
NEK10	FLJ32685		Other	NEK										
NEK11	FLJ23495		Other	NEK		NEK11	NEK11	NEK11		NEK11				NEK11
NEK2	NEK2		Other	NEK		NEK2	NEK2	NEK2	NEK2	NEK2	NEK2	NEK2	NEK2	NEK2
NEK3	NEK3		Other	NEK		NEK3	NEK3	NEK3		NEK3				NEK3
NEK4	STK2		Other	NEK			NEK4	NEK4	NEK4	NEK4	NEK4	NEK4		
NEK5	TIG36334		Other	NEK				NEK5						NEK6
NEK6	NEK6		Other	NEK		NEK6	NEK6	NEK6	NEK6	NEK6	NEK6	NEK6		
NEK7	tar7718		Other	NEK		NEK7	NEK7	NEK7	NEK7	NEK7	NEK7	NEK7		NEK7
NEK8	LOC255185		Other	NEK										
NEK9	tar7709		Other	NEK			NEK9	NEK9	NEK9	NEK9	NEK9	NEK9		NEK9
NIK	MAP3K14		STE	STE-Unique			NIK/MAP3K14			NIK	MAP3K14 (NIK)			
NIM1	tar7699		CAMK	CAMKL	NIM1			NIM1				MGC42105		

Appendix. (*Continued*)

Kinase Name	Gene Name	Alias	Group	Family	Sub-family	Millipore Assay	RBC Assay	Ambit Assay	Nanosyn Assay	ProQinase Assay	InVitrogen Assay	Carna Assay	Caliper Assay Kit	Signalchem
NLK	NLK		CMGC	MAPK	nmo	NLK	NLK	NLK		NLK	NLK			
NRBP1	NRBP		Other	NRBP										
NRBP2	TIG36417		Other	NRBP										
NuaK1	KIAA0537	ARK5	CAMK	CAMKL	NuaK	ARK5	ARK5/NUAK1	ARK5	ARK5	ARK5	NUAK1 (ARK5)	NuaK1	NuaK1	
NuaK2	DKFZ-P434J037		CAMK	CAMKL	NuaK		SNARK/NUAK2	SNARK	NUAK2	SNARK		NuaK2		NUAK2
Obscn	tar7694		CAMK	Trio										
OSR1	OSR1		STE	STE20	FRAY		OSR1/OXSR1	OSR1						
p38a	MAPK14	Sapk2a	CMGC	MAPK	p38	SAPK2a	P38a/MAPK14	p38-alpha	P38á	p38-alpha	MAPK14 (p38 alpha)	p38á(MAPK14)	p38a	p38á
p38a (T106M)						SAPK2a (T106M)	p38a (T106M)						p38a (T106M)	
p38a Direct											MAPK14 (p38 alpha) Direct			
p38b	MAPK11	Sapk2b	CMGC	MAPK	p38	SAPK2b	P38b/MAPK11	p38-beta	P38â	p38-beta	MAPK11 (p38 beta)	p38â(MAPK11)	p38b2	p38â
p38d	MAPK13	SAPK4	CMGC	MAPK	p38	SAPK3	P38d/MAPK13	p38-delta	P38ä	p38-delta	MAPK13 (p38 delta)	p38ä(MAPK13)	p38d	p38ä
p38g	MAPK12	SAPK3, ERK6	CMGC	MAPK	p38	SAPK4	P38g/MAPK12	p38-gamma	P38ã	p38-gamma	MAPK12 (p38 gamma)	p38ã(MAPK12)	p38g	p38ã
p70S6K	RPS6KB1		AGC	RSK	p70	p70S6K	p70S6K/RPS6KB1	S6K1	P70S6K1	S6K	RPS6KB1 (p70S6K)	p70S6K(RPS6-KB1)	P70S6K	p70S6K
p70S6Kb	RPS6KB2		AGC	RSK	p70		p70S6Kb/RPS6KB2		P70S6K2	S6K-beta		p70S6Ká(RPS6-KB2)		p70S6Kb
PAK1	PAK1		STE	STE20	PAKA		PAK1	PAK1	PAK1	PAK1	PAK1	PAK1		PAK1/CDC42
PAK2	PAK2		STE	STE20	PAKA	PAK2	PAK2	PAK2	PAK2	PAK2	PAK2 (PAK65)	PAK2	PAK2	PAK2
PAK3	PAK3		STE	STE20	PAKA	PAK3	PAK3	PAK3	PAK3	PAK3	PAK3	PAK3	PAK3	PAK3
PAK4	PAK4		STE	STE20	PAKB	PAK4	PAK4	PAK4	PAK4	PAK4	PAK4	PAK4	PAK4	PAK4
PAK5	PAK5	PAK7	STE	STE20	PAKB	PAK5	PAK5	PAK7	PAK5(7)		PAK7 (KIAA1264)	PAK5(PAK7)	PAK5 (PAK7)	PAK7
PAK6	PAK6		STE	STE20	PAKB	PAK6	PAK6	PAK6		PAK6	PAK6	PAK6	PAK6	PAK6
PASK	KIAA0135		CAMK	CAMKL	PASK	PASK	PASK		PASK	PASK	PASK	PASK	PASK	PASK
PBK	TOPK		Other	TOPK			PBK/TOPK		PBK	PBK		PBK		TOPK
PCTAIRE1	PCTK1		CMGC	CDK	TAIRE			PCTK1	PCTK1	PCTAIRE1	PCTAIRE1			
PCTAIRE2	PCTK2		CMGC	CDK	TAIRE			PCTK2	PCTK2					
PCTAIRE3	PCTK3		CMGC	CDK	TAIRE			PCTK3	PCTK3					
PDGFRa	PDGFRA		TK	PDGFR	RTK	PDGFRá	PDGFRa	PDGFRA	PDGFRá	PDGFR-alpha wt	PDGFRA (PDGFR alpha)	PDGFRá(PDGFRA)	PDGFRa	PDGFRá (h)
PDGFRa (T674I)							PDGFRa (T674I)		PDGFRa (T674I)	PDGFR-alpha T674I	PDGFRA T674I	PDGFRá(PDGFRA) [T674I]		
PDGFRb	PDGFRB		TK	PDGFR	RTK	PDGFRb	PDGFRb	PDGFRB	PDGFRá	PDGFR-beta	PDGFRB (PDGFR beta)	PDGFRá(PDGFRB)	PDGFRb	PDGFRá
PDGFRá (D842V)						PDGFRá (D842V)	PDGFRa (D842V)		PDGFRa (D842V)	PDGFR-alpha D842V	PDGFRA D842V		PDGFRa (D842V)	

Appendix. (*Continued*)

Kinase Name	Gene Name	Alias	Group	Family	Sub-family	Millipore Assay	RBC Assay	Ambit Assay	Nanosyn Assay	ProQinase Assay	InVitrogen Assay	Carna Assay	Caliper Assay Kit	Signalchem
PDGFRá (V561D)						PDGFRá (V561D)	PDGFRa (V561D)		PDGFRa (V561D)	PDGFR-alpha V561D	PDGFRA V561D	PDGFRá(PDG-FRA) [V651D]	PDGFRa [V561-D]	
PDHK1	PDK1		Atypical	PDHK										
PDHK2	PDK2		Atypical	PDHK								PDHK2(PDK2)		
PDHK3	PDK3		Atypical	PDHK					PDHK4					
PDHK4	PDK4		Atypical	PDHK			PDK4/PDHK4 (Pyruvate Dehydrogenase Kinase)					PDHK4(PDK4)		
PDK1	PDPK1		AGC	PKB		PDK1	PDK1/PDPK1	PDPK1	PDK1	PDK1	PDK1	PDK1(PDPK1)		PDK1
PDK1 Direct											PDK1 Direct			
PEK	EIF2AK3		Other	PEK	PEK		EIF2AK3/PERK			EIF2AK3		PEK(EIF2AK3)		EIF2AK3
PFTAIRE1	PFTK1		CMGC	CDK	TAIRE			PFTK1						
PFTAIRE2	tar7701		CMGC	CDK	TAIRE			PFTAIRE2						
PHKg1	PHKG1	PBK at Dundee	CAMK	PHK			PHKg1	PHKG1	PHKg1	PHKG1	PHKG1	PHKG1	PHKG1	PHKG1
PHKg2	PHKG2		CAMK	PHK		PhKá2	PHKg2	PHKG2	PHKá2	PHKG2	PHKG2	PHKG2	PHKG2	PHKG2
PI3K alpha								PIK3CA	PI3Ka					
PI3K gamma						PI3Ká	PI3Kg (p120g)	PIK3CG			PIK3CG (p110 gamma)			PI3K (p120á)
PI3K-C2á						PI3K-C2á	PI3KC2a				PIK3C2A (PI3K-C2 alpha)			
PI3K-C2ã						PI3K-C2ã		PIK3C2G						
PI3K-C3							PI3KC3				PIK3C3 (hVPS34)			PI3K (Vps34/ Vps15)
PI3Ká (C420R)								PIK3CA(C420R)						
PI3Ká (E542K)						PI3Ká (E542K)/ p85á		PIK3CA(E542K)						
PI3Ká (E545A)								PIK3CA(E545A)						
PI3Ká (E545K)						PI3Ká (E545K)/ p85á		PIK3CA(E545K)						PI3K (p110á(E545K)/ p85á)
PI3Ká (H1047L)								PIK3CA(H1047L)						
PI3Ká (H1047R)						PI3Ká (H1047R)/p85á								
PI3Ká (H1047Y)								PIK3CA(H1047Y)						
PI3Ká (I800L)								PIK3CA(I800L)						
PI3Ká (M1043I)								PIK3CA(M1043I)						
PI3Ká (Q546K)								PIK3CA(Q546K)						
PI3Ká/p65á						PI3Ká/p65á								PI3K (p110á/ p65á)

Appendix. *(Continued)*

Kinase Name	Gene Name	Alias	Group	Family	Sub-family	Millipore Assay	RBC Assay	Ambit Assay	Nanosyn Assay	ProQinase Assay	InVitrogen Assay	Carna Assay	Caliper Assay Kit	Signalchem
PI3Ká/p85á						PI3Ká/p85á	PI3Ka (p110a/p85a)				PIK3CA/PIK3R1 (p110 alpha/p85 alpha)			PI3K (p110á/p85á) (h)
PI3Kâ								PIK3CB						
PI3Ká/p85á						PI3Ká/p85á	PI3Kb (p110b/p85a)							PI3K (p110â/p85á)
PI3Ká/p85â						PI3Ká/p85â								
PI3Kä								PIK3CD						
PI3Kä/p85á						PI3Kä/p85á	PI3Kd (p110d/p85a)				PIK3CD/PIK3R1 (p110 delta/p85 alpha)			PI3K (p110ä/p85á)
PI3Kä/p85á						PI3Kä/p85á								
PI4Ka											PI4KA			
PI4Kb								PIK4CB	PI4Kb		PI4KB (PI4K beta)			
PIK3-C2â								PIK3C2B			PIK3C2B (PI3K-C2 beta)			
PIK3R4	PIK3R4		Other	VPS15										
PIM1	PIM1		CAMK	PIM		Pim-1	PIM1	PIM1	PIM-1	PIM1	PIM1	PIM1	PIM1	PIM1
PIM2	PIM2		CAMK	PIM		Pim-2	PIM2	PIM2	PIM-2	PIM2	PIM2	PIM2	PIM2	PIM2
PIM3	tar7710		CAMK	PIM		Pim-3	PIM3	PIM3	PIM-3	PIM3		PIM3	PIM3	
PINK1	PINK1		Other	NKF2										
PIP4K2á						PIP4K2á								
PIP5K1C								PIP5K1C						
PIP5K1á							PIP5K1á	PIP5K1A						
PIP5K1ä							PIP5K1ä							
PIP5K2B								PIP5K2B						
PIP5K2C								PIP5K2C						
PITSLRE	CDC2L1		CMGC	CDK	PITSLR-E			CDC2L1						
PKACa	PRKACA		AGC	PKA		PKA	PKA	PKAC-alpha	PKA	PKA	PRKACA (PKA)	PKACá(PRKA-CA)	PKACa	PKAcá
PKACb	PRKACB		AGC	PKA				PKAC-beta				PKACâ(PRKA-CB)		PKAcâ
PKACg	PRKACG		AGC	PKA			PKAcg					PKACä(PRKA-CG)		PKAcä
PKCa	PRKCA		AGC	PKC	Alpha	PKCá	PKCa		PKCa	PKC-alpha	PRKCA (PKC alpha)	PKCá(PRKCA)	PKCa	PKCá (h)
PKCb	PRKCB1		AGC	PKC	Alpha	PKCâI	PKCb1		PKCb1	PKC-beta1	PRKCB1 (PKC beta I)	PKCâ1(PRKC-B1)	PKCb	PKCâ I
PKCd	PRKCD		AGC	PKC	Delta	PKCä	PKCd	PRKCD		PKC-delta	PRKCD (PKC delta)	PKCä(PRKCD)	PKCd	PKCä

Appendix. (Continued)

Kinase Name	Gene Name	Alias	Group	Family	Sub-family	Millipore Assay	RBC Assay	Ambit Assay	Nanosyn Assay	ProQinase Assay	InVitrogen Assay	Carna Assay	Caliper Assay Kit	Signalchem
PKCe	PRKCE		AGC	PKC	Eta	PKCå	PKCepsilon	PRKCE		PKC-epsilon	PRKCE (PKC epsilon)	PKCå(PRKCE)	PKCe	PKCå
PKCg	PRKCG		AGC	PKC	Alpha	PKCã	PKCg		PKCg	PKC-gamma	PRKCG (PKC gamma)	PKCã(PRKCG)	PKCg	PKCã
PKCh	PRKCH		AGC	PKC	Eta	PKCç	PKCeta	PRKCH	PKCç	PKC-eta	PRKCH (PKC eta)	PKCç(PRKCH)	PKCh	PKCç
PKCi	PRKCI		AGC	PKC	Iota	PKCé	PKCiota	PRKCI	PKCi	PKC-iota	PRKCI (PKC iota)	PKCé(PRKCI)	PRKCI (PKCi)	PKCé
PKCt	PRKCQ		AGC	PKC	Delta	PKCè	PKCtheta	PRKCQ	PKCè	PKC-theta	PRKCQ (PKC theta)	PKCè(PRKCQ)	PKCq	PKCè
PKCz	PRKCZ		AGC	PKC	Iota	PKCz	PKCzeta			PKC-zeta	PRKCZ (PKC zeta)	PKCæ(PRKCZ)		PKCæ
PKCåII						PKCåII	PKCb2		PKCå2	PKC-beta2	PRKCB2 (PKC beta II)	PKCå2(PRKC-B2)		PKCå II
PKD1	PRKCM		CAMK	PKD		PKCm	PKCmu/PRKD1	PRKD1	PRKD1	PKC-mu	PRKD1 (PKC mu)	PKD1(PRKD1)	PKD1	PKCi
PKD2	PKD2		CAMK	PKD		PKD2		PRKD2	PRKD2	PKD2	PRKD2 (PKD2)	PKD2(PRKD2)	PKD2	PKD2
PKD3	PRKCN		CAMK	PKD			PKCnu/PRKD3	PRKD3	PRKD3	PKC-nu	PRKCN (PKD3)	PKD3(PRKD3)	PKD3	PKCï
PKG1	PRKG1		AGC	PKG		PKG1á	PKG1a	PRKG1	PKG1 (PRKG1)	PKG1	PRKG1	PGK(PRKG1)	PKG1a	PRKG1
PKG1â						PKG1â	PKG1b						PKG1b	PRKG2
PKG2	PRKG2		AGC	PKG			PKG2/PRKG2	PRKG2	PKG2 (PRKG2)	PRKG2	PRKG2 (PKG2)	CGK2(PRKG2)		
PKN1	PRKCL1		AGC	PKN			PKN1/PRK1	PKN1		PRK1	PKN1 (PRK1)	PKN1		PKN1/PRK1
PKN2	PRKCL2	PRK2	AGC	PKN		PRK2	PKN2/PRK2	PKN2		PRK2				PKN2/PRK2
PKN3	pknbeta		AGC	PKN										
PKR	PRKR		Other	PEK			EIF2AK2	PRKR		EIF2AK2		PKR(EIF2AK2)		EIF2AK2
PLK1	PLK		Other	PLK		Plk1	PLK1	PLK1	PLK1	PLK1	PLK1	PLK1		PLK1
PLK2	SNK		Other	PLK		SNK	PLK2	PLK2		SNK	PLK2	PLK2		PLK2
PLK3	CNK		Other	PLK		PLK3	PLK3	PLK3		PLK3	PLK3	PLK3		
PLK4	STK18		Other	PLK				PLK4		SAK		PLK4		PLK4
PRKX	PRKX		AGC	PKA		PrKX	PRKX	PRKX	PRKX	PRKX	PRKX	PRKX	PRKX	PRKX
PRKY	PRKY		AGC	PKA										
PRP4	PRP4		CMGC	DYRK	PRP4				PRP4					
PRPK	C20orf64	TP53RK	Other	Bud32										
PSKH1	PSKH1		CAMK	PSK										
PSKH2	PSKH2		CAMK	PSK										
PYK2	PTK2B		TK	Fak		Pyk2	PYK2	PYK2	PTK2 (FAK2)	PYK2	PTK2B (FAK2)	PYK2(PTK2B)	PYK2 (PTK2B)	PYK2
QIK		SIK2	CAMK	CAMKL	QIK		SIK2	SIK2	QIK (SNF1L-K2)	SNF1LK2	SNF1LK2	QIK(SNF1LK2)		QIK
QSK	tar7719		CAMK	CAMKL	QIK				QSK					
RAF1	RAF1		TKL	RAF		c-RAF	RAF1	RAF1	cRAF			RAF1	Raf-1	RAF1
RAF1 Y340D/Y341D										RAF1 Y340D/ Y341D	RAF1 (cRAF) Y340D Y341D			

Appendix. (*Continued*)

Kinase Name	Gene Name	Alias	Group	Family	Sub-family	Millipore Assay	RBC Assay	Ambit Assay	Nanosyn Assay	ProQinase Assay	InVitrogen Assay	Carna Assay	Caliper Assay Kit	Signalchem
RAF1(EE)														RAF1(EE)
RET	RET		TK	Ret	RTK	Ret	RET	RET	RET	RET wt	RET	RET	RET	RET
Ret (V804L)						Ret (V804L)	RET (V804L)	RET(V804L)	RET (V804L)	RET V804L	RET V804L		RET (V804L)	
Ret (V804M)						Ret (V804M)	RET (V804M)	RET(V804M)	RET (V804M)					
RET (Y791F)									RET (Y791F)	RET Y791F	RET Y791F	RET [Y791F]	RET (Y791F)	
RET [G691S]												RET [G691S]		
RET [S891A]												RET [S891A]		
RET(M918T)								RET(M918T)				RET [M918T]		
RHOK	RHOK		AGC	GRK	GRK			GRK1						
RIOK1	AD034		Atypical	RIO	RIO1			RIOK1						
RIOK2	FLJ11159		Atypical	RIO	RIO2			RIOK2						
RIOK3	SUDD		Atypical	RIO	RIO3			RIOK3						
RIPK1	RIPK1		TKL	RIPK				RIPK1						
RIPK2	RIPK2	RICK	TKL	RIPK		RIPK2	RIPK2	RIPK2		RIPK2	RIPK2			RIPK2
RIPK3	RIPK3		TKL	RIPK										
RNAseL	RNASEL		Other	Other-Unique										
ROCK1	ROCK1		AGC	DMPK	ROCK	ROCK-I	ROCK1	ROCK1	ROCK1	ROCK1	ROCK1	ROCK1	ROCK1	ROCK1
ROCK2	ROCK2		AGC	DMPK	ROCK	ROCK-II	ROCK2	ROCK2	ROCK2	ROCK2	ROCK2	ROCK2	ROCK2	ROCK2
RON	MST1R		TK	Met	RTK	Ron	RON/MST1R	MST1R	RON	RON	MST1R (RON)	RON(MST1R)	MST1R	RON
ROR1	ROR1		TK	Ror	RTK									
ROR2	ROR2		TK	Ror	RTK				ROR2					ROR2
ROS	ROS1		TK	Sev	RTK	Ros	ROS/ROS1	ROS1	ROS	ROS	ROS1	ROS(ROS1)	ROS (ROS1)	ROS1
RPS6KA4(Kin.Dom.1-N-terminal)								RPS6KA4(Kin.Dom.1-N-terminal)						
RPS6KA4(Kin.Dom.2-C-terminal)								RPS6KA4(Kin.Dom.2-C-terminal)						
RPS6KA5(Kin.Dom.1-N-terminal)								RPS6KA5(Kin.Dom.1-N-terminal)						
RPS6KA5(Kin.Dom.2-C-terminal)								RPS6KA5(Kin.Dom.2-C-terminal)						
RSK1	RPS6KA2	MAPKAP-K1a, RPS6KA1	AGC	RSK	RSK	Rsk1	RSK1		RSK1	RPS6KA1	RPS6KA1 (RSK1)	RSK1(RPS6K-A1)	RSK1	RSK1
Rsk1 - S637N, G697A						Rsk1								
RSK1(Kin.Dom.1-N-terminal)								RSK1(Kin.Dom.1-N-terminal)						
RSK1(Kin.Dom.2-C-terminal)								RSK1(Kin.Dom.2-C-terminal)						

Appendix. (*Continued*)

Kinase Name	Gene Name	Alias	Group	Family	Sub-family	Millipore Assay	RBC Assay	Ambit Assay	Nanosyn Assay	ProQinase Assay	InVitrogen Assay	Carna Assay	Caliper Assay Kit	Signalchem
RSK2	RPS6KA3	MAPKAP-K1b, RPS6KA3	AGC	RSK	RSK	Rsk2	RSK2		RSK2	RPS6KA3	RPS6KA3 (RSK2)	RSK2(RPS6K-A3)	RSK2	RSK2
RSK2(Kin.Dom.1-N-terminal)								RSK2(Kin.Dom.1-N-terminal)						
RSK3	RPS6KA1	MAPKAP-K1c, RPS6KA2	AGC	RSK	RSK	Rsk3	RSK3		RSK3	RPS6KA2	RPS6KA2 (RSK3)	RSK3(RPS6K-A2)	RSK3	RSK3
RSK3(Kin.Dom.1-N-terminal)								RSK3(Kin.Dom.1-N-terminal)						
RSK3(Kin.Dom.2-C-terminal)								RSK3(Kin.Dom.2-C-terminal)						
RSK4	RPS6KA6		AGC	RSK	RSK	Rsk4	RSK4		RSK4	RPS6KA6	RPS6KA6 (RSK4)	RSK4(RPS6K-A6)	RSK4	RSK4
RSK4(Kin.Dom.1-N-terminal)								RSK4(Kin.Dom.1-N-terminal)						
RSK4(Kin.Dom.2-C-terminal)								RSK4(Kin.Dom.2-C-terminal)						
RSKL1	RPS6KC1		AGC	RSKL										
RSKL2	MGC11287		AGC	RSKL										
RYK	RYK		TK	Ryk	RTK									
SBK	TIG31440		Other	NKF1				SBK1						
SCYL1	NTKL		Other	SCY1										
SCYL2	FLJ10074		Other	SCY1										
SCYL3	LOC57147		Other	SCY1										
SGK	SGK		AGC	SGK		SGK	SGK1		SGK1	SGK1	SGK (SGK1)	SGK	SGK	SGK1
SgK069	tar7723		Other	NKF1										
SgK071	MGC43306		Other	Other-Unique										
SgK085	LOC221757	MYLK4	CAMK	MLCK				MYLK4						
SgK110	LOC284299		Other	NKF1				SgK110						
SgK196	FLJ23356		Other	Other-Unique										
SGK2	SGK2		AGC	SGK		SGK2	SGK2		SGK2	SGK2	SGK2	SGK2	SGK2	SGK2
SgK223	DKFZ-p761P0423		Other	NKF3										
SgK269	FLJ21140		Other	NKF3										
SGK3	SGKL	CISK	AGC	SGK		SGK3	SGK3/SGKL	SGK3	SGK3	SGK3	SGKL (SGK3)	SGK3(SGKL)	SGK3	SGK3
SgK307	TEX14		Other	NKF5										
SgK396	STK31		Other	Other-Unique										
SgK424	SK707		Other	NKF5										

Appendix. (Continued)

Kinase Name	Gene Name	Alias	Group	Family	Sub-family	Millipore Assay	RBC Assay	Ambit Assay	Nanosyn Assay	ProQinase Assay	InVitrogen Assay	Carna Assay	Caliper Assay Kit	Signalchem
SgK493	LOC91461		Other	Other-Unique										
SgK494	FLJ25006		AGC	RSK										
SgK495	MGC4796		CAMK	CAMK-Unique										
SgK496	KIAA0472	RIPK5	Other	Other-Unique			RIPK5	RIPK5		RIPK5				RIPK5
SIK	SNF1LK		CAMK	CAMKL	QIK	SIK		SIK	SIK			SIK(SNF1LK)		SIK
skMLCK	MYLK2		CAMK	MLCK			MLCK2/MYLK2	MYLK2		MYLK2	MYLK2 (skMLCK)	skMLCK(MY-LK2)		
SLK	SLK		STE	STE20	SLK		SLK/STK2	SLK	SLK	SLK	SLK	SLK		SLK
Slob	FLJ20335		Other	Slob										
SMG1	SMG1		Atypical	PIKK	SMG1									
smMLCK	MYLK	MLCK	CAMK	MLCK		MLCK	MLCK/MYLK	MYLK	MLCK	MYLK	MYLK (MLCK)			MLCK
SNRK	SNRK		CAMK	CAMKL	SNRK			SNRK						
SPEG	tar7777		CAMK	Trio										
SPHK1							SPHK1		SPHK1		SPHK1			SPHK1
SPHK2							SPHK2		SPHK2		SPHK2			SPHK2
SRC	SRC		TK	Src		cSRC	c-Src	SRC	SRC	SRC	SRC	SRC	SRC	SRC (h)
Src (1-530)						Src (1-530)								
SRC N1											SRC N1 (neuronal splice variant of SRC)			
SRM	SRMS		TK	Src			SRMS	SRMS	SRM	SRMS	SRMS (Srm)	SRM(SRMS)	SRM (SRMS)	
SRPK1	SRPK1		CMGC	SRPK		SRPK1	SRPK1	SRPK1	SRPK1	SRPK1	SRPK1	SRPK1		SRPK1
SRPK2	SRPK2		CMGC	SRPK		SRPK2	SRPK2	SRPK2	SRPK2	SRPK2	SRPK2	SRPK2		SRPK2
SSTK	SSTK		CAMK	TSSK										
STK33	STK33		CAMK	CAMK-Unique		STK33	STK33	STK33		STK33	STK33			STK33
STLK3	PASK	STK39	STE	STE20	FRAY		STK39/STLK3	STK39		STK39				
STLK5	FLJ90524		STE	STE20	STLK									
STLK6	ALS2CR2		STE	STE20	STLK									
SuRTK106	DKFZ-p761P1010		TK	TK-Unique										
SYK	SYK		TK	Syk		Syk	SYK	SYK	SYK	SYK	SYK	SYK	SYK	SYK
TAF1	TAF1		Atypical	TAF1										
TAF1L	TAF1L		Atypical	TAF1										
TAK1	MAP3K7		TKL	MLK	TAK1	Tak1	TAK1	TAK1				TAK1-TAB1(MAP3-K7)		TAK1-TAB1
TAO1	KIAA1361		STE	STE20	TAO	TAO1	TAOK1	TAOK1						TAOK1
TAO2	TAO1		STE	STE20	TAO	TAO2	TAOK2/TAO1	TAOK2		TAOK2	TAOK2 (TAO1)	TAOK2		TAOK2

Appendix. (*Continued*)

Kinase Name	Gene Name	Alias	Group	Family	Sub-family	Millipore Assay	RBC Assay	Ambit Assay	Nanosyn Assay	ProQinase Assay	InVitrogen Assay	Carna Assay	Caliper Assay Kit	Signalchem
TAO3	JIK		STE	STE20	TAO	TAO3	TAOK3/JIK	TAOK3		TAOK3	TAOK3 (JIK)			TAOK3
TBCK	MGC16169		Other	TBCK										
TBK1	TBK1		Other	IKK		TBK1	TBK1	TBK1	TBK1	TBK1	TBK1	TBK1		TBK1
TEC	TEC		TK	Tec			TEC	TEC	TEC	TEC	TEC	TEC	TEC	TEC
Tec, activated						Tec, activated								
TESK1	TESK1		TKL	LISK	TESK			TESK1	TESK1					TESK1
TESK2	TESK2		TKL	LISK	TESK									
TGFbR1	TGFBR1	Alk5	TKL	STKR	Type1	TGFBR1		TGFBR1		TGFB-R1	TGFBR1 (ALK5)			
TGFbR2	TGFBR2		TKL	STKR	Type2		TGFBR2	TGFBR2		TGFB-R2				TGFäR2
TIE1	TIE		TK	Tie	RTK			TIE1	TIE1					
TIE2	TEK		TK	Tie	RTK	Tie2	TIE2/TEK	TIE2	TIE2	TIE2	TEK (Tie2)	TIE2(TEK)		TIE 2
Tie2 (R849W)						Tie2 (R849W)								
Tie2 (Y897S)						Tie2 (Y897S)								
TIF1a	TIF_V3		Atypical	TIF1										
TIF1b	TRIM28		Atypical	TIF1										
TIF1g	TRIM33		Atypical	TIF1										
TLK1	TLK1		Other	TLK				TLK1	TLK1					
TLK2	TLK2		Other	TLK		TLK2	TLK2	TLK2	TLK2					TLK2
TNK1	TNK1		TK	Ack				TNK1	TNK1			TNK1		
Trad	TRAD		CAMK	Trio										
Trb1	C8FW		CAMK	Trbl										
Trb2	GS3955		CAMK	Trbl										
Trb3	LOC57761		CAMK	Trbl										
Trio	TRIO		CAMK	Trio										
TRKA	NTRK1		TK	Trk	RTK	TrkA	TRKA	TRKA	TRKA	TRK-A	NTRK1 (TRKA)	TRKA(NTR-K1)		TRKA
TRKB	NTRK2		TK	Trk	RTK	TrkB	TRKB	TRKB	TRKB	TRK-B	NTRK2 (TRKB)	TRKB(NTRK2)	NTRK2 (TRKB)	TRKB
TRKC	NTRK3		TK	Trk	RTK		TRKC	TRKC	TRKC	TRK-C	NTRK3 (TRKC)	TRKC(NTR-K3)	TRKC (NTRK3)	TRKC
TRRAP	TRRAP		Atypical	PIKK	TRRAP									
TSSK1	FKSG81		CAMK	TSSK		TSSK1	STK22D/TSSK1	TSSK1B	TSSK1	TSSK1	STK22D (TSSK1)	TSSK1	TSSK1	TSSK1
TSSK2	STK22B	TSK2	CAMK	TSSK		TSSK2	TSSK2		TSSK2	TSK2	STK22B (TSSK2)	TSSK2	TSSK2	TSSK2
TSSK3	STK22C		CAMK	TSSK								TSSK3		
TSSK4	LOC283629		CAMK	TSSK										
TTBK1	KIAA1855		CK1	TTBK										
TTBK2	TTBK		CK1	TTBK										
TTK	TTK		Other	TTK			TTK	TTK	TTK	TTK	TTK	TTK		TTK
TTN	TTN		CAMK	MLCK			TXK							
TXK	TXK		TK	Tec		Txk		TXK	TXK	TXK	TXK	TXK	TXK	TXK
TYK2	TYK2		TK	JakA			TYK2		TYK2	TYK2	TYK2	TYK2		TYK2

Appendix. *(Continued)*

Kinase Name	Gene Name	Alias	Group	Family	Sub-family	Millipore Assay	RBC Assay	Ambit Assay	Nanosyn Assay	ProQinase Assay	InVitrogen Assay	Carna Assay	Caliper Assay Kit	Signalchem
TYK2(JH1domain-catalytic)								TYK2(JH1domain-catalytic)						
TYK2(JH2domain-pseudokinase)								TYK2(JH2domain-pseudokinase)						TYRO3
TYRO3	TYRO3		TK	Axl	RTK	Rse	TYRO3/SKY	TYRO3	TYRO3	TYRO3	TYRO3 (RSE)	TYRO3	TYRO3	
ULK1	ULK1		Other	ULK			ULK1	ULK1						ULK1
ULK2	ULK2		Other	ULK		ULK2	ULK2	ULK2						ULK2
ULK3	tar7763		Other	ULK		ULK3	ULK3	ULK3						ULK3
ULK4	FLJ20574		Other	ULK										
VACAMKL	MGC8407		CAMK	CAMK-Unique										
VRK1	VRK1		CK1	VRK			VRK1			VRK1				
VRK2	VRK2		CK1	VRK		VRK2		VRK2						
VRK3	LOC51231		CK1	VRK										
Wee1	WEE1		Other	WEE			WEE1	WEE1		WEE1	WEE1	WEE1		WEE1
Wee1B	tar7751		Other	WEE				WEE2						
Wnk1	PRKWN-K1		Other	Wnk								WNK1		
Wnk2	tar7707		Other	Wnk		WNK2	WNK2		WNK2	WNK2	WNK2	WNK2		
Wnk3	PRKWN-K3		Other	Wnk		WNK3	WNK3		WNK3	WNK3		WNK3		
Wnk4	PRKWN-K4		Other	Wnk										
YANK1	TIG45206		AGC	YANK				YANK1						
YANK2	HSA250839		AGC	YANK				YANK2						
YANK3	tar7697		AGC	YANK				YANK3						
YES	YES1		TK	Src		Yes	YES/YES1	YES	YES	YES	YES1	YES(YES1)	YES	Yes1
YSK1	SOK1		STE	STE20	YSK		STK25/YSK1	YSK1		STK25	STK25 (YSK1)			SOK1
ZAK	ZAK		TKL	MLK	MLK		ZAK/MLTK	ZAK		ZAK	ZAK			ZAK
ZAP70	ZAP70		TK	Syk		ZAP-70	ZAP70	ZAP70	ZAP70	ZAP70	ZAP70	ZAP70		ZAP70
ZC1/HGK	MAP4K4		STE	STE20	MSN		HGK/MAP4K4	MAP4K4	MAP4K4 (HGK)	MAP4K4	MAP4K4 (HGK)	HGK(MAP4-K4)	HGK	HGK
ZC2/TNIK	KIAA0551		STE	STE20	MSN			TNIK				TNIK		
ZC3/MINK	MINK		STE	STE20	MSN	MINK	MINK/MINK1	MINK	MINK	MINK1	MINK1	MINK(MINK1)		MINK1
ZC4/NRK	NRK		STE	STE20	MSN									

CHAPTER 2

Contemporary Approaches to Kinase Lead Generation

IAIN SIMPSON AND RICHARD A. WARD

AstraZeneca, Oncology Innovative Medicines, Alderley Park, Macclesfield, SK10 4TG, UK

2.1 Introduction

Historically, finding hits and leads against kinase targets has predominantly been achieved through high throughput or directed screening. In more recent years fragment based approaches have come to the fore as a viable way of finding chemical starting points for kinase inhibition and are becoming an embedded practise for hit finding. As kinase drug discovery, especially in oncology research, has been a major focus for pharmaceutical companies for the past 20 years or so, it is reasonable to assume that corporate compound collections are well populated with agents able to provide hits against this target class. So why is there a need to continually refine our approaches to kinase inhibition? There are several reasons for this, one of the most commonly cited is the abundance of ATP competitive agents known, where selectivity against other kinases can be a challenge. In this chapter we present some contemporary approaches to finding novel hits and leads against kinase targets. Of special interest are the approaches where novel inhibition mechanisms have been discovered through screening in alternative ways and–or by utilising structural insights.

RSC Drug Discovery Series No. 19
Kinase Drug Discovery
Edited by Richard A. Ward and Frederick Goldberg
© Royal Society of Chemistry 2012
Published by the Royal Society of Chemistry, www.rsc.org

2.2 Isoform Selective, PH Domain Dependent Akt Inhibitors

Merck and Co. screened $\sim$270 000 compounds against the 3 human isoforms of Akt (Akt1, 2 and 3). The screening approach used protein constructs containing both the pleckstrin homology (PH) domain and the kinase domain of the proteins. The PH domain directs Akt translocation from the cytosol to the plasma membrane by binding to membrane phosphatidylinositides. Akt is subsequently phosphorylated resulting in kinase activation. Given the high structural homology in the kinase domains ($\sim$ 85%, active sites are identical) of the Akt isoforms it was expected that finding isoform selective compounds by screening against the kinase domain alone would be difficult. By screening something more akin to the biologically relevant protein they identified two compounds (Akt-I-1 (**1**) and Akt-I-1,2 (**2**)) with interesting selectivity profiles (Figure 2.1). Akt-I-1 (**1**) was shown to be a selective inhibitor of Akt1 and was not active when the PH domain was removed from the protein (ΔPH). Akt-I-1,2 (**2**) was shown to be active against Akt1 and 2 and, like Akt-I-1 (**1**), required the PH domain for activity. The compounds were also shown to be selective inhibitors relative to 16 other tyrosine and serine–threonine kinases (IC_{50} values against the 16 kinases tested were > 40 μM).[1] A model to explain the mechanism of inhibition by these agents was postulated where the inhibitor bound outside of the active site and interacts with the PH domain and–or the hinge region. More recently, a related PH domain dependent inhibitor (Inhibitor VIII (**3**), Figure 2.2) has been shown, by X-ray crystallography in complex with Akt1, to bind at the interface between the PH domain and the kinase domain confirming the original binding mode hypothesis.[2]

Name	Structure	IC_{50} (μM)				
		Akt1	ΔPHAkt1	Akt2	ΔPHAkt2	Akt3
Akt-I-1 (**1**)	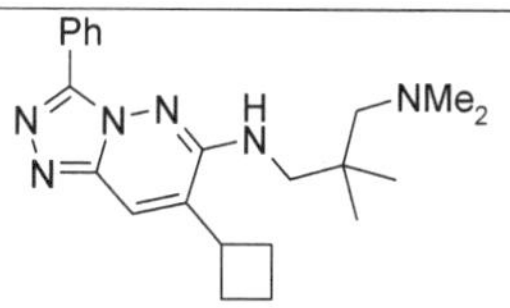	4.6	> 250	> 250	> 250	> 250
Akt-I-1,2 (**2**)	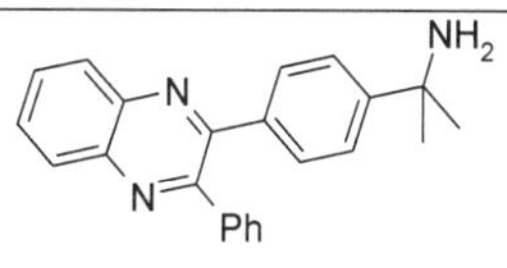	2.7	> 250	21	> 250	> 250

Figure 2.1 The structures of two PH domain dependent and isoform selective Akt inhibitors.

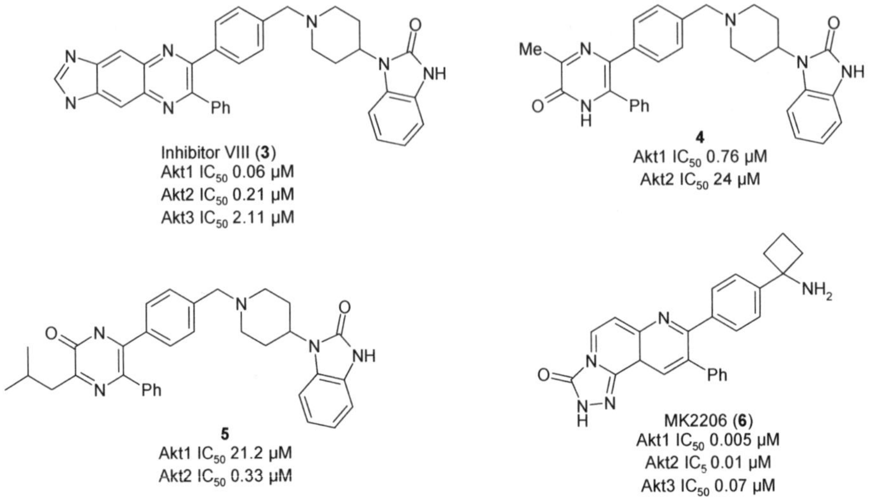

Inhibitor VIII (**3**)
Akt1 IC$_{50}$ 0.06 µM
Akt2 IC$_{50}$ 0.21 µM
Akt3 IC$_{50}$ 2.11 µM

4
Akt1 IC$_{50}$ 0.76 µM
Akt2 IC$_{50}$ 24 µM

5
Akt1 IC$_{50}$ 21.2 µM
Akt2 IC$_{50}$ 0.33 µM

MK2206 (**6**)
Akt1 IC$_{50}$ 0.005 µM
Akt2 IC$_{5}$ 0.01 µM
Akt3 IC$_{50}$ 0.07 µM

Figure 2.2 Merck and Co. PH domain dependent Akt inhibitors.

From these initial observations the researchers at Merck and Co. developed the Akt-I-1,2 (**2**) inhibitor scaffold into sub-series where potency and selectivity between Akt1 and 2 could be achieved. A notable example of this is with the ring truncated pyrazinone compounds (**4**) and (**5**) (Figure 2.2).[3] Compound (**4**) is a potent inhibitor of Akt1 (IC$_{50}$ 0.76 µM) and is a weaker inhibitor of Akt2 (24 µM). By inverting the orientation of the pyrazinone and modifying the nature of the alkyl sidechain, compound (**5**) was discovered. This compound is a potent inhibitor of Akt2 (IC$_{50}$ 0.33 µM) and a weaker inhibitor of Akt1 (IC$_{50}$ 21.2 µM). Importantly, for both these compounds their potency is dependent on the presence of the PH domain of the protein and they are non-competitive with ATP.

Merck and Co. have published extensively on the optimisation of compounds that had their origins in the identification of Akt-I-1,2 (**2**).[4–8] To review all of this work would be beyond the scope of this chapter. It is noteworthy however that the Phase II clinical candidate MK2206 (**6**) (Figure 2.2) has its origins in the discovery of the PH domain dependent Akt inhibitors discussed above. MK2206 has been optimised to be a very potent inhibitor of Akt1, 2 and 3 and is highly selective *versus* other AGC kinases.[9] This work exemplifies the potential benefits of understanding the opportunities for inhibiting kinases in alternative ways. This research was dependent on the establishment of biochemical screens capable of findings hits with a different mechanism of action. As well as generating isoform selective Akt inhibitors, excellent selectivity against related kinases was achieved and most significantly this approach led to the identification of a compound that is currently in Phase II clinical evaluation.

2.3 Exploitation of Inactive Kinase Conformations

The majority of kinase inhibitors which are in clinical development target the active site of the enzyme, directly competing with ATP. Although targeting this region can be highly effective there are potential disadvantages. All protein kinases bind ATP, thereby targeting the ATP site of a kinase makes achieving high levels of selectivity very challenging. This is particularly problematic if the active conformation of a kinase is being targeted. With the significant chemistry investment across the pharmaceutical industry in ATP-competitive compounds, the chemistry landscape of kinase inhibitors can be crowded and it can be a challenge to find a suitable starting point for new inhibitors. An additional issue with a number of kinase inhibitors, which target the active form of a kinase, is that they can be vulnerable to large drop-offs in activity when measuring inhibition in cellular models due to the competition with high concentrations of endogenous ATP. As a result of these factors the inactive conformations of kinases have been targeted by drug discovery scientists for a number of years. The two most commonly referred to inactive conformations are 'DFG-out' and '*c*-helix-out'. The DFG motif is a highly conserved region on the activation loop, and in a subset of kinases (including the most commonly studied P38-MAP Kinase),[10] the activation loop exists in equilibrium between the active conformation (DFG-in) and the inactive DFG-out form. Phosphorylation of the activation loop often triggers activation of a kinase through a mechanism of driving this equilibrium towards the active form. A further subset of kinases, such as EGFR, show a *c*-helix-out inactive form,[11] whereas the *c*-helix is referred to as 'in' when active. This conformational change is often driven by the activation of these kinases through dimerisation. An additional advantage which could be exploited by such a mechanism is if

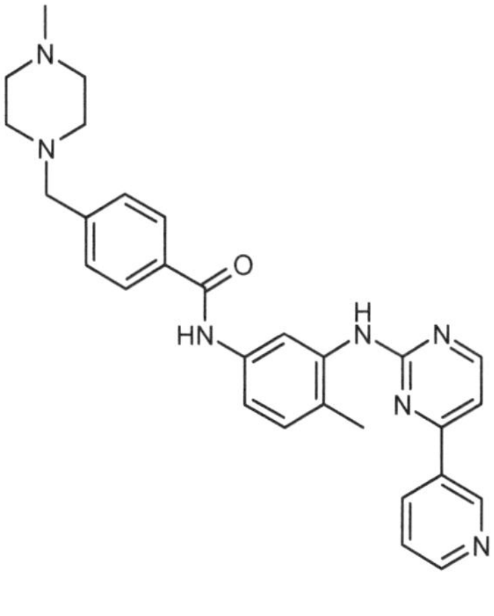

7 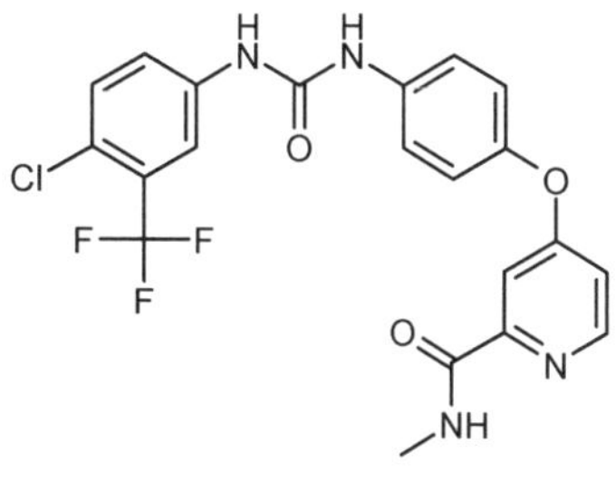   **8**

Figure 2.3 Chemical structures of imatinib (**7**) and sorafenib (**8**).

these conformational changes lead to inhibitors with slower off-rates which may offer additional advantages in terms of duration of cover when used *in vivo*.[12] It should be pointed out that although targeting the inactive conformations has enabled the identification of inhibitors with high levels of selectivity (such as imatinib (**7**)),[13] not all compounds targeting inactive forms have such impressive selectivity profiles including the FDA approved sorafenib (**8**),[14] structures shown in Figure 2.3.

Although these inactive conformations have been studied for a number of years, recent approaches have been published which look to exploit these inactive forms from different perspectives, two such examples are discussed in this section.

2.3.1 Switch Pocket Inhibitors

The concept of 'Switch Pocket' inhibitors has recently been published by Deciphera Pharmaceuticals (http://www.deciphera.com). This term describes the pocket which is created when a kinase converts between its active and inactive conformation(s). Targeting the inactive form of the kinase, specifically the region around the switch pocket, is designed to block conformational activation of the kinase. Deciphera have recently published their work on the design of switch pocket inhibitors of P38-MAP Kinase,[15] exploiting the pyrazole urea scaffold shared with the P38 inhibitor BIRB-796 (**9**),[10] structure shown in Figure 2.4.

BIRB-796 binds to the inactive DFG-out conformation of P38 and has been tested in Phase II clinical trials for rheumatoid arthritis, Crohn's disease and psoriasis but was withdrawn, possibly due to elevation in liver enzymes.[16] A key aim of Deciphera's work was to synthesise compounds which interacted more specifically with the switch pocket amino acids to remove the need for the

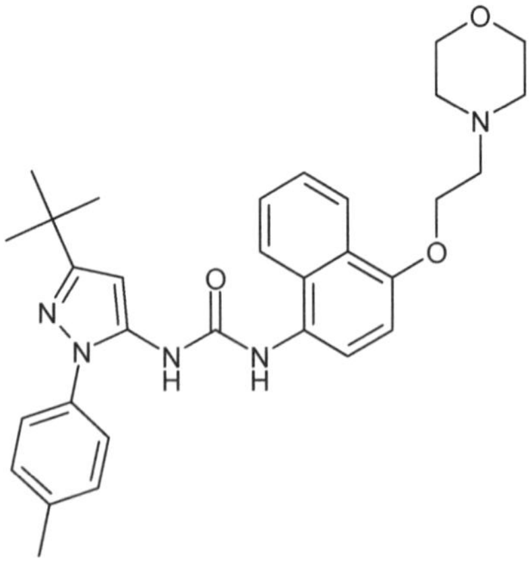

9

Figure 2.4   Chemical structures of BIRB-796 (**9**).

inhibitor to interact with the kinase hinge region, as observed with BIRB-796. Inspection of the ATP-bound structure of doubly-phosphorylated P38 gamma (PDB 1CM8),[17] shows the kinase in its active DFG-in conformation, whereas the crystal structure of unphosphorylated P38 with BIRB-796 bound (PDB 1KV2),[10] shows the activation loop in the inactive DFG-out conformation (Figure 2.5). In this figure the cyan ribbon is the ATP-bound structure (with ATP removed for clarity) with the brown loop from the BIRB-796 crystal structure. The movement of the Phe-169 from the DFG-in and DFG-out crystal structure can clearly be seen, using the residue numbering from the 1KV2 structure. An important observation is how in the DFG-in conformation, the DFG-loop occupies the region of the active site where BIRB-796 binds, thereby BIRB-796 is able to prevent the activation of the kinase.

Deciphera were particularly interested in exploiting the 'switch control' amino acids, which they define as the positively charged arginine residues which play key roles in co-ordinating the phosphorylated warhead of phospho-Thr-180, which stabilises the active conformation. These arginine residues are positioned on the *c*-helix (Arg-67 and Arg-70) and on the activation loop (Arg-149). Arg-67 and Arg-70 can both be seen on Figure 2.5, but Arg-149 is not visible as this region is disordered in the 1KV2 crystal structure. Inhibitors were therefore synthesised to target Arg-67 or Arg-70 to exploit the specific switch pocket residues of P38, binding affinities were measured against the unphosphorylated protein using the fluoroprobe SKF-86002 (**10**) (Figure 2.6).

The synthesised compound, DP-802 (**11**), demonstrated potent binding to the unphosphorylated protein with an IC_{50} of 9 nM, with thermal melt studies

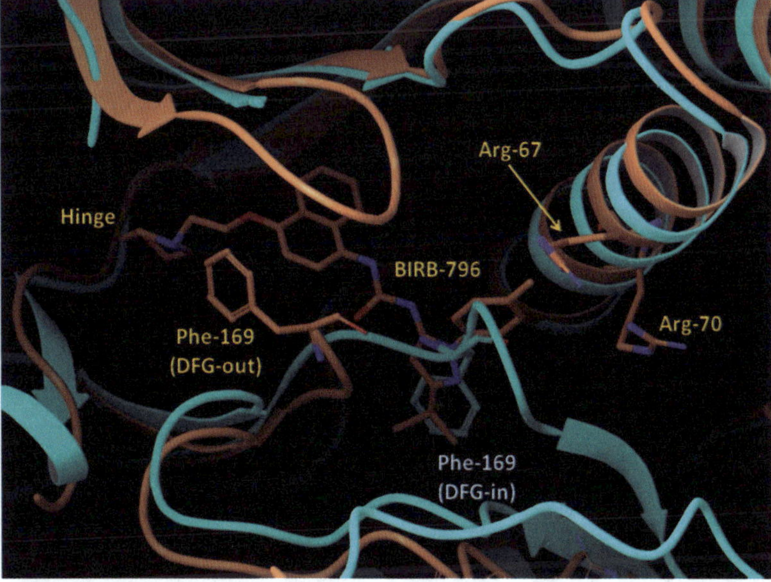

Figure 2.5 Overlay of P38 structure crystallised with ATP (cyan) and BIRB-796 (brown).

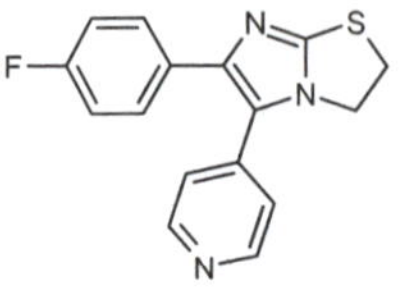

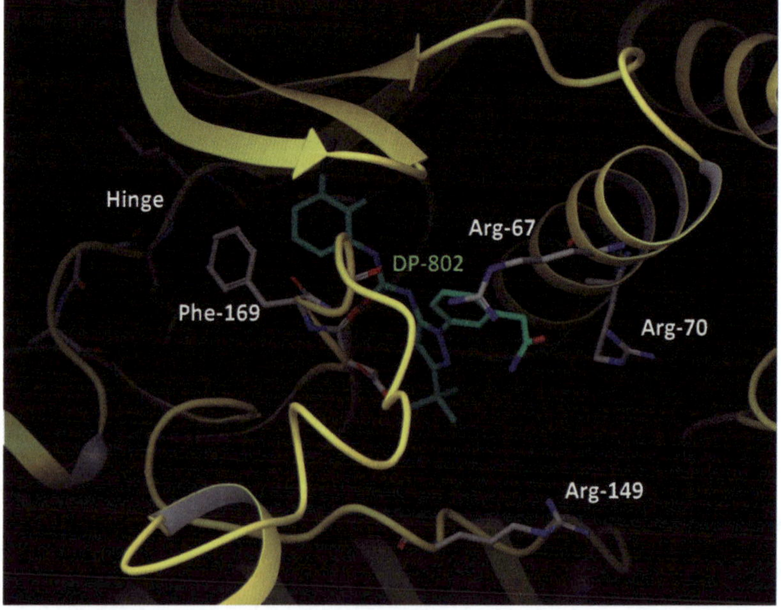

10 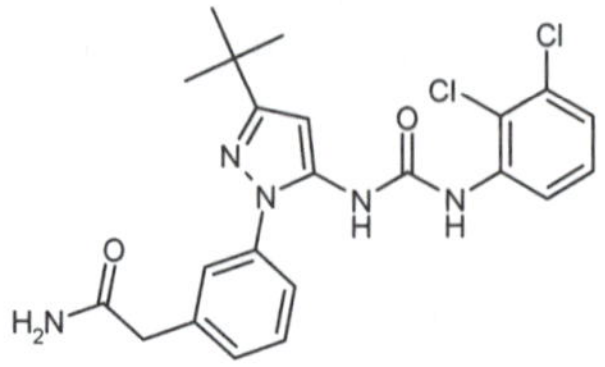   **11** 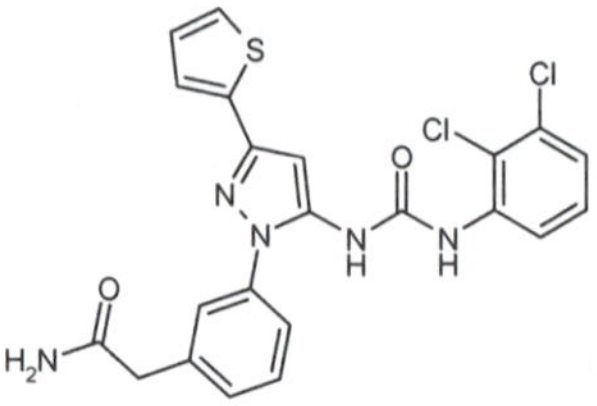   **12**

Figure 2.6   Chemical structures of SFK-86002 (**10**), DP-802 (**11**) and DP-1376 (**12**).

showing significant stabilisation of the kinase structure (ΔT_M = 12.8 °C). Furthermore, DP-802 exhibited slow off-rates (289 min) when evaluated against the active protein, suggesting that this inhibitor binds to a conformation of the kinase which requires time dependant fluxing to release the inhibitor. The selectivity of this compound was assessed against a kinase panel and apart from P38-*alpha* and P38-*beta*, only BRAF (17 nM), CRAF (11 nM) and EPHB2 (82 nM) were significantly inhibited. It is inferred that this selectivity is in part driven by the specific interactions with Arg-70. An X-ray crystal structure was solved for DP-802 in unphosphorylated P38-*alpha* (PDB 3NNW), where the entire activation loop was visible, unlike the 1KV2 structure shown in Figure 2.5. Figure 2.7 shows the binding mode of DP-802 along with key switch pocket amino acid residues in the active site.

Figure 2.7   Binding mode of DP-802 in unphosphorylated P38-*alpha*.

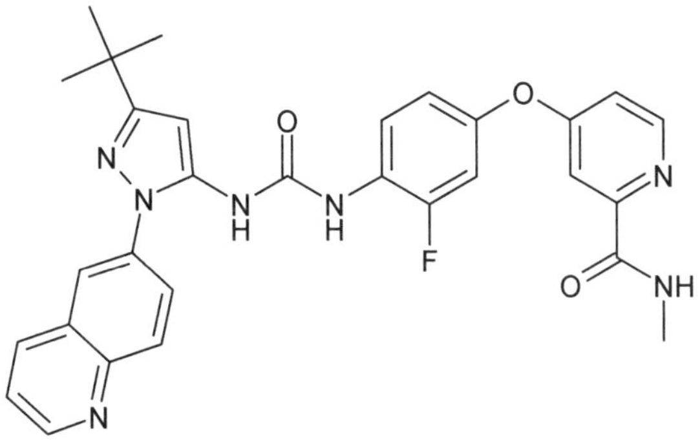

13

Figure 2.8 Chemical structure of DCC-2036 (**13**).

To gain further selectivity over kinases such as BRAF and CRAF, it was observed that Ile-182 at the base of the DFG hydrophobic pocket was typically a larger residue such as leucine. The *tert*-butyl group of DP-802 was therefore replaced with a thiophene group to push further into this region, producing compound DP-1376 (**12**) which showed exquisite selectivity across a panel of kinases. Additionally, Deciphera have exploited equivalent switch pocket residues in BCR-ABL in an effort to overcome the T315I resistance mechanism to which a number of ATP competitive inhibitors are vulnerable to.[18] Their lead compound DCC-2036 (**13**), structure shown in Figure 2.8, is currently in Phase I evaluation for use in patients with imatinib-refractory CML.

It will be interesting to observe how these reported benefits from the exploitation of these switch pocket residues may translate into advantages in the clinic.

2.3.2 Hydrophobic Motifs

Arqule (www.arqule.com) have recently published work on the potential advantages of targeting autoinhibited, inactive kinase conformations. First reported was the discovery of ARQ 197 (**14**), Figure 2.9, a cMET inhibitor which was shown to bind to an inactive kinase conformation and is currently in Phase III studies.[19]

The inhibitor has been described as exquisitely selective across a panel of over 200 kinases due to the specific nature of the interactions it makes with the inactive form of cMET. ARQ 197 has been extensively characterised as non-ATP competitive by biochemical, biophysical and structural studies. X-Ray crystal structure studies of ARQ 197 bound to cMET (PDB 3RHK) provides specific rationale for the selectivity profile of the compound, where the protein is seen in a conformation which has not been observed for other protein kinases. Figure 2.10 shows the overlay with ARQ 197 bound in cMET (brown) with a structure of cMET bound with ATP (3DKC,[20] cyan). For ease of viewing only key side chains have been included and ATP has been removed.

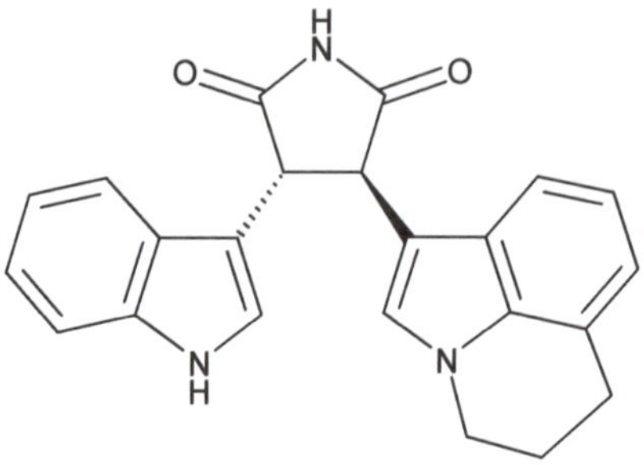

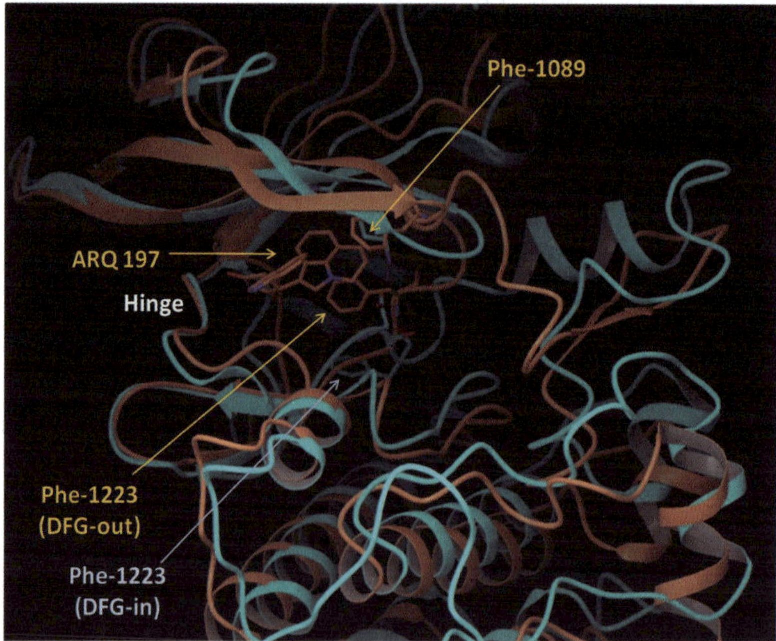

14

Figure 2.9   Chemical structure of ARQ 197 (**14**).

ARQ 197 binds to cMET in the inactive DFG-out conformation, interacting with the hinge region but with significant movements of the glycine-rich loop, activation loop and *c*-helix are required to accommodate the inhibitor. This allows the tri-cyclic group of the inhibitor to interact with centrally located hydrophobic pocket (formed partly by Phe-1089 on the glycine-rich loop and Phe-1223 on the DFG-loop), disrupting the ion-pair interaction of the catalytic residues. It can also be observed how significant the movement of the DFG-

Figure 2.10   Overlay of cMET structure crystallised with ATP (cyan) and ARQ 197 (brown). The structure of ATP has been removed for clarity.

loop movement is between the DFG-out and DFG-in conformations. Arqule have subsequently exploited the learning from their work on cMET with the aim of using this approach in a systematic way against other kinase targets. In a recent paper,[21] they describe their approach to identify and exploit a novel 'hydrophobic motif' in Fibroblast Growth Factor 2 (FGFR2). They used the previously described ARQ 198 bound cMET crystal structure along with kinases crystallised in their autoinhibited form to build a model of the key structural features required for this pocket to be present in a kinase. These features were a relatively conserved phenylalanine in the glycine-rich loop, the phenylalanine of the DFG loop and the aliphatic side-chain of the arginine residue from the activation loop. Their paper also reports on their work to construct homology models of a number of kinase targets and use virtual screening of compound databases in an effort to find chemical leads. This approach identified ARQ 523 (**15**), which bound to the unphosphorylated protein (using mass spectrometric analysis) with K_d 16 μM and was able to reduce the rate of autophosphorylation in a concentration-dependant manner (IC$_{50}$ 18 μM). Efforts were made to improve the affinity of ARQ 523 *via* the introduction of a phenyl group at the 6-position, producing the enantiomers ARQ 068 (**16**) and ARQ 069 (**17**). The chemical structures of these compounds are shown in Figure 2.11.

ARQ 068 showed no affinity to FGFR2 protein in a tryptophan quench assay (K_d > 100 μM) whereas ARQ 069 showed a moderate increase in affinity, compared to ARQ 523, with K_d 5.2 μM. Although the addition of the phenyl group in ARQ 069 did increase affinity, the ligand efficiency (LE),[22] appears to be reduced by this modification. Monitoring LE can be a useful way of assessing if a chemical change has contributed enough affinity to justify the increase in size and–or lipophilicity to the molecule. ARQ 069 showed activity in biochemical assays, with an IC$_{50}$ of 1.23 μM in unphosphorylated FGFR2 protein whereas it was weaker (IC$_{50}$ 24.8 μM) in phosphorylated FGFR2. Further studies demonstrated that activity was not ATP-dependant and that the compound interacts with FGFR2 to preclude ATP binding. ARQ 069

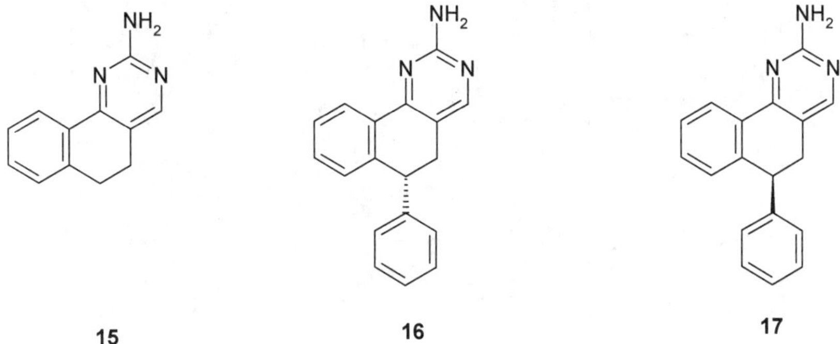

Figure 2.11   Chemical structure of ARQ 523 (**15**), ARQ 068 (**16**) and ARQ 069 (**17**).

appears to have a slow dissociation, which may be due to conformational changes required to release the compound. When tested across a broad panel of 92 kinases, nine of these were inhibited by $> 30\%$ inhibition (at 10 µM compound concentration), however four of these were FGFR family members. Finally, ARQ 069 was found to have an anti-proliferative effect in an FGFR2-dependant human gastric cancer cell line, with an IC_{50} of 9.7 µM. Figure 2.12 highlights crystal structure studies on the binding mode of ARQ 069 (PDB 3RHX, brown) in FGFR1, showing the molecule is binding to a conformation where the side-chain of Phe-489 makes a downward movement and makes contacts with the inhibitor when compared to the *apo*-protein (PDB 1FGK,[23] cyan). In the *apo*-protein structure the Phe-489 side-chain is not visible due to the flexibility of this loop region when unrestrained by compound binding. A further difference in the ARQ 069 bound structure is a small movement of the catalytic lysine (Lys-514) to accommodate the inhibitor. As with ARQ 197, these reported FGFR2 inhibitors all interact with the hinge region of the kinase.

It will be interesting to see how compounds with such mechanisms may be optimised into clinical candidates over the coming years. One particular challenge of compounds exploiting largely hydrophobic regions may be the physicochemical and pharmaceutical properties of the inhibitors and how this may affect their pharmacokinetic (PK) properties *in vivo*.

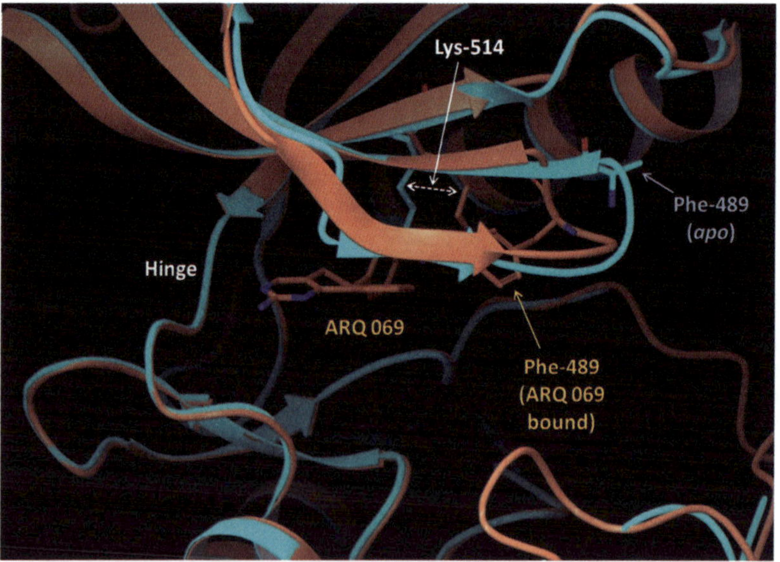

Figure 2.12   Overlay of apo-FGFR1 structure and ARQ 069 bound FGFR1 structure (brown).

2.4 Targeted Kinase Libraries

Although significant challenges have been raised around targeting the active site of kinases, the ever-growing list of FDA-approved kinase inhibitors reinforces the view that this approach should not be ignored. To rise to the challenge of targeting this critical region a number of pharmaceutical companies have enriched their corporate screening collections with compounds targeted specially at kinases. In a recent publication,[24] Abbott discloses their efforts to enrich their kinase screening collection with more than 5000 compounds across chemical 50 series. The focus of their approach was to find novel hinge region binding groups, which is the most crowded area of kinase chemical space and elaborate those cores into chemical libraries to be screened against new targets. Figure 2.13 shows how the key hinge region binding elements of kinase inhibitors can be identified and considered as potential library start points.

Novel hinge region binding groups were identified from NMR screens, biochemical activity screens, virtual screening and through *de novo* design from

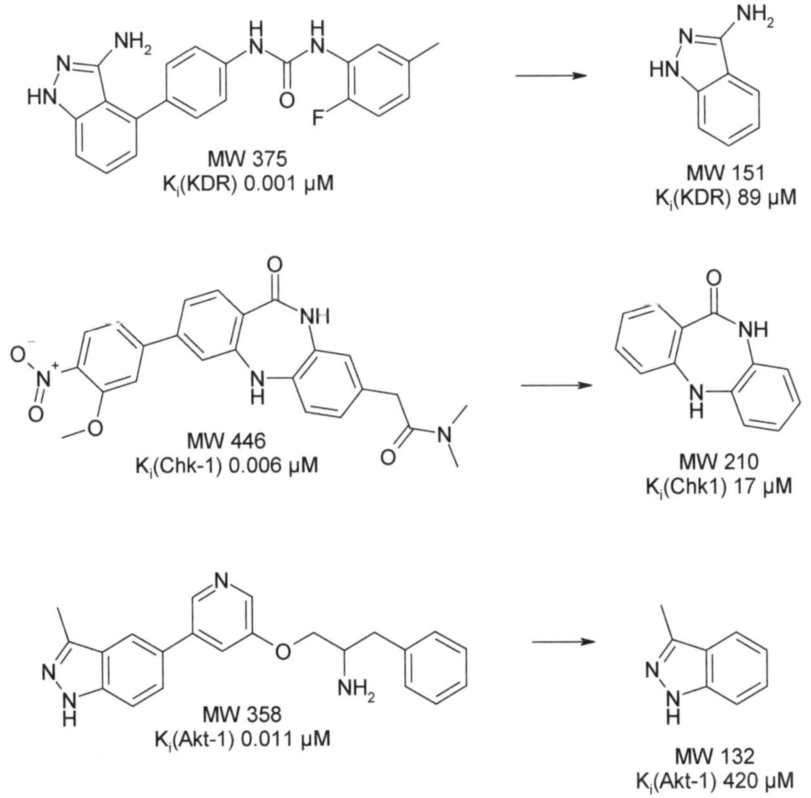

Figure 2.13 Identification of key hinge region binding motifs from kinase inhibitors for consideration as library start points.

knowledge of the critical interactions required for hinge region binding. A complementary approach which we have published at AstraZeneca is the systematic computational enumeration of potential hinge region binding groups around a known binding mode and development of these into synthetic libraries (Figure 2.14).[25]

A total of twenty-nine potential hinge region groups were identified and modified into directed libraries through the addition of substituents which had previously shown kinase activity on alternative scaffolds with the same binding mode. A set of diverse reagents was also added to each scaffold to probe the SAR more widely. An additional observation was that although libraries were targeted to particular binding modes, hinge region binding groups are often able to adopt different binding modes in different kinases, or even against the same kinase.[26] What is also interesting to observe at AstraZeneca, and also reported by Abbott, is that libraries directed at kinases are also observed to bind to other protein targets with enriched hit rates. Often these may be proteins which also bind ATP, or perhaps other co-factors which share some of the key functionality as ATP, but it is encouraging that compounds designed from these principles may have wider application.

2.5 Structure-based Design

Due primarily to the wealth of crystal structures available, along with the high levels of conservation of the ATP pocket, structure-based drug design has made a significant impact in kinase drug discovery.[27] It is beyond the scope of this book to thoroughly review structure-based drug design in its entirety, however, in recent years there have been a number of structure-based approaches published which have proved particularly effective when deployed against kinases. A few such examples are covered in this section.

2.5.1 Scaffold-Hopping and Hybridisation

Kinase chemical space is considered to be very crowded, where it can be challenging to find suitable new inhibitors. As discussed previously this is due to the significant research by a huge number of organisations, against a

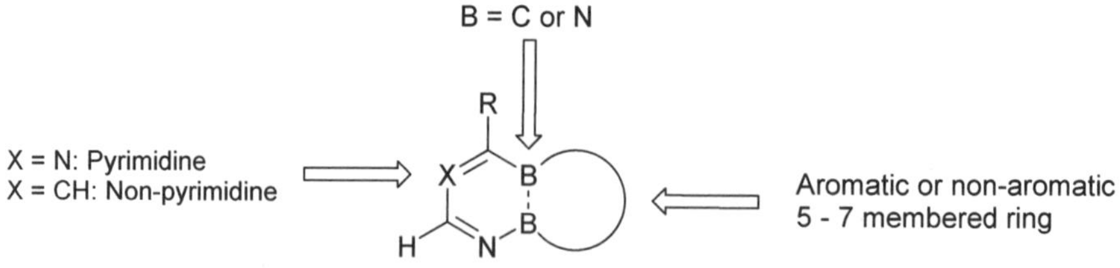

Figure 2.14   Computational enumeration as an approach to design kinase directed libraries. (Figure reproduced in part with permission from J. G. Kettle and R. A. Ward, *J. Chem. Inf. Model.*, 2010, **50** (4), 526 (Table 2)).

number of structurally related kinases. It is therefore no surprise that in the search for new inhibitors, chemistry may diverge out from common chemical start points, or in some cases converge together into similar regions from different chemical start points. Southall *et al.* have published an interesting approach to aid how different patent applications relate to each other from a structural perspective using chemical replacements.[28] In this paper it was shown how the patent applications from particularly crowded areas of chemistry, such as anilino-quinazolines, can be analysed to track scaffold-hopping approaches. For example, it was highlighted that Wyeth had sought to patent 70 molecules across multiple patent applications where the anilino-quinazoline scaffold had been modified to a cyano-quinoline core, Figure 2.15.

Although scaffold-hopping is commonly used as a mechanism of exploring new chemical space, there is a risk that this type of process results in the generation of a chemical series with the same properties (and potential liabilities) as the original series. If the initial compound(s) exploited by scaffold-hopping are significantly further developed, possibly already being evaluated in the clinic, then it is important to consider how a compound less advanced can be differentiated from these. It may be that the leading compound has a known liability that could be improved upon, this may be potency, physico-chemical, or pharmacokinetic properties or perhaps some off-target activity that is seen as a potential issue. One way in which scaffold-hopping approaches have been particularly exploited in kinase drug-discovery is by structural hybridisation. This is where a structural feature of one molecule is hybridised with a structural feature of another. Hybridisation approaches could for example be performed between two inhibitors of the same target to move into new chemical space, and-or remove a liability of one molecule. Alternatively, a feature of an inhibitor of one kinase might be hybridised with a feature from an inhibitor of a different kinase in an aim to gain dual activity. When considering such hybridisation opportunities it is important to not only consider the confidence in the overlay of the ligands, but also whether the hybridised structure could access the required binding mode conformation and–or have suitable electronic properties consistent with the original molecules. The ways in which structural hybridisation can be exploited are varied and allows huge scope for exploitation by drug discovery scientists. Vertex have published an automated method of identifying potential hybridisation opportunities, using

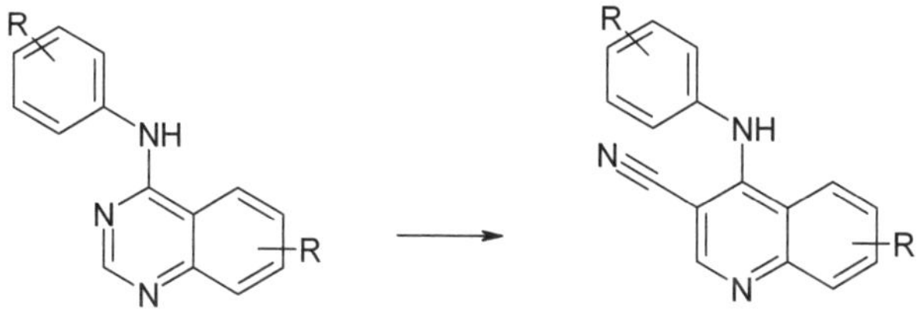

Figure 2.15 Scaffold-hopping from anilino-quinazolines (left) to cyano-quinolines (right).

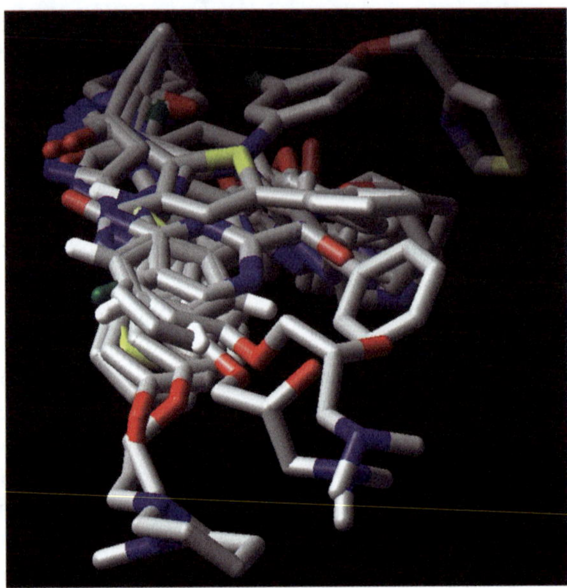

Figure 2.16 Subset of overlayed kinase crystal structures.

an algorithm called BREED.[29] This approach becomes particularly powerful if used in conjunction with a set of proprietary, overlayed, kinase crystal structures (Figure 2.16).

An example of early work around kinase inhibitor hybridisation can be seen from our colleagues at AstraZeneca (Figure 2.17) where a hybridisation approach between two series with a common binding mode (confirmed by X-ray crystallography) lead to the development of potent and selective cyclin-dependant kinase inhibitors.[30] Here they took an imidazo[1,2-a]pyridine series ((**18**), 4 μM in CDK2, 8 μM in CDK4) and a bisanilinopyrimidine series ((**19**), 32 μM in CDK2 and 2 μM in CDK4) and hybridised the solvent channel aniline from compound (**19**) onto compound (**18**) to give compound (**20**) (0.036 μM in CDK2 and 3.6 μM in CDK4). This led to an increase in CDK2 potency of 100-fold, with selectivity over CDK4. Subsequent work led to the identification of potent inhibitors of CDK4, such as compound (**21**) (0.032 μM in CDK2 and 0.15 μM in CDK4). The binding modes of these hybridised inhibitors were also confirmed which validated the hypothesis.

Although it is preferable to have crystal structures of the bound ligands which are to be considered for hybridisation, it is not an absolute requirement, especially when molecular modelling can be exploited to determine reliable binding modes. A more recent publication from AstraZeneca,[31] reports our efforts to use inhibitor hybridisation to rapidly identify Transforming Growth Factor-β Type 1 Receptor (TGFβR1) inhibitors. In this work, a range of reported TGFβR1 inhibitors, along with examples from in-house screening were modelled into a published crystal structure. The inhibitors were

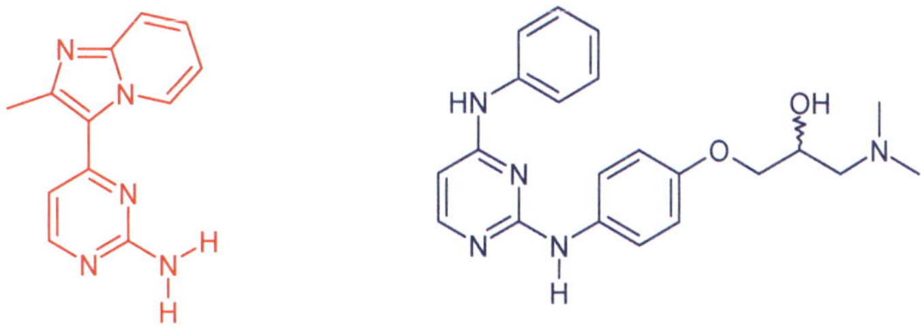

18 4 μM CDK2, 8 μM CDK4

19 32 μM CDK2, 2 μM CDK4

20 0.036 μM CDK2, 3.6 μM CDK4

21 0.032 μM CDK2, 0.15 μM CDK4

Figure 2.17 Hybridisation approach applied to CDK2 and CDK4. The red and blue colours show the regions of each molecule included in the synthesised hybrid molecules.

22 42 nM (enzyme), 55 nM (cell)

23 72 nM (enzyme), 22 nM (cell)

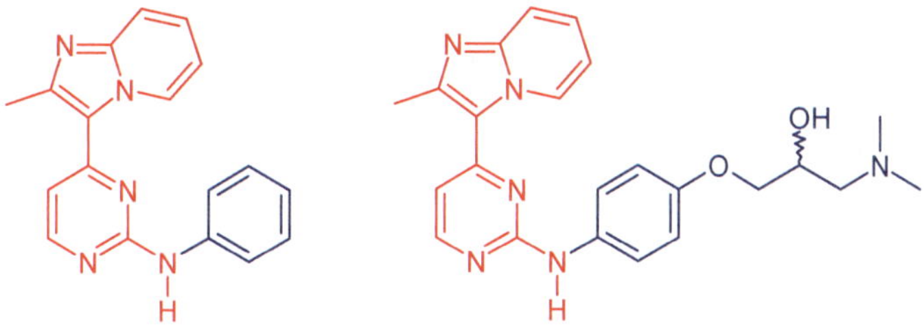

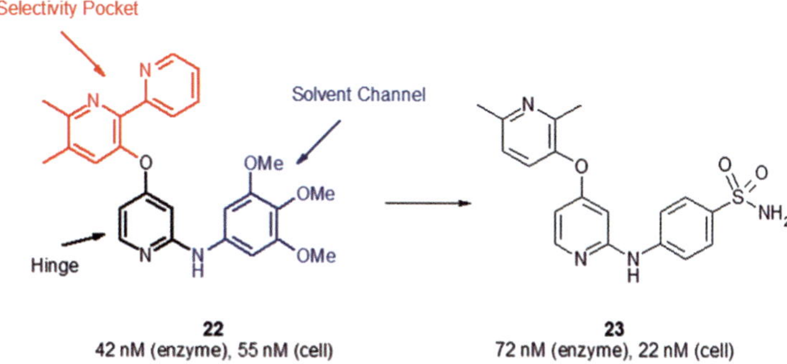

Figure 2.18 Identified inhibitors of TGFβR1 identified through inhibitor hybridisation. The red region shows the selectivity pocket fragment, the hinge region binding element is black and the solvent channel unit is coloured blue.

fragmented into hinge, selectivity pocket and solvent channel units and hybridised together exhaustively so each proposed molecule contained one example of each feature. These potential inhibitors were prioritised by further protein-ligand docking, novelty and synthetic tractability. The process led to the identification of the hit molecule **(22)**, which was subsequently optimised into **(23)** (Figure 2.18). Molecule **(22)** showed potent enzyme and cell inhibition (44 and 55 nM respectively), with **(23)** being broadly similar in terms of potency (72 and 22 nM) but with significantly improved oral bioavailability.

2.5.2 Fragment-Based Lead Generation

Fragment-based lead generation (FBLG) has made significant impact more recently in drug discovery.[32] The hypothesis is that a relatively small number of fragments are screened, which although bind only weakly, fit well into the pocket and can then be elaborated (usually guided by X-ray crystal structures) into lead series. This often requires specialist assays capable of detecting weakly binding compounds with good reproducibility. These fragments are then grown into other regions of the pocket, or alternatively they can be linked with other fragments,[33] which may be identified in other pockets. This compares with a high throughput screening (HTS) approach where a large number of more elaborated compounds are screened most routinely in a biochemical assay, often finding more potent chemical start points, but these compounds may not be elaborated ideally for your target of interest and as a result suffer from poor ligand efficiency (LE) and physicochemical properties. These more elaborated hits may therefore not be ideal chemical leads to develop into a clinical candidate. Although by no means used only against kinases, kinases have proved to be particularly tractable to FBLG approaches.[34] This may be in part due to the availability of X-ray crystal structures, the abundance of known SAR to elaborate fragments into lead series and the hinge region of a kinase which has been shown to be a 'hot-spot' for identifying highly ligand efficient (LE) fragments. An additional benefit of fragment-based approaches is that it removes the requirement for a large compound screening collection which limits HTS approaches to organisations with access to such compound databases. Therefore fragment-based approaches have proved popular with biotech companies and academia as constructing suitable fragment screening libraries is more accessible. There are a number of very good reviews, book chapters and publications around FBLG approaches and successes against kinases in particular, therefore, we have made the decision not to duplicate these here.[35–38] A novel extension of fragment based lead generation has been reported by Erlanson *et al.* to develop potent and selective PDK1 inhibitors.[39] The technique applied is known as "tethering with extenders" which is an evolution of the tethering approach which was disclosed in 2000.[40] The tethering with extenders approach to PDK1 is schematically represented in Figure 2.19. Firstly, the glutamic acid residue at position 166 of the protein was mutated to a cysteine residue. This cysteine residue was then used to

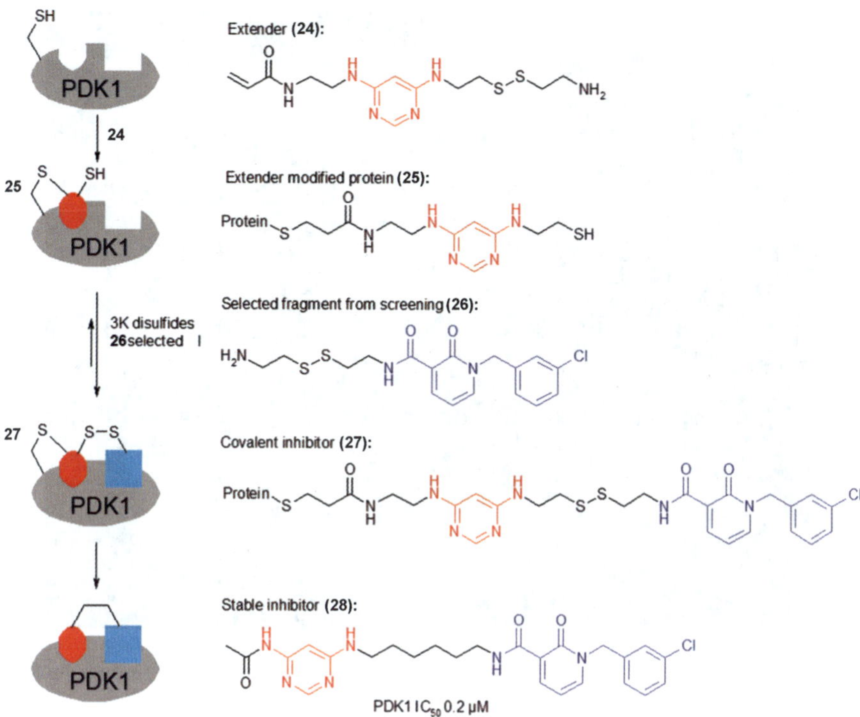

Figure 2.19 Tethering with extenders approach (Figure reproduced in part from D. A. Erlanson *et al.*, *Bioorg. Med. Chem. Lett.*, 2011, **21**, 3079 (Scheme 1) with permission from Elsevier).

covalently tether a small molecule to the protein close the hinge region. The acrylamide functionality of (**24**) was used to form the covalent bond at Cys-166 and the bound molecule also contained the diaminopyrimidine moiety and a thiol linker. It was shown, by X-ray crystallography of the active, DFG-in conformation, that compound (**24**) had covalently reacted with the cysteine 166 residue and the diaminopyrimidine group was interacting with the hinge region of PDK1, through hydrogen bonding with serine 160 (Figure 2.20). Under the covalent labelling conditions the disulfide bond of compound (**24**) was also cleaved to form the extender modified protein structure (**25**). Around 3000 disulfide containing compounds were then screened against the extender modified protein structure (**25**), with the unphosphorylated (inactive) form of the protein, to find fragments that are preferentially to bind *e.g.* (**26**) to the protein and react with the thiol of (**25**) to generate a covalent complex, (**27**). The pyridinone (**26**) was one of the most potent selected fragments and this was used as the basis of further optimisation. Interestingly, this fragment bound less well to the phosphorylated form of the protein, indicating a potential preference for inhibition of the unphosphorylated (inactive) form of the protein. The last stage in the lead finding was to remove the disulfide linker and the acrylamide to provide a stable inhibitor (**28**) that had good potency

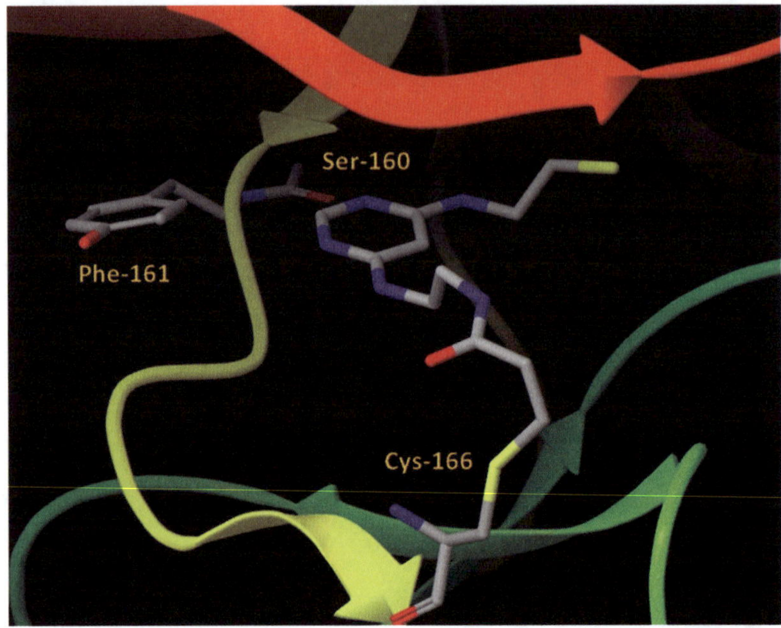

Figure 2.20 X-Ray structure of PDK1 showing the covalent binding of the molecular tether.

against PDK1 in an autophosphorylation assay (IC$_{50}$ 0.2 µM). Compound **(28)** was less active when PDK1 was pre-activated by phosphorylation. This indicates that **(28)** has a preference to bind to the inactive form of the protein.

Compound **(28)** was then optimised, through more conventional structure-activity driven medicinal chemistry, to discover compound **(29)** (Figure 2.21). This compound is a very potent inhibitor of PDK1 (IC$_{50}$ < 0.013 µM against the unphosphorylated protein; IC$_{50}$ 0.002 µM against the phosphorylated protein) and has cellular activity (IC$_{50}$ 0.4 µM). The agent is also highly selective, when screened against a panel of 241 kinases at 10 µM, only one target (other than PDK1) was inhibited (Musk, 98% inhibition) by more than 80%. An X-ray structure of the complex of compound **(29)** bound to PDK1 was also obtained and confirmed the DFG-out binding mode. This is the first example of a DFG-out inhibitor of PDK1 that has been disclosed. Along with being an example of FBLG being used successfully against a kinase target, this specific publication also shows the exploitation of inactive kinase conformations as discussed earlier in this chapter.

This work is a good example of where modification of a protein can be used to anchor a molecular starting point. This approach may be of particular value when targeting less tractable kinases or indeed other targets classes. Additionally, the approach utilised different forms of the target protein to look for different mechanisms of action and in this case uncovered a novel inhibitor series that preferentially binds to the unphosphorylated form of the

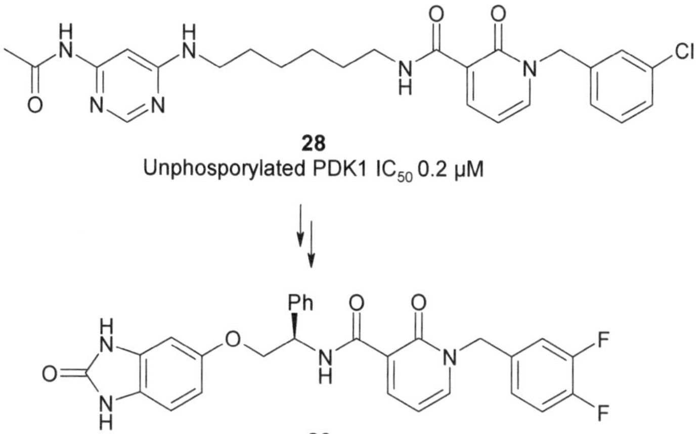

28
Unphosporylated PDK1 IC$_{50}$ 0.2 µM

29
Unphosporylated PDK1 IC$_{50}$ <0.013 µM
Phosphorylated PDK1 IC$_{50}$ 0.002 µM
Cell EC$_{50}$ 0.4 µM

Figure 2.21 Optimisation of the stable lead (**28**) to a potent PDK1 inhibitor (**29**).

protein. These compounds were also shown to be very selective against other kinases.

2.5.3 Virtual Screening

Virtual screening is a term which covers a number of approaches in which computational methodology can be used to identify a subset of compounds to test against a target, the aim being the enrichment in activity of these compounds compared to a random subset.[41] Similarly to FBLG, an attraction of these approaches is that large compound collections are not required, libraries of commercially available compounds,[42] can be assessed computationally, a subset purchased and then tested in a suitable assay. Specific approaches which are often used are protein-ligand docking,[43] and pharmacophore-based methods,[44] using known inhibitors or similarity methods.[45] Although there are features of kinases which may initially appear unsuitable for approaches such as protein-ligand docking,[46] (large active site, significant protein flexibility),[46] experience with this target class has allowed computational chemists to use their knowledge to refine and post-filter docking output quite successfully.[47,48]

Lyne *et al.* at AstraZeneca have published their virtual screening approach to identify checkpoint kinase-1 (CHK1) inhibitors using a combined pharmacophore and docking protocol.[48] A pharmacophore model was constructed from an analysis of kinase protein crystal structures, insisting that molecules had a hydrogen bond donor and acceptor pair within a distance of 1.35–2.40 Å apart to enable bi-dentate coordination to the hinge region. Although there are examples of kinase inhibitors forming single hydrogen bonds to the hinge

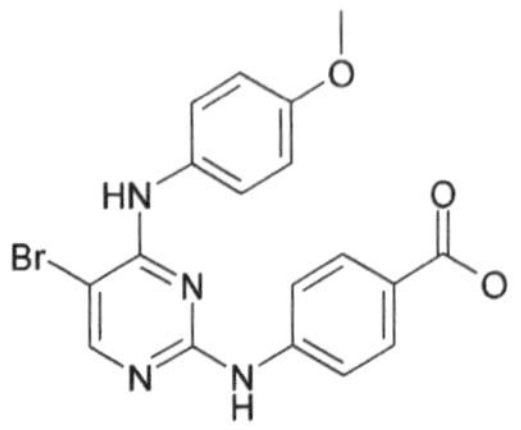

30
IC$_{50}$ 450nM

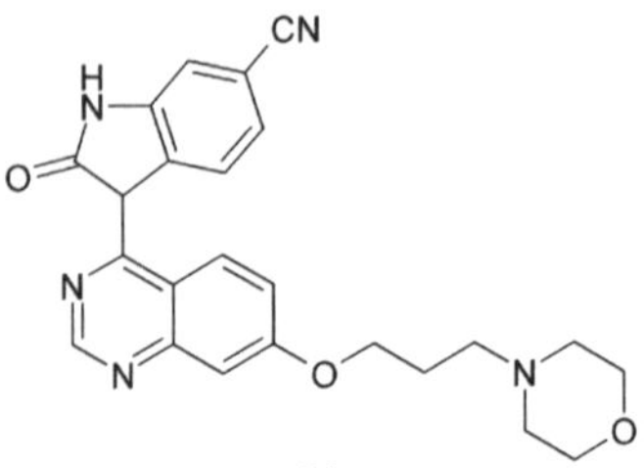

31
IC$_{50}$ 4µM

Figure 2.22   Actives from checkpoint kinase-1 virtual screening (**30** and **31**).

region it was felt that insisting on two hydrogen bonds produced a reasonable number of hits to further analyse by protein-ligand docking. Out of the 103 compounds screened in their biochemical assay, they identified an impressive 36 hits. The activity values for these compounds varied from 110 nM (**30**) to 68 µM (**31**). The chemical structures of these compounds are shown in Figure 2.22.

Additionally, fragments are commonly screened by computational means against kinases, this has proved particularly successful when using constraints to ensure suitable interactions with the hinge region. Further work at AstraZeneca has recently been published which used fragment docking to identify PI3K (p110β isoform) binders.[49] As a PI3K p110β isoform crystal structure was not available, a homology model was constructed using the published p110γ crystal structure (PDB 2CHW). A fragment library of 183 330 unique structures were docked using Glide and post-filtered based on their hydrogen bonds and key interactions with the kinase hinge. Screening of 210 fragments was undertaken, of which 18 showed activity in the biochemical assay, an 8.6% hit rate. Figure 2.23 shows compounds (**32**) and (**33**), the most potent fragment binders from the virtual screening.

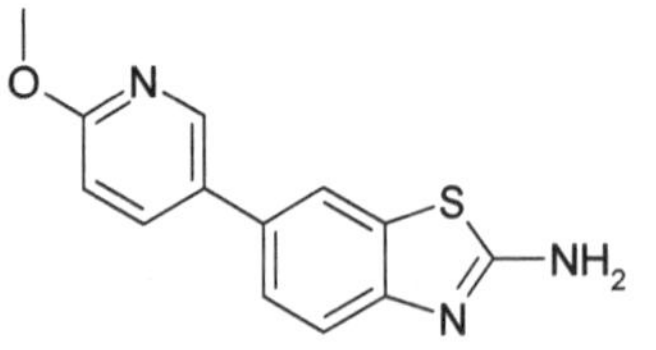

32
IC$_{50}$ 3.6µM

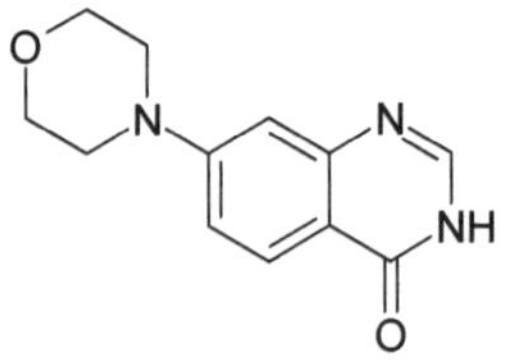

33
IC$_{50}$ 5.8µM

Figure 2.23   Actives from PI3K (p110β isoform) virtual fragment screening (**32** and **33**).

2.6 Summary

In this chapter we have attempted to summarise some of the more recently published approaches to kinase lead-generation which have either been specifically developed to target kinases, or approaches which have been successful against other target classes but have been tailored towards kinases. We believe that in an effort to overcome some of the discussed challenges (and opportunities) in kinase lead generation, many more examples of innovative and imaginative techniques will be published over the coming years. As increasing amounts of data becomes published year on year, whether it is protein structural data or compound activity data, drug discovery scientists will need to constantly respond and rethink how this new knowledge can be effectively exploited. As we gain more experience in kinase lead generation these approaches will be tuned and further improved to maximise their value in future uses. We therefore believe that kinase drug discovery will continue to be a vibrant and important area of drug discovery research for many years to come.

References

1. S. F. Barnett, D. Defeo-Jones, S. Fu, P. J. Hancock, K. M. Haskell, R. E. Jones, J. A. Kahana, A. M. Kral, K. Leander, L. L. Lee, J. Malinowski, E. M. McAvoy, D. D. Nahas, R. G. Robinson and H. E. Huber, *Biochem. J.,* 2005, **385**, 399–408.
2. W. Wu, W. C. Voegtli, H. L. Sturgis, F. P. Dizon, G. P. A. Vigers, and B. J. Brandhuber, *PLoS One,* 2010, **5**, e12913.
3. C. W. Lindsley, Z. Zhao, W. H. Leister, R. G. Robinson, S. F. Barnett, D. Defeo-Jones, R. E. Jones, G. D. Hartman, J. R. Huff, H. E. Huber and M. E. Duggan, *Bioorg. Med. Chem. Lett.,* 2005, **15**, 761–764.
4. J. C. Hartnett, S. F. Barnett, M. T. Bilodeau, D. Defeo-Jones, G. D. Hartman, H. E. Huber, R. E. Jones, A. M. Kral, R. G. Robinson and Z. Wu, *Bioorg. Med. Chem. Lett.,* 2008, **18**, 2194–2197.
5. Z. Wu, R. G. Robinson, S. Fu, S. F. Barnett, D. Defeo-Jones, R. E. Jones, A. M. Kral, H. E. Huber, N. E. Kohl, G. D. Hartman and M. T. Bilodeau, *Bioorg. Med. Chem. Lett.,* 2008, **18**, 2211–2214.
6. M. T. Bilodeau, A. E. Balitza, J. M. Hoffman, P. J. Manley, S. F. Barnett, D. Defeo-Jones, K. Haskell, R. E. Jones, K. Leander, R. G. Robinson, A. M. Smith, H. E. Huber and G. D. Hartman, *Bioorg. Med. Chem. Lett.,* 2008, **18**, 3178–3182.
7. T. Siu, J. Liang, J. Arruda, Y. Li, R. E. Jones, D. Defeo-Jones, S. F. Barnett and R. G. Robinson, *Bioorg. Med. Chem. Lett.,* 2008, **18**, 4186–4190.
8. T. Siu, Y. Li, J. Nagasawa, J. Liang, L. Tehrani, P. Chua, R. E. Jones, D. Defeo-Jones, S. F. Barnett and R. G. Robinson, *Bioorg. Med. Chem. Lett.,* 2008, **18**, 4191–4194.

9. H. Hirai, H. Sootome, Y. Nakatsuru, K. Miyama, S. Taguchi, K. Tsujioka, Y. Ueno, H. Hatch, P. K. Majumder, B. Pan and H. Kotani, *Mol. Cancer Ther.*, 2010, **9**, 1956–1967.
10. C. Pargellis, L. Tong, L. Churchill, P. F. Cirillo, T. Gilmore, A. G. Graham, P. M. Grob, E. R. Hickey, N. Moss, S. Pav, and J. Regan, *Nat. Struct. Biol.*, 2002, **9**, 268–272.
11. E. R. Wood, A. T. Truesdale, O. B. McDonald, D. Yuan, A. Hassell, S. H. Dickerson, B. Ellis, C. Pennisi, E. Horne, K. Lackey, K. J. Alligood, D. W. Rusnak, T. M. Gilmer, and L. Shewchuk, *Cancer Res.* 2004, **64**, 6652–6659.
12. S. E. Szedlacsek and R. G. Duggleby, *Methods Enzymol.*, 1995, **249**, 144–180.
13. R. Capdeville, E. Buchdunger, J. Zimmermann and A. Matter, *Nat. Rev. Drug Discovery*, 2002, **1**, 493–502.
14. D. Strumberg, *Drugs Today*, 2005, **41**, 773–784.
15. Y. M. Ahn, M. Clare, C. L. Ensinger, M. M. Hood, J. W. Lord, W. Lu, D. F. Miller, W. C. Patt, B. D. Smith, L. Vogeti, M. D. Kaufman, P. A. Petillo, S. C. Wise, J. Abendroth, L. Chun, R. Clark, M. Feese, H. Kim, L. Stewart and D. L. Flynn, *Bioorg. Med. Chem. Lett.*, 2010, **20**, 5793–5798.
16. L. A. J. O'Neill, *Nat. Rev. Drug Discovery*, 2006, **5**, 549–563.
17. S. Bellon, M. J. Fitzgibbon, T. Fox, H. M. Hsiao and K. P. Wilson, *Structure* 1999, **7**, 1057–1065.
18. C. A. Eide, L. T. Adrian, J. W. Tyner, M. MacPartlin, D. J. Anderson, S. C. Wise, B. D. Smith, P. A. Petillo, D. L. Flynn, M. W. N. Deininger, T. O'Hare and B. J. Druker, *Cancer Res.* 2011, **71**, 3189–3195.
19. S. Eathiraj, R. Palma, E. Volckova, M. Hirschi, D. S. France, M. A. Ashwell, and T. C. K. Chan, *J. Biol. Chem.* **2011**, *286,* 20666–20676.
20. S. G. Buchanan, J. Hendle, P. S. Lee, C. R. Smith, P. Bounaud, K. A. Jessen, C. M. Tang, N. H. Huser, J. D. Felce, K. J. Froning, M. C. Peterman, B. E. Aubol, S. F. Gessert, J. M. Sauder, K. D. Schwinn, M. Russell, I. A. Rooney, J. Adams, B. C. Leon, T. H. Do, J. M. Blaney, P. A. Sprengeler, D. A. Thompson, L. Smyth, L. A. Pelletier, S. Atwell, K. Holme, S. R. Wasserman, S. Emtage, S. K. Burley and S. H. Reich, *Mol. Cancer Ther.*, 2009, **8**, 3181–3190.
21. S. Eathiraj, R. Palma, M. Hirschi, E. Volckova, E. Nakuci, J. Castro, C. Chen, T. C. K. Chan, D. S. France and M. A. Ashwell, *J. Biol. Chem.*, 2011, **286**, 20677–20687.
22. A. L. Hopkins, C. R. Groom and A. Alex, *Drug Discovery Today*, 2004, **9**, 430–431.
23. M. Mohammadi and J. Schlessinger, S. R. *Cell*, 1996, **86**, 577–587.
24. I. Akritopoulou-Zanze and P. J. Hajduk, *Drug Discovery Today*, 2009, **14**, 291–297.
25. J. G. Kettle and R. A. Ward, *J. Chem. Inf. Model.*, 2010, **50**, 525–533.
26. K. L. Constantine, L. Mueller, W. J. Metzler, P. A. McDonnell, G. Todderud, V. Goldfarb, Y. Fan, J. A. Newitt, S. E. Kiefer, M. Gao,

D. Tortolani, W. Vaccaro and J. Tokarski, *J. Med. Chem.*, 2008, **51**, 6225–6229.

27. L. M. Toledo, N. B. Lydon and D. Elbaum, *Curr. Med. Chem.*, 1999, **6**, 775–805.

28. N. T. Southall, *J. Med. Chem.*, 2006, **49**, 2103–2109.

29. A. C. Pierce, G. Rao and G. W. Bemis, *J. Med. Chem.*, 2004, **47**, 2768–2775.

30. M. Anderson, J. F. Beattie, G. A. Breault, J. Breed, K. F. Byth, J. D. Culshaw, R. P. A. Ellston, S. Green, C. A. Minshull, R. A. Norman, R. A. Pauptit, J. Stanway, A. P. Thomas and P. J. Jewsbury, *Bioorg. Med. Chem. Lett.*, 2003, **13**, 3021–3026.

31. F. W. Goldberg, R. A. Ward, S. J. Powell, J. E. Debreczeni, R. A. Norman, N. J. Roberts, A. P. Dishington, H. J. Gingell, K. F. Wickson and A. L. Roberts, *J. Med. Chem.*, 2009, **52**, 7901–7905.

32. J. S. Albert, Fragment-based lead discovery in *Lead generation approaches in drug discovery*, John Wiley, 2010; pp 105–139.

33. P. J. Hajduk, J. R. Huth and C. Sun, *Methods Princ. Med. Chem.*, 2006, **34**, 181–192.

34. A. Gill, *Mini-Rev. Med. Chem.*, 2004, **4**, 301–311.

35. K. M. Grimshaw, L. K. Hunter, T. A. Yap, S. P. Heaton, M. I. Walton, S. J. Woodhead, L. Fazal, M. Reule, T. G. Davies, L. C. Seavers, V. Lock, J. F. Lyons, N. T. Thompson, P. Workman and M. D. Garrett, *Mol. Cancer Ther.*, 2010, **9**, 1100–1110.

36. J. Lanter, X. Zhang and Z. Sui, *Methods Enzymol.*, 2011, **493**, 421–445.

37. J. Leonard and N. Thorsteinson, in *Novel scaffold replacement methodology applied to the discovery of P38 MAP kinase inhibitors,* 2010, ACS Fall Meeting, abstract COMP-252.

38. A. J. Woodhead, H. Angove, M. G. Carr, G. Chessari, M. Congreve, J. F. Coyle, J. Cosme, B. Graham, P. J. Day, R. Downham, L. Fazal, R. Feltell, E. Figueroa, M. Frederickson, J. Lewis, R. McMenamin, C. W. Murray, M. A. O'Brien, L. Parra, S. Patel, T. Phillips, D. C. Rees, S. Rich, D. Smith, G. Trewartha, M. Vinkovic, B. Williams and A. J. Woolford, *J. Med. Chem.*, 2010, **53**, 5956–5969.

39. D. A. Erlanson, J. W. Arndt, M. T. Cancilla, K. Cao, R. A. Elling, N. English, J. Friedman, S. K. Hansen, C. Hession, I. Joseph, G. Kumaravel, W. Lee, K. E. Lind, R. S. McDowell, K. Miatkowski, C. Nguyen, T. B. Nguyen, S. Park, N. Pathan, D. M. Penny, M. J. Romanowski, D. Scott, L. Silvian, R. L. Simmons, B. T. Tangonan, W. Yang and L. Sun, *Bioorg. Med. Chem. Lett.*, 2011, **21**, 3078–3083.

40. D. A. Erlanson, A. C. Braisted, D. R. Raphael, M. Randal, R. M. Stroud, E. M. Gordon and J. A. Wells, *Proc. Natl. Acad. Sci. U. S. A.*, 2000, **97**, 9367–9372.

41. M. J. Stoermer, *J. Med. Chem.*, 2006, **2**, 89–112.

42. www.emolecules.com.

43. W. M. Rockey and A. H. Elcock, *Curr. Protein Pept. Sci.*, 2006, **7**, 437–457.

44. G. Wolber and T. Langer, *J. Chem. Inf. Comput. Sci.*, 2005, **45**, 160–169.
45. J. Hert, P. Willett, D. J. Wilton, P. Acklin, K. Azzaoui, E. Jacoby and A. Schuffenhauer, *Org. Biomol. Chem.*, 2004, **2**, 3256–3266.
46. C. N. Cavasotto and R. A. Abagyan, *J. Mol. Biol.*, 2004, **337**, 209–225.
47. C. Chuaqui and Z. Deng, J. Singh, *J. Med. Chem.*, 2005, **48**, 121–133.
48. P. D. Lyne, P. W. Kenny, D. A. Cosgrove, C. Deng, S. Zabludoff, J. J. Wendoloski and S. Ashwell, *J. Med. Chem.*, 2004, **47**, 1962–1968.
49. F. Giordanetto, B. Kull, and A. Dellsen, *Bioorg. Med. Chem. Lett.*, 2011, **21**, 829–835.

The Learning and Evolution of Medicinal Chemistry against Kinase Targets

MARTIN E. SWARBRICK

Cancer Research Technology Discovery Laboratories, 12 Rosemary Lane, Cambridge, CB1 3LQ, United Kingdom

3.1 Introduction

More than 500 kinases have been identified in the human genome and, given the importance of phosphoryl transfer processes in cellular signal transduction pathways, it is no surprise that the kinome has been extensively investigated as a family of potential drug targets.[1,2] A large number of small molecule kinase inhibitors have been studied in the context of human diseases, with an overwhelming focus in the field of cancer where several protein kinase inhibitors have progressed through development and onto the market as approved therapeutic agents.[3,4]

As a result of this demonstrated therapeutic potential the identification and optimisation of drug-like small molecule inhibitors for kinase targets has attracted significant attention over the past few decades, and protein kinases have become the second most extensively prosecuted drug target family after G-protein coupled receptors.[3,5] From the medicinal chemist's perspective, kinase drug discovery consistently presents the same key challenges regardless of the intended target. For example, and by no means uniquely to this target class, the challenges of achieving adequate selectivity for the target protein

RSC Drug Discovery Series No. 19
Kinase Drug Discovery
Edited by Richard A. Ward and Frederick Goldberg
© Royal Society of Chemistry 2012
Published by the Royal Society of Chemistry, www.rsc.org

over other family members, of translating enzyme inhibitory activity to efficacy against the native protein and of addressing developability properties such as solubility are typically encountered. Furthermore, the level of interest in kinase targets has created a crowded intellectual property landscape such that achieving and protecting novel chemical space is becoming increasingly challenging. As kinase inhibitor drug discovery has matured a wealth of structural information has been generated and, although again by no means uniquely to this target class, it is not unusual for the insight available from crystallography and/or modelling studies to be at the centre of a kinase inhibitor medicinal chemistry strategy. In this chapter the impact of understanding the structural basis of inhibitor-target interactions, as well as the key characteristics of the inhibitors themselves, on the typical challenges of kinase inhibitor medicinal chemistry will be discussed.

3.2 Structural Basis of Kinase Inhibition and its Effects on Selectivity

A large number of kinase family members have been successfully crystallised and solved at high resolution, generating a wealth of important structural information; at the end of 2010 there were more than 2300 kinase crystal structures deposited in the Protein Data Bank,[6] representing more than 100 unique kinase family members. As every kinase mediates a similar phosphoryl transfer process it is unsurprising that a degree of sequence and structural homology exists, particularly in active site regions, across all family members. This is further exaggerated within subclasses of kinases where the catalytic activity of each member is identical. For example, each member of the protein kinase family catalyses the transfer of a phosphate group from ATP to an acceptor present in an amino acid of a substrate protein, and as a result key structural features are common to all family members; in particular the nucleotide binding domains demonstrate a high degree of structural similarity.

With such structural similarity across family members, the identification and optimisation of selective small molecule inhibitors of specific kinases is clearly a challenging task and one which was believed to be virtually impossible in the early days of kinase drug discovery. Indeed, most early inhibitors exhibited poor selectivity, and protein kinases were considered by some to be an undruggable class of targets due to the complexity of minimising potential side effects by achieving selective inhibition. There have been a number of reports of screening exercises which neatly illustrate the scale of this challenge: for example, researchers at GlaxoSmithKline screened a selection of almost 600 compounds, intended to represent kinase inhibitor chemical space, against 203 protein kinases and found that over 60% of them bound to more than 10 kinases.[7] Despite such evidence the intervening years have seen a large number of successful kinase drug discovery programmes deliver drug-like compounds with a trend towards ever more potent and selective inhibitors.

The challenge of identifying selective kinase inhibitors is further compounded by the scale of the task of assessing cross-reactivity at the other > 500 family members. Beyond choosing related isoforms, close homologues and certain kinases believed or known to be associated with undesirable biological effects, the difficulty of building and running large numbers of assays to measure kinase inhibition activity can severely limit the amount of selectivity data that can realistically be generated. Recognising this, researchers at Ambit Biosciences developed an experimental approach for assessing the specificity of kinase inhibitors based on ATP site-dependent competition binding assays, introducing the kinase dendogram-based interaction map as a depiction of kinase inhibitor selectivity.[8] Briefly, human kinases are expressed as fusions to T7 bacteriophage which renders the attached protein amplifiable and amenable to exquisitely sensitive detection. The binding of test compounds is assessed relative to a small set of immobilised probe ligands, which bind to the ATP site of one or more kinases, by quantifying the amount of fusion protein bound to the solid support. The potential of this competition binding assay methodology was further demonstrated by the evaluation of 38 known kinase inhibitors against a panel of 317 kinases.[9] The depiction of kinase inhibitor selectivity was also developed into the selectivity score (S), initially defined and calculated by dividing the number of kinases found to bind with a $K_d < 3\ \mu M$ by the total number of distinct kinases tested, excluding mutant variants. The selectivity score is intended to enable quantitative comparison between compounds, and scores were found to range from 0.01 for lapatinib (Tykerb®, **7**) to 0.87 for staurosporine. Marketed under the KINOME*scan*® brand, this platform for the broad assessment of kinase inhibitor selectivity against more than 400 kinases is now well established and very widely used, with the original report having received close to 600 literature citations in the six years since it first appeared.[8]

For rationalisation and design of kinase inhibitor selectivity through structural and computational studies, understanding the mode of binding to the enzyme can be considered crucial.[10] ATP-competitive kinase inhibitors typically display a hydrogen bond donor-acceptor motif which interacts with key amino acid residues in the adenine binding pocket. The majority of kinase inhibitors bind to the ATP binding site of the enzyme in its active, or 'DFG-in' state, characterised by the position of the conserved aspartic acid-phenylalanine-glycine (DFG) motif at the beginning of the activation loop in its open conformation. Such compounds are referred to as type I inhibitors and, as they effectively mimic the interaction of the adenine ring of ATP with the hinge region of the kinase, inhibition of multiple kinase family members is not uncommon. Type I inhibitors include the marketed agents gefitinib (Iressa®, **1**), erlotinib (Tarceva®, **2**), sunitinib (Sutent®, **3**), dasatinib (Sprycel®, **4**) and vandetanib (Caprelsa®, **5**, Figure 3.1).

In contrast, a second group of inhibitors bind to the enzyme in an inactive form, the majority of which are characterised by a 'DFG-out' state, with the activation loop in a closed conformation. As well as preventing binding of the

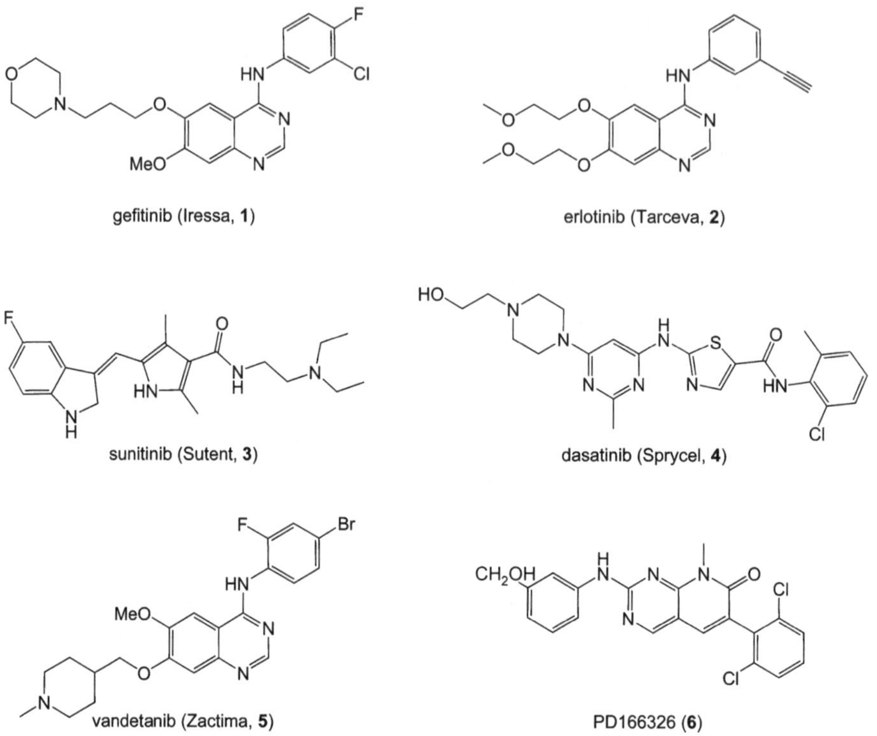

gefitinib (Iressa, **1**)

erlotinib (Tarceva, **2**)

sunitinib (Sutent, **3**)

dasatinib (Sprycel, **4**)

vandetanib (Zactima, **5**)

PD166326 (**6**)

Figure 3.1 Example type I kinase inhibitors.

enzyme substrates, the change in disposition of the DFG motif reveals an additional allosteric hydrophobic pocket which these compounds, known as type II inhibitors, typically interact with in addition to the ATP binding site. Another major inactive form for kinases such as EGFR, where dimerisation causes activation rather than the phosphorylation which drives the DFG-in–out conformational change, is 'c-helix-out'. Although many type II inhibitors contain a hinge binding motif, this is not essential and by specifically targeting the allosteric site they have the potential to be more selective than type I inhibitors, although this is by no means certain. Type II inhibitors include the marketed agents imatinib (Gleevec®, **7**), sorafenib (Nexavar®, **8**), nilotinib (Tasigna®, **9**) and lapatinib (Tykerb®, **10**) and are shown in Figure 3.2.

Type I and type II kinase inhibitor binding modes are exemplified in Figure 3.3, which shows PD166326 (**6**, a) and imatinib (**7**, b) bound to ABL, represented by ribbon structures (PDB 1OPK,[11] and 1IEP,[12] respectively). In the latter, the 'DFG-out' conformation of the activation loop reveals the allosteric site occupied by the benzylpiperazine moiety of imatinib, **7**.

The c-helix in–out conformational change is illustrated in Figure 3.4 which shows lapatinib-bound (brown, PDB 1XKK,[13]) and ATP-bound (blue, PDB 2ITX,[14] ATP removed) EGFR. The movement of the activation loop is also indicated.

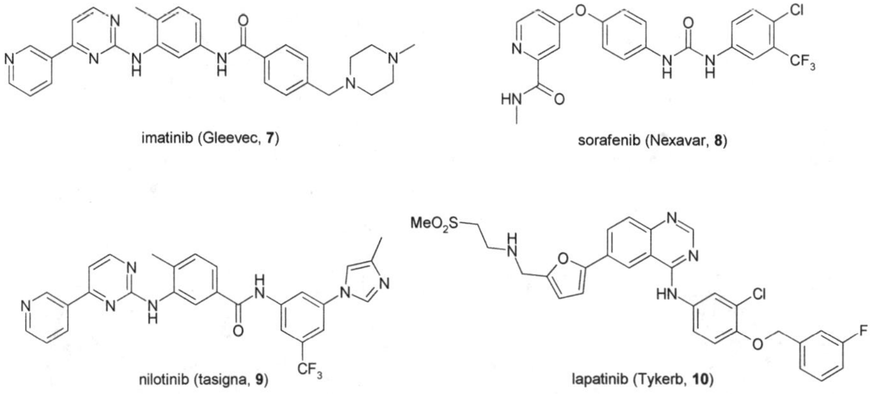

imatinib (Gleevec, **7**)

sorafenib (Nexavar, **8**)

nilotinib (tasigna, **9**)

lapatinib (Tykerb, **10**)

Figure 3.2 Example marketed type II kinase inhibitors.

The KINOME*scan* selectivity scores for a selection of the marketed kinase inhibitors demonstrate the potential for the type II inhibitors imatinib **7**, sorafenib **8** and lapatinib **10** to display higher selectivity than the type I inhibitors sunitinib **3** and dasatinib **4**, especially when only higher affinity off-target interactions are considered (S(100 nM) scores). It can also be seen that, despite their potential to inhibit multiple kinase family members, it is possible to achieve good levels of selectivity with type I inhibitors (*e.g.* gefitinib **1** and erlotinib **2**),[9] (Table 3.1).

The levels of selectivity achieved with type I inhibitors such as gefitinib **1** and erlotinib **2** is thought to arise from interactions with the hydrophobic back cavity of the ATP binding site, often referred to as the 'selectivity pocket' as differences in structure between kinases can often be found in this region. The ATP back cavity was originally recognised by researchers at Boehringer Ingelheim who solved the crystal structure of an inhibitor-bound p38 MAP kinase complex.[15] The size of, and access to, this pocket is controlled by the first residue of the hinge region, known as the gatekeeper residue: for example this key residue is much smaller in B-Raf (threonine gatekeeper) than in CDK-2 (phenylalanine) (Figure 3.5). Targeting this pocket can offer the potential to achieve good levels of selectivity in kinases with an appropriately small gatekeeper residue. The binding mode of a promiscuous compound which does not extend into this region is illustrated in Figure 3.6, where sunitinib **3** bound to KIT kinase (PDB 2G0E, protein removed) has been overlayed with the ATP-bound structure (PDB 1PKG). In contrast, the overlay of gefitinib **1** bound to EGFR kinase (PDB 1ITY, protein removed) with the requisite ATP bound structure (PDB 1ITX) in Figure 3.7 shows it extending towards the gatekeeper–selectivity pocket region.

With kinases so well established as protein targets for drug discovery, it is inevitable that binding modes will emerge which are subtly different to the three broad classes outlined above. Following structural analysis of published kinase inhibitors, researchers at Nerviano recently identified a class of

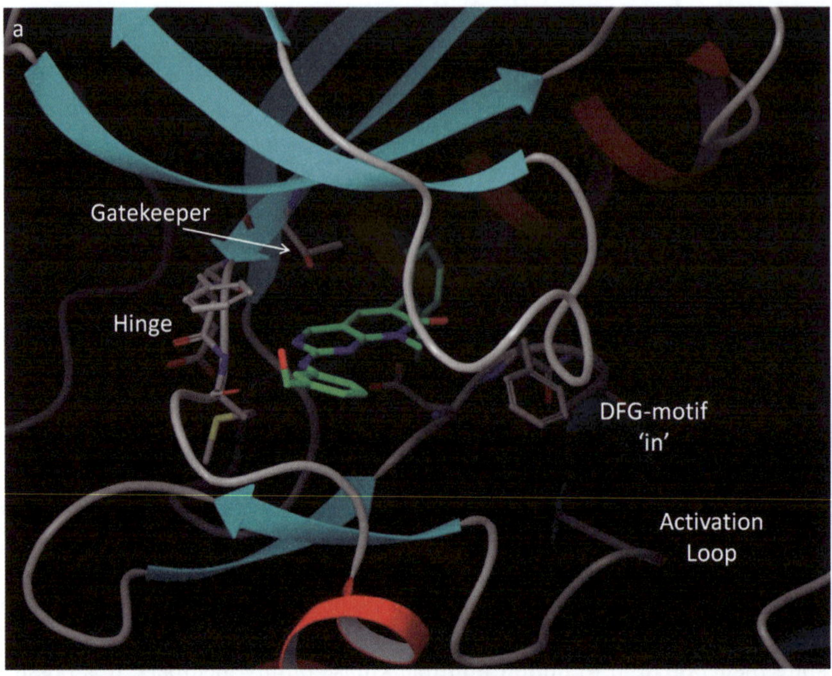

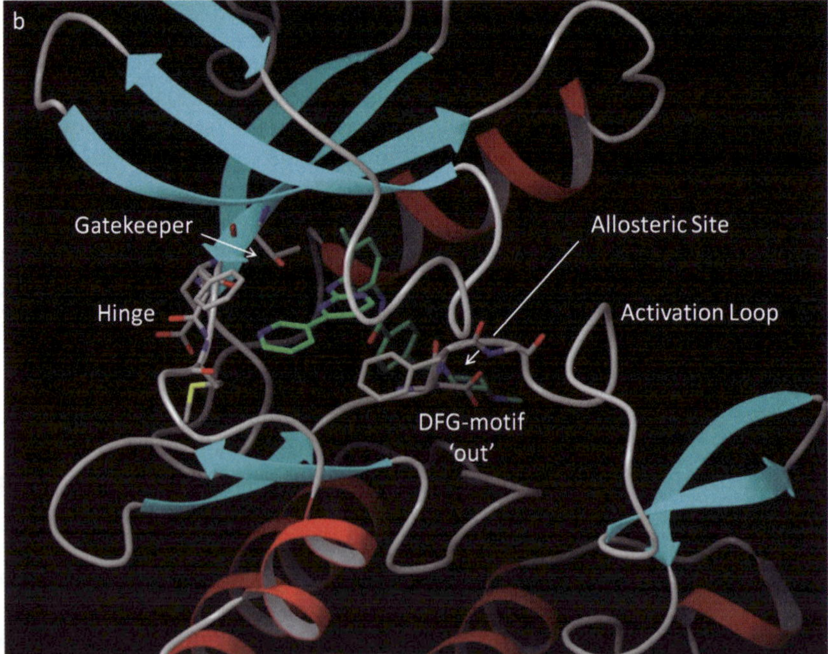

Figure 3.3 Comparison of Type I (a) and Type II (b) kinase inhibitor binding modes.

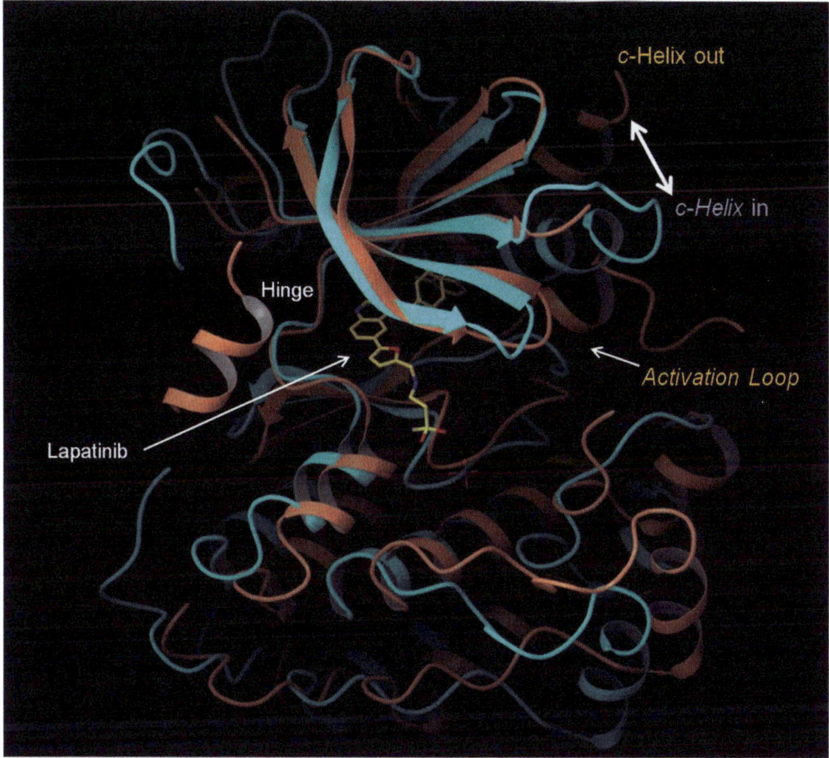

Figure 3.4   Active (*c*-helix-in) and inactive (*c*-helix-out) forms of EGFR.

compounds which target the hydrophobic back cavity of the ATP binding site in *either* the catalytically active (DFG-in) or inactive (DFG-out) conformations.[16] Viewing this novel pharmacophore as a hybrid of types I and II, such compounds have been described as type I½ inhibitors.

Table 3.1   KINOME*scan* selectivity scores (S) for selected marketed kinase inhibitors.

Inhibitor	$S(3\ \mu M)^a$	$S(100\ nM)^b$
gefitinib **1**	0.072	0.007
erlotinib **2**	0.152	0.014
sunitinib **3**	0.569	0.183
dasatinib **4**	0.283	0.159
imatinib **5**	0.066	0.031
sorafenib **6**	0.179	0.048
lapatinib **7**	0.010	0.010

[a]S(3 μM) calculated by dividing the number of kinases found to bind with a $K_d < 3\ \mu M$ by the total number of distinct kinases tested, excluding mutant variants. [b]S(100 nM) calculated by dividing the number of kinases found to bind with a $K_d < 100\ nM$ by the total number of distinct kinases tested, excluding mutant variants.

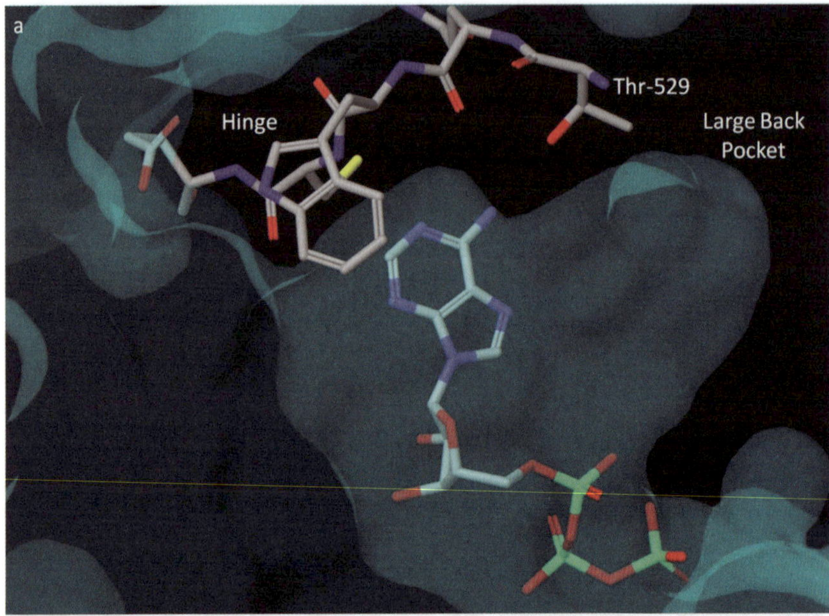

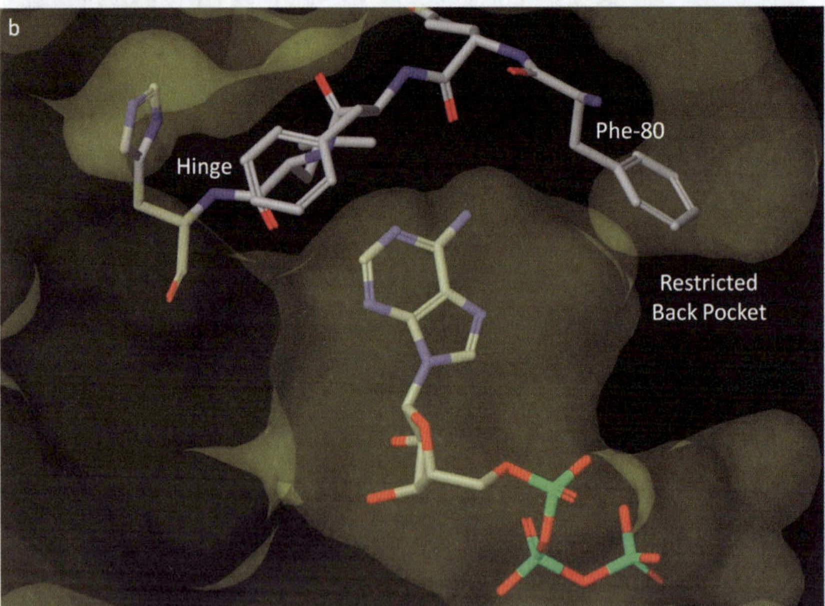

Figure 3.5   Gatekeeper residues in B-Raf (a, PDB 2FB8) and CDK-2 (b, PDB 1B39).

A less well-populated class of compounds, thought to have the potential to exhibit the highest levels of selectivity, modulate kinase activity in a non-ATP-competitive manner.[10] Known as type III inhibitors, these avoid the highly

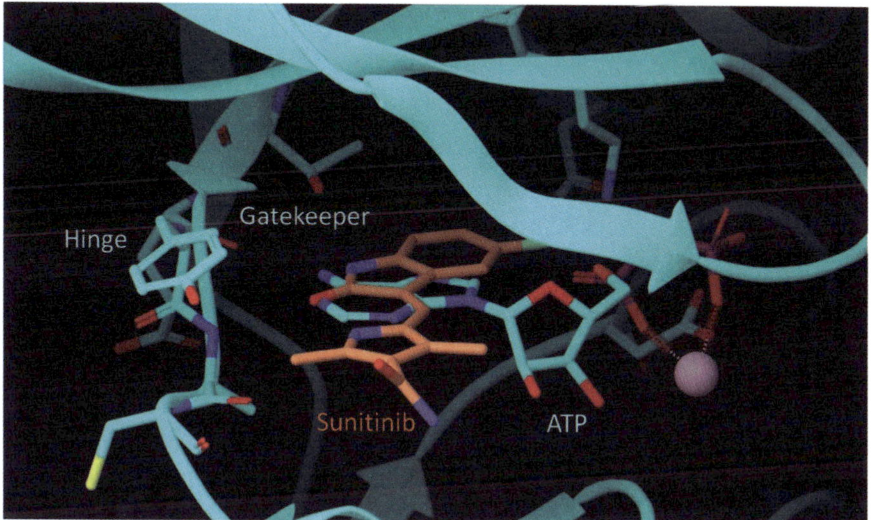

Figure 3.6 Overlays of sunitinib **3** and ATP bound to KIT kinase.

conserved ATP binding site and interact with allosteric sites which are, in principle, unique to a particular kinase. Type III inhibitors include the highly selective MEK inhibitors,[18] such as PD334581 **11** and ARRY-142886 (AZD-6244, **12**) (Figure 3.8). ARRY-142886 **12** is described as having no effect on more than 40 other kinases and a minimal inhibitory effect on MEK5.[17]

As with any other drug target family, no discussion on the challenges of achieving highly selective inhibition of kinases to treat complex human diseases

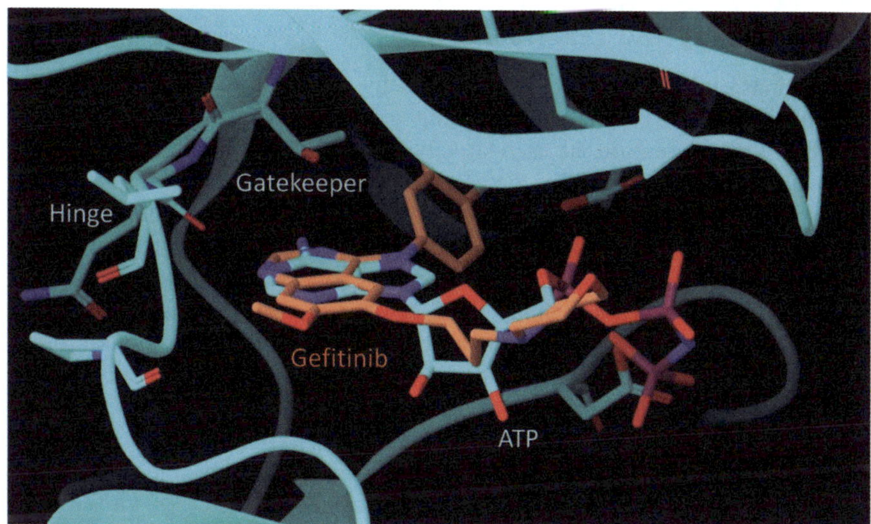

Figure 3.7 Overlays of gefitinib **1** and ATP bound to EGFR kinase.

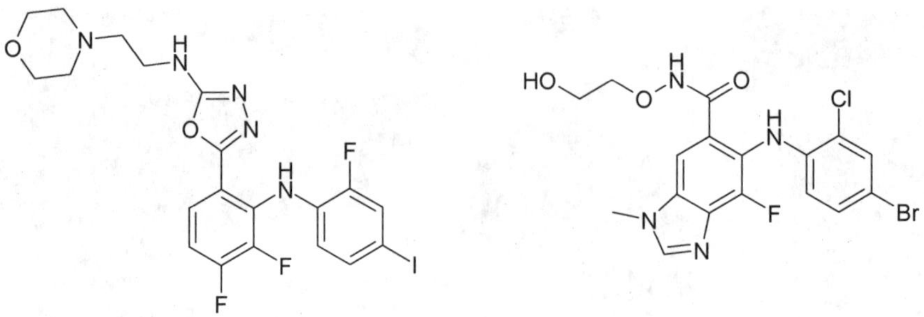

PD334581 (**11**)   ARRY-142886 (AZD-6244, **12**)

Figure 3.8   Example type III kinase inhibitors.

would be complete without also questioning the merits of doing so. The trend towards identifying ever more selective kinase inhibitors has clearly been driven by a desire to minimise the risk unwanted, off-target biological effects. This strategy, though, carries a concomitant risk that efficacy could be compromised by redundancy and bypass mechanisms in signalling pathways. In reality, few clinically efficacious drugs are highly selective for a single target and in a recent review,[19] the case for 'selectively nonselective' inhibition of kinases is argued. Indeed, the first marketed kinase inhibitor imatinib **7** inhibits other tyrosine kinases such as c-Kit and BCR-ABL in addition to its originally intended target PDGFR and displays clinical efficacy against cancers dependent on all three targets.

3.3 Translating Isolated Enzyme Inhibition to Efficacy Against the Native Kinase

Many kinase inhibitor leads are identified by high throughput screening, hit validation and early optimisation using assays based on inhibition of the catalytic kinase domain at a relatively low ATP concentration. As lead optimisation progresses, it becomes increasingly important to measure the potency of compounds against the target in a more native situation, and cell-based assays are used to inform the medicinal chemistry strategy which is aimed at identifying compounds with effects in preclinical models and the potential to treat disease in man. Such cell-based assays typically assess the effects of compounds on, for example, cellular signal transduction using pharmacodynamic (PD) pathway markers or phenotypic effects such as cell proliferation. Although a demonstrably target-specific phenomenon, it is not unusual for there to be a significant drop-off in potency between primary isolated enzyme assays and cellular assays. It is now widely understood that many factors can significantly influence the cellular potency of kinase inhibitors, including high intracellular ATP concentrations, membrane permeability, the presence of phosphatase enzymes and the relative concentrations of the kinase targets and their substrates.[10] High intracellular ATP concentrations are addressed by aiming for high binding affinities, typically in the low nanomolar range, and it is not unusual to attempt to mimic the cellular situation by increasing the ATP concentration in the primary assay as lead optimisation progresses. Cell permeability is typically addressed along with pharmacokinetic (PK) properties in general, where efficient penetration of cells could be expected for compounds achieving a drug-like profile.[20–22] The effects of other complex cellular factors are much more difficult to predict and require detailed understanding of signalling pathways and the coupling between kinase activity and downstream biological effects. Given the intricacy of translating kinase inhibition to cellular potency, it is important to drive lead optimisation and the development of structure-activity relationships using cellular assays where possible.

As well as its effect on the translation of inhibitory potency against the intended target to cellular activity, the complex whole-cell situation can have a profound effect on the observed selectivity of kinase inhibitors. A small selection of data for the type II inhibitor sorafenib **8** demonstrates the potential for selectivity profiles to be different when comparing isolated enzyme or whole cell data, showing it to be a more potent inhibitor of Flt-3 relative to VEGFR-2, PDGFR-β and c-Kit in cells than would be predicted from its kinase inhibitory profile (Table 3.2).[23] Probing the mechanistic basis for cellular kinase inhibitor selectivity, researchers at the University of California demonstrated that differential pathway sensitivity could lead to selective effects in cells.[24]

Cellular assays can also be used in a hit identification–lead generation context and, while this may introduce significant complexity to the understanding of structure-activity relationships and may require extensive and detailed target deconvolution, as an approach it has the advantage of identifying compounds which act on kinases in their native context from the beginning. Cell-based hit identification also offers the potential to identify compounds which may not be active in assays based on inhibition of the catalytic kinase domain, such as type III kinase inhibitors and inhibitors of targets requiring accessory proteins for function.

Table 3.2 Selected isolated enzyme and cellular inhibition data for sorafenib **8**.

Kinase	Enzyme K_i^{app} (nM)[a]	Cellular IC_{50} (nM)[b]
VEGFR-2	4	10 (HUVEC)
PDGFR-β	5	7 (HFF)
c-Kit	15	29 (NCI-H526)
Flt-3	22	1 (RS4;11)

[a]Apparent inhibition constant against purified kinases. [b]Inhibition of ligand-induced receptor autophosphorylation in the cell lines indicated.

3.4   Solubility as a Key Developability Property

Drug dissolution and gastrointestinal permeability are seen as the fundamental parameters that control the rate and extent of drug absorption.[25] Aqueous solubility in particular is a critical attribute in a drug development candidate, as even if permeability is high a drug must be in solution before it can cross the intestinal membrane into the blood. Poor solubility can have a number of significant consequences throughout the drug discovery and development process,[26] ranging from difficulties in obtaining robust and reproducible *in vitro* biological data, to challenges encountered in manufacture and formulation of drug substance, and potentially acute toxicological effects due to crystallisation *in vivo*. Poor absorption, whether due to limiting solubility, limiting permeability or both, is major factor in poor or variable oral bioavailability and exposure. Furthermore, when exposure is limited or difficult to predict and define due to poor absorption, toxicological evaluation can be significantly compromised as well as the assessment of clinical efficacy. While formulation strategies,[27] and the utilisation of prodrug approaches can be very effective in addressing some of the issues caused by sub-optimal solubility and absorption, there is widespread acceptance that the preferred position would be to identify compounds with sufficient aqueous solubility at as early a stage as possible in the drug discovery and development process.

A recent review,[28] describes a general strategy for improving aqueous solubility through decreasing the efficiency of crystal packing as a result of disturbing molecular planarity and symmetry. Several successful examples of improving aqueous solubility are given from discovery programmes spanning the target classes, including representatives of the kinase family. For example, researchers at AstraZeneca found that the introduction of a methyl group at the 2-position of dimethylbenzamide **13** in a series of CDK inhibitors resulted in compound **14** which exhibited > 230-fold higher aqueous solubility at the cost of a very modest decrease in potency.[29] Researchers at Bristol-Myers Squibb also found that the introduction of a substituent, on this occasion an ethoxy group, at the 2-position of a highly planar amide Met kinase inhibitor **15** had an extremely beneficial effect, with compound **16** demonstrating > 40-fold higher solubility (Figure 3.9).[30]

As with most developability properties, the lipophilicity of compounds is an important factor in determining their aqueous solubility,[18] and medicinal chemists are well aware of the many pitfalls associated with improving potency at the desired biological target as a consequence of increased hydrophobic interactions. Modulating lipophilicity is often central to a successful medicinal chemistry programme where a balance must be achieved between potency, which often improves with increasing lipophilicity, and optimal developability properties, which usually require reduced lipophilicity.[20,31,32] The challenge of maintaining the physicochemical properties of kinase inhibitors in particular in drug-like space was illustrated by an analysis of 45 orally bioavailable anticancer protein kinase inhibitors, which found that on average these

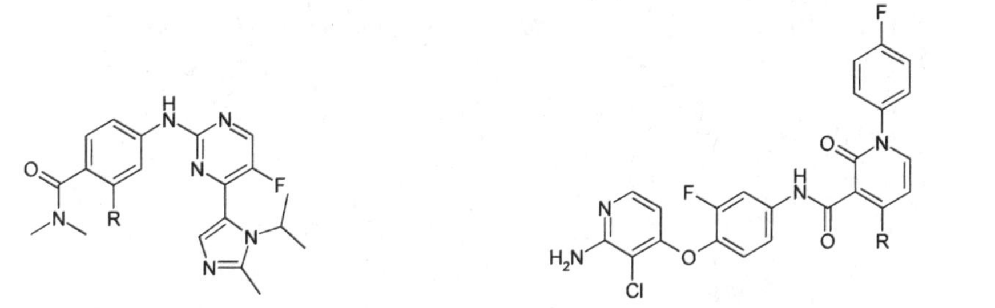

13 (R = H)
CDK2 IC$_{50}$ 2 nM
Solubility 11 µM

14 (R = Me)
CDK2 IC$_{50}$ 9 nM
Solubility >2600 µM

15 (R = H)
Met IC$_{50}$ 1 nM
Solubility (pH 1) <10 µg/mL

16 (R = OEt)
Met IC$_{50}$ 4 nM
Solubility (pH 1) 400 µg/mL

Figure 3.9 Improving solubility by disrupting planarity.

compounds are larger (more than 110 Da) and more lipophilic (more than 1.5 log units) than general oral marketed drugs.[33]

The introduction of particularly hydrophilic functional groups, including those expected to be ionised at physiological pH, is a logical extension of a lipophilicity-lowering approach to improving the aqueous solubility of compounds. Given the important contribution of hydrophobic interactions towards the binding of molecules with their target proteins, it is perhaps not surprising that this approach is not a universally successful medicinal chemistry strategy since the presence of strongly hydrophilic or even charged groups will often interfere with ligand-protein binding. One of the most striking features of protein kinases is that, in contrast to many drug target classes including, for example, nuclear hormone receptors, G-protein coupled receptors and lipid-metabolising enzymes, they are often able to tolerate inhibitors which contain strongly hydrophilic functionality. Among the key structural features of the conserved protein kinase active site (Figure 3.10) is the solvent-exposed front pocket, which is often exploited in order to introduce overall lipophilicity-lowering and solubilising functionality into inhibitors. This has been a successful strategy in the identification of protein kinase inhibitors, and many have been shown to interact with the front pocket including gefitinib **1**, where the morpholine moiety occupies this space (Figure 3.11). In addition to tolerating substituents designed to modulate physicochemical properties, the front pocket can also be exploited to realise significant potency and selectivity gains.

3.5 Intellectual Property Considerations

The identification and optimisation of drug-like small molecule inhibitors for kinase targets is a highly competitive arena and given the high level of structural similarity across family members it is unsurprising that the intellectual property landscape has become crowded, especially around some of the more validated targets. The situation is particularly apparent for type I kinase

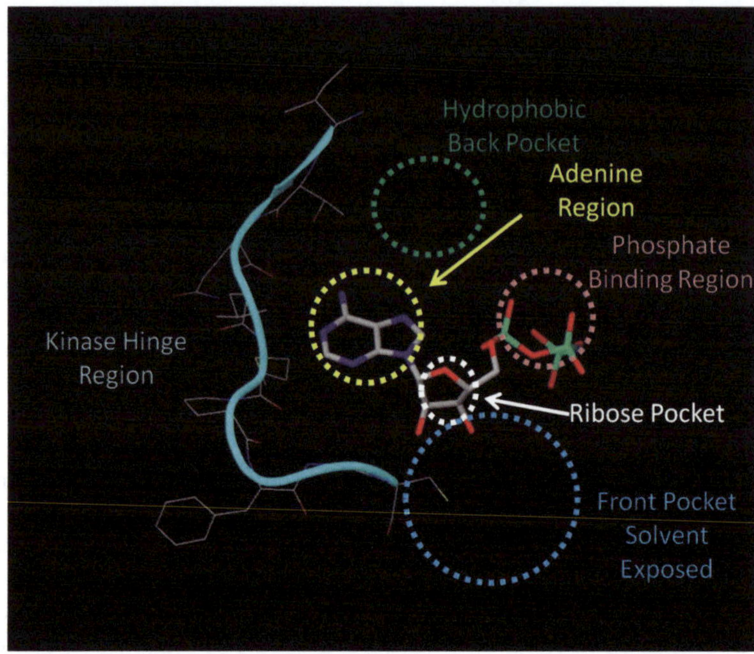

Figure 3.10   ATP-bound EGFR (PDB 2ITX) with key structural features of the protein kinase active site marked.

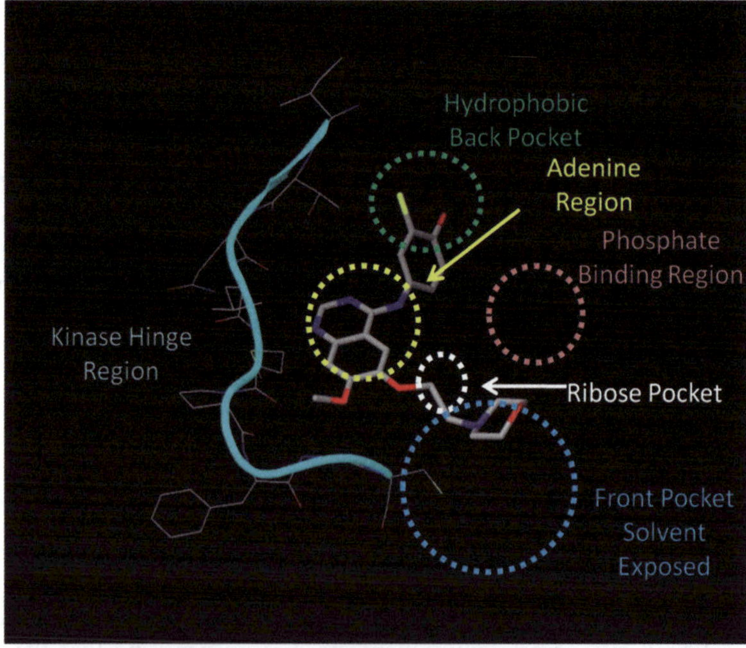

Figure 3.11   Gefitinib-bound EGFR (PDB 2ITY) with the morpholine group occupying the solvent-exposed front pocket.

inhibitors, where it is becoming increasingly difficult to identify novel scaffolds which mimic the interaction of the adenine ring of ATP with the hinge region of the kinase. Although the situation is conceptually more favourable for type II and type III inhibitors due to interactions made with less well-conserved allosteric binding sites, most examples of the former also contain a typical hinge binding motif and pose similar challenges in achieving novelty. As the field has developed intellectual property strategies have varied between the extremes of early and broad filings, aimed at establishing a foothold around a given chemical series, to an approach of late and narrow filings to ensure a secure and competitive position with maximum possible patent term. Competitive advantage is a key factor in any medicinal chemistry programme, and there is a continuing need for medicinal chemists to rise to the challenge of exploring and increasing the boundaries of novelty,[34] within drug-like chemical space.

3.6　Summary

The demonstrated therapeutic potential of kinase inhibitors has led to them becoming one of the most extensively prosecuted drug target families, creating an extremely competitive landscape. Over the past few decades medicinal chemists have developed a thorough understanding of the key challenges presented by this target class, and the approaches to tackling some of these has been discussed. Although the merits of highly selective kinase inhibition as a therapeutic strategy are often debated, there will always be a need to avoid the modulation of kinases believed or known to be associated with undesirable biological effects, and selectivity is likely to remain a central theme of kinase medicinal chemistry strategy. Tackling efficacy as a cause of attrition in drug development continues to drive a need for ever-more thorough understanding of the translation of enzyme inhibitory activity to effects in the contexts of whole cells and whole organisms, and the consideration of developability properties including solubility is certain to remain at the centre of modern medicinal chemistry.

References

1. G. Manning, D. B. Whyte, R. Martinez, T. Hunter and S. Sudarsanam, *Science*, 2002, **298**, 1912.
2. B. D. Manning, *Sci. Signal.*, 2009, **2**(63), pe15.
3. P. Cohen, *Nat. Rev. Drug Discovery*, 2002, **1**, 309.
4. J. Zhang, P. L. Yang and N. S. Gray, *Nat. Rev. Cancer*, 2009, **9**, 28.
5. A. L. Hopkins and C. R. Groom, *Nat. Rev. Drug Discovery*, 2002, **1**, 727.
6. H. M. Berman, K. Henrick and H. Nakamura, *Nat. Struct. Biol.*, 2003, **10**, 980.
7. P. Bamborough, D. Drewry, G. Harper, G. K. Smith and K. Schneider, *J. Med. Chem.*, 2008, **51**, 7898.

8. M. A. Fabian, W. H. Biggs III, D. K. Treiber, C. E. Atteridge, M. D. Azimioara, M. G. Benedetti, T. A. Carter, P. Ciceri, P. T. Edeen, M. Floyd, J. M. Ford, M. Galvin, J. L. Gerlach, R. M. Grotzfeld, S. Herrgard, D. E. Insko, M. A. Insko, A. G. Lai, J. -M. Lélias, S. A. Mehta, Z. V. Milanov, A. M. Velasco, L. M. Wodicka, H. K. Patel, P. P. Zarrinkar and D. J. Lockhart, *Nat. Biotechnol.*, 2005, **23**, 329.

9. M. W. Karaman, S. Herrgard, D. K. Treiber, P. Gallant, C. E. Atteridge, B. T. Campbell, K. W. Chan, P. Ciceri, M. I. Davis, P. T. Edeen, R. Faraoni, M. Floyd, J. P. Hunt, D. J. Lockhart, Z. V. Milanov, M. J. Morrison, G. Pallares, H. K. Patel, S. Pritchard, L. M. Wodicka and P. P. Zarrinkar, *Nat. Biotechnol.*, 2008, **26**, 127.

10. Z. A. Knight and K. M. Shokat, *Chem. Biol.*, 2005, **12**, 621.

11. B. Nagar, O. Hantschel, M. A. Young, K. Scheffzek, D. Veach, W. Bornmann, B. Clarkson, G. Superti-Furga and J. Kuriyan, *Cell*, 2003, **112**, 859.

12. B. Nagar, W. G. Bornmann, P. Pellicena, T. Schindler, D. R. Veach, W. T. Miller, B. Clarkson and J. Kuriyan, *Cancer Res.*, 2002, **62**, 4236.

13. E. R. Wood, A. T. Truesdale, O. B. McDonald, D. Yuan, A. Hassell, S. H. Dickerson, B. Ellis, C. Pennisi, E. Horne, K. Lackey, K. J. Alligood, D. W. Rusnak, T. M. Gilmer and L. Shewchuk, *Cancer Res.*, 2004, **64**, 6652.

14. C. -H. Yun, T. J. Boggon, Y. Li, S. Woo, H. Greulich, M. Meyerson, M. J. Eck, *Cancer Cell*, 2007, **11**, 217.

15. L. Tong, S. Pav, D. M. White, S. Rogers, K. M. Crane, C. L. Cywin, M. L. Brown and C. A. Pargellis, *Nat. Struct. Mol. Biol.*, 1997, **4**, 311.

16. F. Zuccotto, E. Ardini, E. Casale, M. Angiolini, *J. Med. Chem.*, 2010, **53**, 2681.

17. C. -S. Lee and N. S. Duesbery, *Curr. Enzyme Inhib.*, 2010, **6**, 146.

18. T. C. Yeh, V. Marsh, B. A. Bernat, J. Ballard, H. Colwell, R. J. Evans, J. Parry, D. Smith, B. J. Brandhuber, S. Gross, A. Marlow, B. Hurley, J. Lyssikatos, P. A. Lee, J. D. Winkler, K. Koch and E. Wallace, *Clin. Cancer Res.*, 2007, **13**, 1576.

19. R. Morphy, *J. Med. Chem.*, 2010, **53**, 1413.

20. C. A. Lipinski, *J. Pharmacol. Toxicol. Methods*, 2000, **44**, 235.

21. P. D. Leeson and A. M. Davis, *J. Med. Chem.*, 2004, **47**, 6338.

22. P. D. Leeson and B. Springthorpe, *Nat. Rev. Drug Discovery*, 2007, **6**, 881.

23. R. Kumar, M. -C. Crouthamel, D. H. Rominger, R. R. Gontarek, P. J. Tummino, R. A. Levin and A. G. King, *Br. J. Cancer*, 2009, **101**, 1717.

24. C. Kung, D. M. Kenski, K. Krukenberg, H. D. Madhani and K. M. Shokat, *Chem. Biol.*, 2006, **13**, 399.

25. G. L. Amidon, H. Lennernäs, V. P. Shah and J. R. Crison, *Pharm. Res.*, 1995, **12**, 413.

26. S. Stegemann, F. Leveiller, D. Franchi, H. de Jong and H. Lindén, *Eur. J. Pharm. Sci.*, 2007, **31**, 249.

27. M. Palucki, J. D. Higgins, E. Kwong and A. C. Templeton, *J. Med. Chem.*, 2010, **53**, 5897.

28. M. Ishikawa and Y. Hashimoto, *J. Med. Chem.,* 2011, **54**, 1539.
29. C. D. Jones, D. M. Andrews, A. J. Barker, K. Blades, K. F. Byth, M. R. V. Finlay, C. Geh, C. P. Green, M. Johannsen, M. Walker, H. M. Weir, *Bioorg. Med. Chem. Lett.*, 2008, **18**, 6486.
30. G. M. Schroeder, Y. An, Z. -W. Cai, X. -T. Chen, C. Clark, L. A. M. Cornelius, J. Dai, J. Gullo-Brown, A. Gupta, B. Henley, J. T. Hunt, R. Jeyaseelan, A. Kamath, K. Kim, J. Lippy, L. J. Lombardo, V. Manne, S. Oppenheimer, J. S. Sack, R. J. Schmidt, G. Shen, K. Stefanski, J. S. Tokarski, G. L. Trainor, B. S. Wautlet, D. Wei, D. K. Williams, Y. Zhang, Y. Zhang, J. Fargnoli and R. M. Borzilleri, *J. Med. Chem.,* 2009, **52**, 1251.
31. M. J. Waring, *Expert Opin. Drug Discovery,* 2010, **5**, 235.
32. M. P. Gleeson, A. Hersey, D. Montanari and J. Overington, *Nat. Rev. Drug Discovery,* 2011, **10**, 197.
33. A. L. Gill, M. Verdonk, R. G. Boyle and R. Taylor, *Curr. Top. Med. Chem.*, 2007, **7**, 1408.
34. W. R. Pitt, D. M. Parry, B. G. Perry and C. R. Groom, *J. Med. Chem.,* 2009, **52**, 2952.

CHAPTER 4

The Mechanisms and Kinetics of Protein Kinase Inhibitors

WALTER H. J. WARD*

Walter Ward Consultancy & Training, UK

4.1   Introduction

There has been considerable mechanistic research on the inhibition of protein kinases, using a powerful combination of data on protein sequences, 3-D structures, binding and kinetics to yield insight, which not only helps to understand the biological activity of current kinase inhibitors, but also assists in the identification, evaluation and optimisation of future compounds. During drug discovery, enzyme inhibitors often are characterised in isolated protein assays where the only experimental variable is the concentration of test compound. There is a wide awareness that the conditions of these assays influence the potency or IC_{50} (concentration giving 50% inhibition). However, in order to control cost and maximise throughput, assay conditions usually are designed to be quite different to those *in vivo*, so that the results may have limited relevance in drug discovery. We shall see how rates of success may be increased by designing more relevant assays, using full-length (rather than truncated) kinases and authentic substrates at physiological concentrations. The determinants of potency, selectivity, efficacy and safety are complex and multifactorial. In this review, I describe how mechanistic characterisation of kinase inhibitors, by varying substrate concentration, time and enzyme concentration can contribute towards unraveling these complexities to yield information, which has implications for drug discovery and clinical activity.

RSC Drug Discovery Series No. 19

Kinase Drug Discovery

Edited by Richard A. Ward and Frederick Goldberg

© Royal Society of Chemistry 2012

Published by the Royal Society of Chemistry, www.rsc.org

* Email: wal.ward@hotmail.co.uk, Web: www.linkedin.com/walterward

4.2 Mechanisms of Inhibition

Mechanisms may be classified in various ways. Now, I differentiate on the basis of the identity of the binding site, whereas later I describe different kinetic mechanisms. A brief summary of protein kinase 3-D structures helps to explain the observed properties of inhibitors.

4.2.1 Protein Kinase 3-D Structures

The human genome encodes over 500 protein kinases, which exhibit strong similarities in amino-acid sequence at certain positions, which play important roles in structure or function.[1] These enzymes show a high degree of conservation in the 3-D structures of their catalytic domains. A well-studied example is protein kinase A (also known as PKA or cAMP-dependent protein kinase).[2] The catalytic domain consists of two lobes and an intervening linker region. The N-lobe (amino-acid residues 15–120), contains α-helices and β-sheets. The C-helix (residues 85–97) is important in the binding of some kinase inhibitors. Following the N-lobe, is the hinge region (residues 121–127), where main-chain amides form conserved hydrogen bonding interactions with the adenine moiety of ATP in a pocket, sometimes known as the purine site. Next is the C-lobe (residues 128–350), which is largely α-helical. The C-lobe interacts with the bound phosphoacceptor substrate, and contains the activation loop between residues 184 and 208, which is flanked by conserved tripeptide sequences AspPheGly (DFG) and AlaProGlu (APE). The activation loop contains residues (Thr197 in PKA), which may be phosphorylated in many kinases, leading to increases in catalytic activity, and affinity both for ATP and phosphoacceptor substrate. The regulatory phosphate interacts with basic groups in both the N and C-lobes, stabilising an open conformation of the activation loop. In particular, it forms a salt bridge with Arg165, which precedes the catalytic Asp166. This conserved RD sequence is characteristic of kinases regulated by activation loop phosphorylation.[2]

A ubiquitous feature of kinase ATP-binding sites (Figure 4.1) is a hydrogen bond donated by a main-chain amide in the purine site (Val123 in PKA). The α- and β-phosphates of ATP interact with the ammonium group of a conserved Lys (residue 72 in PKA, replaced by a more distant Lys in the WNK family), which is held in place by a charged interaction with a conserved Glu (residue 91 in PKA) from the C-helix. The phosphates of bound ATP are chelated with Mn^{2+} in the crystal structure, which is replaced by Mg^{2+} *in vivo* and, by analogy with other kinases, this is thought to interact with the carboxylate on the side-chain of Asp184 from the DFG-motif.

4.2.2 Binding Modes for Early Kinase Inhibitors

Most kinase inhibitors mimic the interactions of the adenine moiety of ATP, binding in the purine site (Figures 4.2 and 4.3).[3] Between 1 and 3 hydrogen

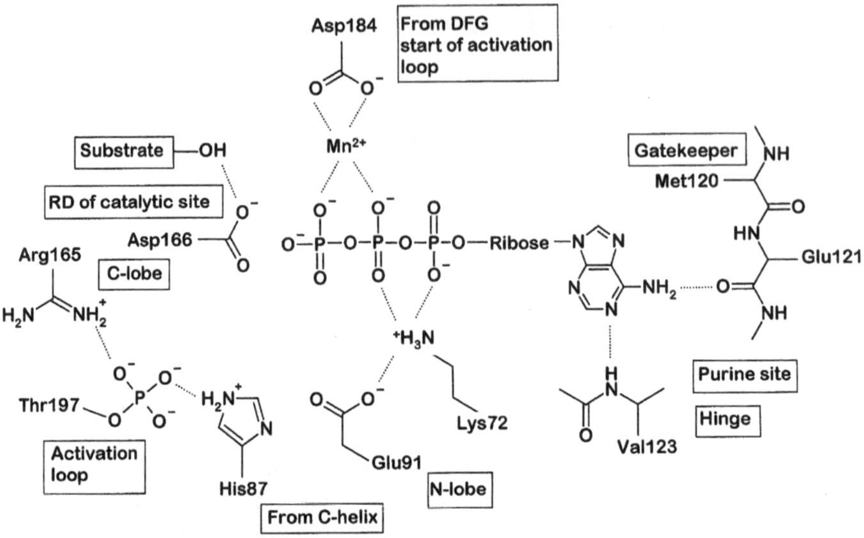

Figure 4.1 The ATP-binding site of protein kinase A. Key polar interactions are shown by dashed lines. Drawn from the crystal structure in Protein Data Bank file 1atp. Mn^{2+} is replaced by Mg^{2+} *in vivo*.

bonds usually are formed with main-chain atoms in the hinge. Ligands making 2 or 3 hydrogen bonds typically have interacting atoms in *cis*, usually alternating with non-hydrogen bonding atoms (up to a 1, 3, 5 interaction pattern). This binding mode can compromise potency in cells due to competition by ATP and ADP. The similarities across kinases may be linked with low selectivity and it can be difficult to secure intellectual property, because many patented inhibitors share related structural features.

This binding mode appears to be used by all small molecule kinase inhibitors currently approved for use as medicines and may reflect how they tend to originate from similar isolated enzyme assays, using activated (and sometimes truncated) kinases and model substrates. Examples of compounds acting in this way include erlotinib and gefitinib (compounds **4.1** and **4.2**), which are used for the treatment of non-small cell lung cancer through the inhibition of EGFR-TK.[4,5] Sunitinib (compound **4.3**), which inhibits many kinases (a so-called "pankinase" inhibitor), binds to KIT using only the purine site.[6] Another pankinase inhibitor, pazopanib, approved for renal cell carcinoma is thought to bind in the same way.[7]

Some compounds extend from a purine site into an adjacent hydrophobic region, which is not thought to be used by natural ligands and is often known as the selectivity pocket (Figure 4.3). This is an allosteric site, because it is not used by substrates of the enzyme. The selectivity pocket is so named because it is not well conserved and inhibitor selectivity often correlates with the identity of a gatekeeper residue at the entrance to the pocket, occurring 3 residues before the main-chain NH donor in the purine site. Most compounds that use

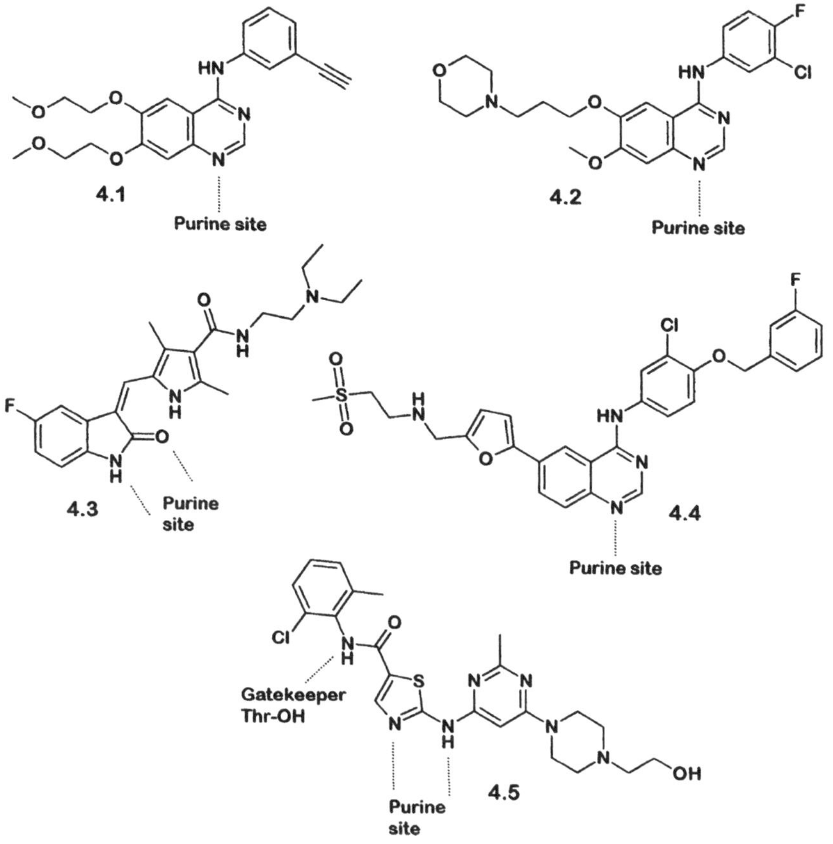

4.1

Purine site

4.2

Purine site

4.3

Purine site

4.4

Purine site

Gatekeeper Thr-OH

Purine site

4.5

Figure 4.2 Binding modes for medicines that use the purine site of protein kinases. Key hydrogen bonds are shown as dotted lines. Lapatinib (**4.4**) extends into an adjacent pocket by displacing the C-helix.

the selectivity pocket form the expected interactions in the purine site and typically contain substituted amide or urea functions, which form hydrogen bonds in *trans* with the carboxylate on the side-chain of the conserved Glu from the C-helix and with the main chain NH of the Asp from the DFG-motif. The conserved Glu-Lys salt-bridge is maintained and if the compounds contain a sufficiently large hydrophobic group, then there is a conformation change from DFG-in to DFG-out, reflecting the displacement of the phenylalanine residue at the start of the activation loop. The phenylalanine side-chain typically moves over 8 Å, so that its interactions with hydrophobic protein side-chains in the selectivity pocket are replaced by hydrophobic interactions with the bound inhibitor. Some authors refer to the selectivity pocket as being a small region adjacent to the gatekeeper, but accessible even in DFG-in conformations, with the allosteric site being an extension of the selectivity pocket accessible only in DFG-out conformations. In this review, I do not distinguish between the two regions.

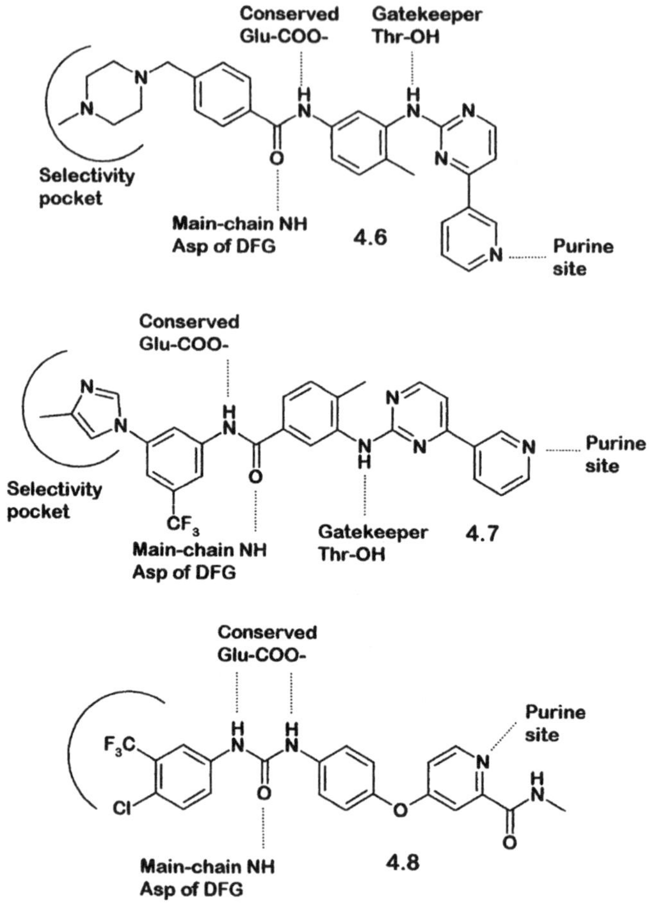

Figure 4.3 Binding modes for medicines that extend from the purine site into the selectivity pocket of protein kinases. Key hydrogen bonds are shown as dotted lines and hydrophobic interactions in the selectivity pocket are shown as arcs.

Imatinib (compound **4.6**) binds in this way to ABL during its use to treat chronic myelogenous leukemia and various other cancers.[8] Nilotinib (compound **4.7**) associates with the same kinase by making analogous interactions in the purine site and selectivity pocket,[9] whereas a pankinase inhibitor, dasatinib (compound **4.5**), uses only the purine site.[10] Sorafenib (compound **4.8**) extends from the purine site and into the selectivity pocket of BRAF.[11] Some diaryl urea-based inhibitors of p38 MAP kinase, use only the selectivity pocket and not the purine site.[12,13] Although these compounds do not use the ATP-binding site, they are competitive with the phosphate donor, because Phe168 from the DFG-motif is displaced into a region, which normally accommodates the ribose of ATP.

A further class of inhibitors induce a different conformation change, which is the displacement of the C-helix. Following disruption of the conserved Glu-Lys salt-bridge, there is a movement from C-helix-in to C-helix-out. For C-helix-in conformations, the distance between the ammonium group of the conserved Lys and the carboxylate of the conserved Glu is 3.2 Å or less. This increases to over 12 Å in C-helix-out conformations. This mechanism is deployed by the MEK inhibitors PD318088 and PD334581,[14] the EGFR-TK / ERBB2 breast cancer drug lapatinib (compound **4**, figure 4.2)[15] and the BRAF inhibitor PLX4032,[16] which has promise for the treatment of metastatic melanoma. Lapatinib and PLX4032 are ATP-competitive, extending from the purine site, whereas the MEK inhibitors bind together with ATP, using an allosteric site.[14] Catalysis requires the DFG-in, C-helix-in conformation, which allows all key groups on the kinase to be in position to bind both ATP and the phosphoacceptor.

There are several published examples of irreversible inhibition of protein kinases, typically involving a Michael addition reaction where there is a nucleophilic addition of a carbanion (such as an enzyme cysteine thiolate in the purine site) to an unsaturated carbonyl compound (Figure 4.4). Compounds working in this way include the EGFR-TK inhibitor, WZ4002 (compound **4.9**).[17]

4.2.3 Diverse Binding Modes for Recent Kinase Inhibitors

A number of ATP-noncompetitive binding modes have been reported or speculated for protein kinase inhibitors (compound structures in Figure 4.5.).

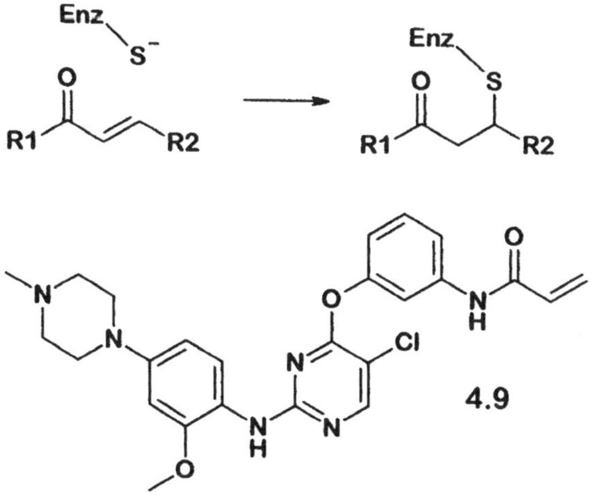

Figure 4.4 Covalent inhibition of protein kinases. Upper part, scheme for a Michael addition reaction. Lower part, structure of a reactive inhibitor, WZ4002 (**4.9**).

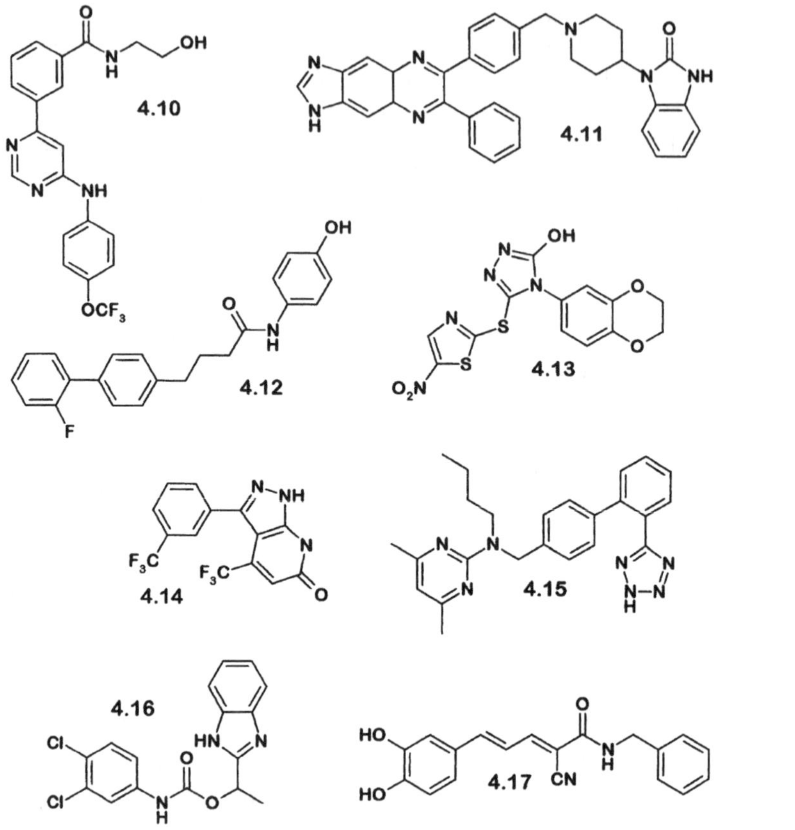

Figure 4.5   Examples of kinase inhibitors, which use diverse binding modes. Further information is in the following references: ABL·GNF-5 (**4.10**),[18] AKT·Inhibitor VIII (**4.11**),[20] p38·biarylbutyranilide (**4.12**),[21] JNK·docking groove binder (**4.13**),[22] p38·allosteric site (**4.14**),[23] JNK·allosteric site (**4.15**),[23] CHK·allosteric site (**4.16**),[25] JAK·substrate competitive (**4.17**).[27]

These inhibitors originate from leads identified in a diverse range of assays, some including physiological substrates and full-length kinase constructs. GNF-5 (compound **4.10**) uses a myristoyl binding site on ABL,[18] and comes from a series, which was identified in cell-based assays. Some AKT inhibitors do not give inhibition *in vitro* if the pleckstrin homology domain is deleted from the enzyme construct (compound **4.11**),[19] with crystal structures subsequently demonstrating association at an interface between the catalytic and pleckstrin homology domains.[20] Deuterium exchange mass spectrometry studies imply that biarylbutyranilide inhibitors of p38 (compound **4.12**) bind adjacent to the catalytic site in the C-lobe of the kinase.[21] Inhibition by these compounds is dependent upon the identity of the phosphoacceptor substrate. Such selectivity is potentially useful if a kinase is at a branch point in signaling networks and therapeutic considerations require inhibition of one pathway more than the other. A further mode of action is reported for certain inhibitors of the JNK and p38 families (compounds **4.13-4.15**). These compounds appear to use the docking groove, which is distal from the catalytic site and involved in the recognition of physiological phosphoacceptor substrates.[22] Compounds using an allosteric site on these kinases were identified in binding assays.[23] Allosteric CHK inhibitors (compound **4.16**) have been identified by their ability to give ATP-noncompetitive inhibition.[24,25]

So far, few inhibitors have been demonstrated to bind in the part of the catalytic site that is used by phosphoacceptor substrates. This is thought to reflect the geometry and physical properties of these sites often being unsuitable for lead-like small molecules. Accordingly, many of these compounds (compound **4.17**) are peptide-like,[26] although a substrate-competitive JAK2 inhibitor reached clinical trials,[27] but has not progressed to launch.

4.3 Kinetics of Inhibition

Inhibitors are evaluated by measuring IC_{50} values, which usually can be estimated by fitting the following equation to dose-response data

$$v = v_0/(1 + [I]/IC_{50}) \tag{4.1}$$

where v_0 is the uninhibited rate.[28,29] Use of percent inhibition is not recommended, because it compounds experimental noise and introduces a bias into the data. More complex equations sometimes are used, but this is inadvisable as a first step, because failure to obtain a good quality of fit with eqn (4.1) indicates that a complex mechanism is occurring and the estimated IC_{50} may be very misleading. A poor fit is indicated by large, nonrandom differences between the experimental data and the best-fit line. Misleading IC_{50} values arise from nonspecific inhibition, slow binding, tight binding, irreversible inhibition or the presence of a mixture of enzyme forms (*e.g.* activation states), which differ in sensitivity to the test compound (See Sections 4.3.2, 4.3.3 and 4.4.1.2.).

The catalytic activity of protein kinases is highly regulated, with both activated and nonactivated states occurring *in vivo*. Many kinases are activated by phosphorylation on the activation loop (see Section 4.2.1). First generation inhibitors were identified in assays, which followed phosphorylation of a substrate by activated forms of the enzyme. Accordingly, these compounds may be described as giving "inhibition of catalysis". Subsequent studies in cells showed that some compounds also decrease levels of phosphorylation of the target protein itself, in addition to the expected effects on proteins downstream in signaling networks. Mechanistic studies revealed that inhibitors of p38 or BRAF, which induce DFG-out inactive conformations, and inhibitors of MEK, which induce C-helix-out inactive conformations, interfere with phosphorylation by upstream activating kinases (sometimes known as "prevention of activation").[13,30] Reported inhibitors give either inhibition of catalysis alone, or both inhibition of catalysis and prevention of activation.

The dose-response equations deployed to analyse most enzyme inhibition data are dependent upon there being a steady state (the degree of inhibition not changing during the assay), approximately equal concentrations of free and total inhibitor, and a one to one stoichiometry for binding and inhibition.[29] Active compounds follow a range of kinetic mechanisms, which may be classified in various ways, for example according to dependence upon ATP or reversibility. Below, I describe some of the most common mechanisms.

4.3.1   ATP-Dependence

The kinetic mechanism of kinase inhibitors often has been investigated by varying the concentration of ATP. The dependence upon concentration of phosphoacceptor substrate is rarely reported. ATP-dependence allows comparison across all kinases as it is the universal phosphate donor, and it gives useful information, because the most inhibitors use the purine site.

The mechanism of inhibition with respect to ATP is assigned on the basis of the potency of the compound after extrapolating to both insignificant and saturating concentrations of the phosphate donor.[29] These concentrations are defined relative to that giving the half-maximal rate, which is the K_m or Michaelis constant for ATP. The value of K_m for ATP may be dependent on several aspects of the assay conditions (See Section 4.4.1.2.). The substrate-dependence of inhibition usually can be characterised using similar rate equations, where [*Substrate*] replaces [*ATP*] and the K_m value refers to that for substrate.

The mechanism of inhibition is defined by the relative values of K_{is} and K_{ii}, which are respectively the inhibition constants at $[ATP] << K_m$ and $>> K_m$.[28,29] Inhibition constants are measures of potency, because they equal the free inhibitor concentration when the rate is reduced by 50%. The mechanism is competitive if inhibition tends to zero when ATP is saturating ($[ATP] >> K_m$). This mechanism is seen if $K_{is} << K_{ii}$. Conversely, the mechanism is uncompetitive if inhibition tends to zero when $[ATP] << K_m$, because $K_{is} >> K_{ii}$. Inhibition is noncompetitive when it occurs both at $[ATP] << K_m$ and $[ATP] >> K_m$. Pure noncompetitive inhibition ($K_{is} = K_{ii}$) arises when potency is independent of ATP-concentration. Mixed noncompetitive inhibition ($K_{is} \neq K_{ii}$) occurs if there is a tendency towards competitive or uncompetitive.

It follows that, for mechanisms other than pure noncompetitive, potency varies according to $[ATP]$ relative to K_m – a phenomenon which needs to be considered when building structure-activity relationships (SAR), measuring selectivity and developing an understanding of the activity observed in cells or *in vivo*. The observed or apparent inhibition constant, K_i', is defined as the concentration of *free* compound giving 50% inhibition at any specified concentration of ATP. In many cases it approximates to the *total* concentration of inhibitor, IC_{50}, which is fortunate because it can be calculated from the amount added, whereas the free concentration is difficult to measure. Drug discovery projects may encounter difficulties, which compromise success rates, if K_i' does not approximate to IC_{50}. This situation often is encountered,

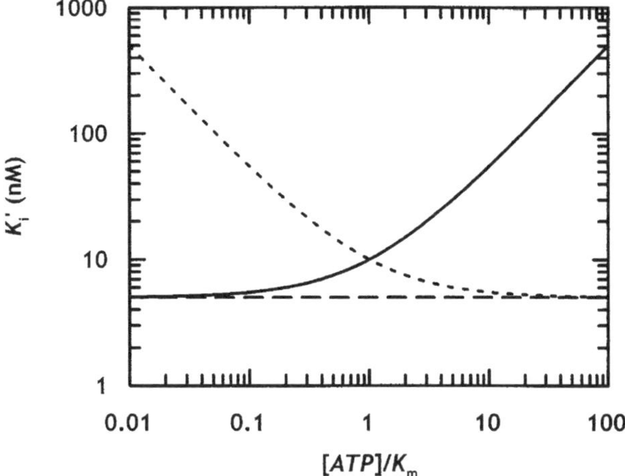

Figure 4.6 Apparent inhibition constant as a function of ATP-concentration relative to K_m for competitive inhibition (solid line), pure noncompetitive inhibition (dashed line) and uncompetitive inhibition (dotted line). The inhibition constant is 5 nM for each mechanism.

resulting from slow or tight binding during the optimisation of potent inhibitors (see below). The relationship between K_i' and substrate concentration ([S], *e.g.* [ATP]) was derived by Cheng and Prusoff.[28]

$$K_i' = \frac{K_{is}K_{ii}(K_m + [S])}{K_m K_{ii} + [S]K_m} \tag{4.2}$$

Figure 4.6 shows the relationship between K_i' and [ATP] for various mechanisms of inhibition. For a competitive inhibitor, when ATP is present much below its K_m value, K_i' approximates to the inhibition constant, K_{is}. At [ATP] = K_m, K_i' = 2K_{is}. When [ATP] $>>$ K_m, K_i' increases linearly with [ATP], K_i' $\approx$ [ATP]K_{is}/K_m. The extent of competition by ATP, therefore, is a key factor in determining potency and selectivity (See Section 4.4.2.1.). The selectivity at physiological [ATP] is influenced by K_m, which is an intrinsic property of the enzyme determining the relative ease of competitive inhibition. Uncompetitive inhibition may be barely detectable when [ATP] $<<$ K_m, because K_i' $\approx$ $K_{ii}K_m$/[ATP]. When [ATP] = K_m, K_i' = 2K_{ii}, and at [ATP] $>>$ K_m, K_i' $\approx$ K_{ii}. For pure noncompetitive inhibitors, K_i' = K_i at all concentrations of ATP. It is noteworthy that when [ATP] = K_m, K_i' is between 1 and 2-times the value of the inhibition constant for each of these mechanisms.

4.3.2 Rates for Onset and Reversal of Inhibition

IC_{50} values are used routinely to monitor the biological activity of kinase inhibitors. However, multiple factors influence the magnitude of IC_{50} and may make it unreliable in terms of relevance and comparison of compounds. In some cases, the confounding factor is that the IC_{50} changes during the course of the assay. Equilibration of inhibitor binding often is effectively instantaneous during isolated enzyme assays. However, around 30% of the enzyme inhibitors approved as drugs give "slow binding" inhibition, where the rate of reaction may change during the assay, so that the magnitude of IC_{50} becomes dependent upon time.[31] Compounds with slow on rates must also have slow off rates in order to accumulate on the enzyme and give inhibition. Consequently, their potencies may be overestimated if preincubated with enzyme prior to starting the assay (*e.g.* by addition of substrate), or underestimated if there is no preincubation. Special methods are required to characterise slowly reversible and irreversible inhibitors.[32]

Most mechanisms for slow binding inhibition involve one or two kinetic steps during association of the inhibitor with the enzyme.[32] In the one step mechanism, a high affinity complex ($E \cdot I^*$, apparent inhibition constant $K_i^{*'}$), is generated directly, without any detectable intermediates. In the two-step process, there is an initial complex ($E \cdot I$, apparent inhibition constant K_i'), in equilibrium with uninhibited enzyme, followed by a subsequent tightening to give the final complex ($E \cdot I^*$) with an overall steady state apparent inhibition constant $K_i^{*'}$.

In order to understand SAR, it can be important to determine whether there is a second step. In the one step mechanism, inhibition is governed by the on and off rate constants, say k_3 and k_4. Whereas in the two-step mechanism, potency is controlled by the affinity of the initial inhibited complex and the values of the rate constants k_5 ($E \cdot I$ to $E \cdot I^*$) and k_6 ($E \cdot I^*$ to $E \cdot I$). Inhibition is irreversible if k_4 or k_6 tends to zero. A second step may involve a conformation change or formation of a covalent bond.

Although on and off rates can be measured in binding assays (Section 4.3.4), project teams may prefer to avoid developing a new assay and instead use the conventional activity assay. First, slow binding compounds may be identified as those giving different IC_{50} values with and without preincubation. On rates can be measured by following the time course of product accumulation in assays started by addition of enzyme to a mixture of substrates and inhibitor. Off rates can be determined by following time courses where enzyme is preincubated with inhibitor and then assays are started by addition of substrate. The drug-target residence time, τ, is defined as $1/k_{off}$ and is a measure of the duration of inhibition after complex formation.

Irreversible, covalent inhibition has been explored as an approach to overcome competition by ATP and ADP. Such competition is thought to contribute to the large drop in potency often seen on moving from isolated enzyme assays, typically at around 10 µM ATP, into cells where are ATP and ADP each are present around mM concentrations.

4.3.3 Tight Binding Inhibition

Many irreversible inhibitors and potent reversible compounds give $IC_{50} \approx$ 50% binding sites. Such IC_{50} values, therefore, help in the qualitative recognition of potent compounds, but are misleading if used as quantitative estimates of potency. These compounds are known as "tight binding" inhibitors and can be recognised because the IC_{50} increases linearly with enzyme concentration. When the accumulation of product is linear with time (known as "steady state") and there is 50% inhibition, then

$$IC_{50} = K_i' + [E]/2 \tag{4.3}$$

where IC_{50} and K_i' are, respectively, the total and free concentrations of inhibitor and $[E]$ is the total enzyme concentration. Under tight binding conditions, $K_i' << [E]/2$ (K_i' tends to zero for irreversible inhibitors), so that $IC_{50} \approx [E]/2$ and under non-tight binding conditions, $K_i' >> [E]/2$, so that $IC_{50} \approx K_i'$.

The measured IC_{50} values for tight binding and irreversible inhibitors are of limited value in determining SAR, because further increases in affinity do not decrease the IC_{50}. It has been suggested that it is not important to measure affinity for such compounds, because it is sufficient to know that they have high potency. However, there are examples where different tight binding inhibitors are approximately equipotent in isolated enzyme assays and have very different activities inside cells. It can be important to know if this effect is related to differences in affinity for the desired target, or if they are off target effects. Thus, it may be desirable to resolve between the isolated enzyme potencies for different tight binding inhibitors. Various approaches may be suitable. It may be informative to fit a dose-response equation, which allows for tight binding by including an $[E]$ term to account for depletion of free inhibitor.[33] Alternatively it may help to decrease the enzyme concentration, perhaps using a longer assay duration or more sensitive method to follow rate of reaction. The apparent potency also may be moved into a measurable range by increasing the concentration of competing substrate (often ATP). Binding assays may provide an alternative approach to rank the affinities of tight binding inhibitors.

4.3.4 Binding Assays

The equilibrium dissociation constant, K_d, is a measure of binding affinity, defined as the free ligand concentration when 50% of the binding sites are occupied. It is equivalent to the ratio of the rate constants for dissociation and association, $K_d = k_{off}/k_{on}$. Accordingly, the magnitude of K_d can be measured either at equilibrium, or as k_{off}/k_{on}. Several approaches have been used to measure these parameters in kinase drug discovery.[34] Below, I outline some of the widely used methods.

Binding assays usually offer a potentially important advantage over assays that require turnover of substrate—they can characterise the effects of test compounds against kinases in non-activated states, which frequently have low or undetectable levels of catalytic activity. However, binding alone may not give the desired biological activity. There are examples of compounds, which bind to non-activated kinases, but do not give prevention of activation.[13] Several binding assays involve competition with reference or reporter molecules, and such experiments usually are designed to ensure that there is no significant perturbation of binding of the test compound. It can be advisable to confirm that different methods give consistent values, or to compare with a direct binding assay.

Ambit Biosciences published competition binding assays, involving an immobilised ATP-competitive "bait" compound and bacteriophages, which are coated by engineered proteins containing kinase sequences.[35] The concentration of test compound is varied to determine the level at which there is 50% inhibition of the bait-phage interaction. This approach has been used to estimate large numbers of K_d values and has given considerable insight into kinase inhibitor selectivity and the effects of mutations and activation state on binding affinity.[35–37]

Biosensor instruments such as Biacore[TM] (General Electric) exploit the sensitivity of a surface plasmon resonance response to the mass localised near the surface of a sensor chip. Various approaches can be used for kinases. Inhibition in solution assays involve immobilisation of a "target definition compound" (TDC) on the sensor surface.[13,30] A buffer containing the kinase is flowed over the surface so that the protein is able to bind to the immobilised TDC, giving a signal. When test compounds are included in the buffer, they can compete for the TDC-kinase interaction, allowing estimation of K_d.

Inhibition in solution assays can give precise, reliable estimates of K_d values,[13] but the approach has relatively low throughput, consumes moderate amounts of protein, requires a suitable TDC and does not give rate constants. These issues can be overcome using direct binding assays.[38] Here, the target kinase is immobilised on the surface of the sensor chip and the association, dissociation or equilibrium binding of the test compounds can be measured. It can be challenging to obtain sufficient signal from small molecule binding and there may be difficulties in retaining authentic binding characteristics after immobilisation of the kinase. However, rigorous quality control methods are available and there has been progress in overcoming these technical issues, especially for compounds of Mol Wt > 500.

Specific binding of a ligand increases the thermal stability of the target protein. This phenomenon has been exploited in experiments, which involve determination of the melting temperatures for kinases in the presence and absence of test compounds. The approach generally works well for potent compounds (K_d < 0.1 μM), but is less reliable for weak binders.[39] Indeed, protein stabilisation may be the only method suitable for measuring affinity for very tight binders, where there often is an approximate correlation between shift in melting temperature and the free energy of binding. This approach was used to determine SAR for slow, tight binding inhibitors of p38α.[40]

Pyridinyl imidazoles have been used as fluorescent reporters of binding to p38α MAP kinase.[12,13] To measure k_{on} for test compounds, the kinase was preincubated with the reporter and then the decrease in fluorescence was followed as the test compound displaced the reporter. In order to determine the value of k_{off}, the kinase was preincubated with the test compound and then the increase in fluorescence monitored when it was displaced by the reporter. Some binding assays use kinases, which are derivatized with a fluorescent reporter in order to help identify compounds that induce specific conformation changes.[41] However, such proteins may not be fully reliable models for the authentic enzyme.

NMR has been used extensively in the identification and evaluation of enzyme inhibitors.[42] Various methods have been deployed, which may be divided into 2 classes, according to whether NMR follows signals from the protein or the test compound. Two-dimensional protein-observe NMR may be an excellent method to identify binders, because it can give information on the interaction site if resonances have been assigned. K_d values may be estimated from dose-response studies. Ligand-observe NMR is suitable for identification of weak binding fragments, but binding by compounds with K_ds < 1 μM is not detected. Compounds which form molecular aggregates—a common mechanism of nonspecific inhibition,[43]—are false positives using this approach.

Isothermal titration calorimetry gives precise measurements of affinity and stoichiometry.[39] It follows the uptake or release of heat, when aliquots of test compound are injected into a solution of the target enzyme. The heat change is a widely applicable signal and allows estimation of the enthalpy change upon binding. Large shifts in enthalpy of binding between closely related compounds suggest a possible change in binding mode. Experiments may be performed in the presence or absence of substrates, products or regulators, giving insight into binding mechanisms. However, throughput is low and the method consumes significant quantities of purified protein.

4.3.5 Relationships Between IC$_{50}$, K_i and K_d

The discussion above has included several different measures of potency. It may be helpful to clarify the relationships between them. When the test compound gives 50% inhibition, its total concentration is IC$_{50}$ and the free concentration is K_i'. K_i' is a steady state value, whereas the IC$_{50}$ may or may not be at steady state. So, $K_i' \approx$ IC$_{50}$ if the assays are at steady state and if binding leads to no significant depletion of free ligand. The inhibition constant, K_i, K_{is} or K_{ii}, is the concentration of free compound giving 50% inhibition at steady state under particular assay conditions (K_{is} if $[S] << K_m$, K_{ii} if $[S] >> K_m$, and K_i if K_i' does not vary with $[S]$). K_i^* is the inhibition constant at steady state for a slow binding inhibitor and reflects the equilibria between free E, free I, E·I complex and E·I* complex. K_d is the equilibrium dissociation constant measured in binding assays and approximates to an inhibition constant measured under conditions where all of the enzyme is in a form capable of binding the inhibitor.

4.4 Implications of Mechanisms and Kinetics

4.4.1 Identification and Evaluation of Inhibitors

4.4.1.1 Design of Hit Identification Assays

Many kinases interact with their physiological protein substrates using grooves or pockets in addition to the required binding at the catalytic site. However, physiological substrates only rarely have been used in hit identification, with common model substrates including poly(Glu-Ala-Tyr), myelin basic protein, or oligopeptides derived from the natural substrate. This is because the physiological substrate may be unknown, or difficult to obtain in sufficient purity and quality for use in large screening campaigns. The use of small model substrates often biases the assay away from detecting compounds that act at distal substrate recognition sites (*e.g.* docking groove binders, see Figure 4.5), and favours active site binders.

Section 4.3 describes how the magnitude of IC_{50} depends upon the assay conditions. Test compounds must be present at concentrations approaching the IC_{50} value in order to give detectable inhibition. Accordingly, the compounds identified as active depend upon the assay conditions. IC_{50} is influenced by affinity and the relative abundance of the enzyme forms bound by the inhibitor. If the relevant form (*e.g.* free kinase) is only 1% of the total (*e.g.* because $[ATP] >> K_m$), then $IC_{50} = 100\ K_i$. Fragment-based lead generation typically uses small compounds (Mol Wt $\leqslant$ 300) in binding assays against the free enzyme,[44] however such assays tend to favour compounds which will suffer from competition by ATP and ADP *in vivo*. Binding assays in the absence of ATP also fail to detect compounds that follow an uncompetitive mechanism. Such a mode of action could be useful, because it would avoid displacement by ATP or ADP *in vivo*. Assays designed to give a "balanced" distribution of enzyme forms (complexes with substrates, products and intermediates each present at similar abundance) are capable of identifying a greater diversity of hits with different mechanisms of inhibition.[45] Such compounds would be diverse in chemical and biological properties, giving differentiated options for further progression. However, a more focused distribution of enzyme forms may be desirable, for example to favour compounds active under physiological conditions, or to avoid previously explored areas of medicinal chemistry space. Favouring enzyme forms, which are common during the disease state *in vivo* could be an approach to identify hits where potency is less compromised on moving from isolated enzyme assays. Efficient transfer of potency *in vivo* may result in improved efficacy, and the administration of less compound may enhance tolerability.

Most kinase screens follow inhibition of catalysis by enzymes in their activated states. An alternative assay format is required to identify and characterise compounds that function by prevention of activation. Here, the target kinase is used as a substrate for an upstream activating kinase. This assay accordingly also identifies compounds, which inhibit the upstream activator, and so mechanistic deconvolution assays are required to understand the SAR. Careful assay design and data analysis are required to identify and evaluate prevention of activation compounds.[13] The substrate should be present at or below a K_m concentration, if not then its depletion by binding of inhibitor has little effect on the rate of reaction. A further requirement is that the concentration of substrate should be less than that of the test compound. For example, to identify a compound giving inhibition at 100 nM usually requires less than 100 nM substrate in the assay. This may mean that sensitive methods are required to the monitor rate of reaction.[34]

Some kinases have low levels of catalytic activity prior to activation, so that compounds may act by inhibiting non-activated kinases. Characterisation of such compounds can be difficult, because affinity may depend on kinase activation state and the kinase may be capable of autoactivation. Binding assays can be useful.[30,37]

4.4.1.2 Evaluation of Inhibitors

Having identified compounds that inhibit protein kinases, the first step in characterisation is to determine whether inhibition is specific. It is not unusual for well over 10% of the inhibitors from activity-based hit identification screens to be nonspecific, although they may not rapidly be recognised as such and so lead to considerable confusion and waste of resource.[46] Non-specific inhibition is different to non-selective inhibition, which is a specific effect on multiple target proteins. Non-selective kinase inhibitors may be clinically useful (*e.g.* dasatininb and sorafenib), because simultaneous suppression of multiple kinases may be required for beneficial therapeutic effect. Specific compounds utilise particular binding sites and bind at a defined stoichiometry (usually 1 : 1). Non-specific compounds interact in many different ways with the same enzyme.[43,46] Inhibition may involve many hundreds of inhibitor molecules interacting with each enzyme molecule. Such compounds typically inhibit the activity of diverse, unrelated enzymes. Nonspecific inhibition is unlikely to be of any value in drug discovery, although some drugs appear nonspecific at high multiples of their target IC_{50} values (over 1 μM). There are many physical mechanisms for nonspecific inhibition, although most such compounds often can be identified as they give several of the following characteristics: steep dose-response curves (derived from the high stoichiometry), sensitivity to detergent (which prevents formation of molecular aggregates), non-competitive kinetics, time-dependent inhibition, "flat" SAR (potency changes little amongst a series of structurally similar compounds.).

Having removed nonspecific active compounds, it may be advantageous to determine if inhibition is reversible. Standard approaches in the interpretation of IC_{50} values assume that inhibition is reversible on a time-scale much shorter than the duration of the assay, but this assumption may not hold and around one in three clinically useful enzyme inhibitors are not rapidly reversible.[31] Reversibility often is most easily determined by preincubating a mixture of enzyme plus inhibitor (at 10-times IC_{50}) and then measuring regain of catalytic activity after extensive (*e.g.* 100-fold) dilution into an assay.

The Cheng-Prusoff relationship (see Section 4.3.1) describes how potency varies according to the substrate concentration and the K_m of the target enzyme. Additional factors may affect the potency of kinase inhibitors: activation state, choice of phosphoacceptor (which can change the K_m for ATP), design of protein construct and choice of counterion for ATP. The physiological counterion is Mg^{2+}, although Mn^{2+} sometimes has been used due to its ability to increase kinase activity and so improve signal to noise in the assay.[47] Data collected using Mn^{2+} can be misleading, because it often affects the K_m for ATP by 10-fold or more and can change additional aspects of the kinetics. Experiments to characterise kinase inhibitors should be designed specifically to measure the desired parameters. For example, ranking at $[ATP] = K_m$ to assess selectivity in binding *in vitro* may be quite different to ranking at physiological $[ATP]$ to help understand efficacy *in vivo*.

In some cases, it is important to characterise the effects of compounds on different activation states of the same kinase. Compounds may give prevention of activation by binding to non-activated kinase in addition to inhibition of catalysis by association the activated state. Some compounds show marked changes in affinity according to the kinase activation state. The magnitude of any effect varies considerably between kinases and tends to be similar for inhibitors that bind the same kinase conformation.[13,30,37]

Multiple factors affect activity and selectivity in cells and even more influence efficacy *in vivo*.[48,49] Properties of the kinase, the inhibitor and the target cell all contribute to the observed effects of the compound. Activity in cells does not always correlate with that in isolated protein assays.

4.4.1.3 Mechanisms and SAR

Following hit identification, compounds are clustered by structural similarity in order to predict activity and generate ideas for synthesis of improved inhibitors. SAR should be similar within, not between, clusters. SAR is more likely to be consistent if compounds follow the same mechanism of inhibition, because compounds with the same mechanism bind to the same enzyme forms. Enzyme forms differ in accessibility and conformation of binding sites, and in presence or absence of bound substrates, intermediates or products, which may contribute directly to an inhibitor binding site or indirectly influence inhibitor binding. Knowledge of mechanism of inhibition, therefore, increases confidence in SAR clustering. It helps to identify less obvious clusters and when substitution on the compound changes the identity of the target forms. Mechanistic information is valuable in addition to experimentally determined 3-D structures of complexes. Such structures typically relate to only one enzyme form, which may have limited physiological relevance, and the implications for molecular design are only qualitative because it is difficult to predict affinity from structure. For some MEK inhibitors, small changes in structure can affect whether inhibitors require the prior binding of ATP.[30] In a series of anthranilic acid derivatives, neither an *N*-alkoxy amide nor the parent acid require prior binding of ATP in order to associate with non-activated MEK. Conversely, a closely related *N*-alkyl amide derivative does not bind unless ATP is present at sufficient concentrations. These effects were not expected from considering only the structures of the compounds. This work also reveals large shifts in potency dependent upon the activation state of MEK.

Similarly, Okram *et al.*,[50] highlight that adding a 3-trifluorobenzamide onto four distinct ABL inhibitor scaffolds (each binding only in the purine site) generates compounds, which extend into the selectivity pocket and induce DFG-out. The substitution increases activity and selectivity in enzyme and cell assays, and gives greater potency against non-activated ABL relative to the activated enzyme. *N*-methylation of the amide or urea causes a large decrease in potency, presumably by interfering with the ability to hydrogen bond with the conserved Glu involved in accessing the selectivity pocket.

4.4.1.4 Structure-Based Design

Crystal structures may be generated by co-crystallisation of enzymes and ligands or by soaking ligands into previously formed crystals of the free enzymes. Although soaking has given the structures of many kinase-inhibitor complexes, co-crystallisation sometimes may be required, because the enzyme in previously formed crystals may have insufficient mobility to permit access to the authentic binding mode.

The value of K_d is a measure of affinity in the absence of substrates. As such, it often can be used to interpret 3-D structures determined by X-ray crystallography, because it relates to the form of the enzyme in the structural model. However, most drug discovery projects do not measure K_d values, so that structure-based design usually is dependent upon measurements of IC_{50}. Under certain circumstances (see Section 4.3.1), the magnitude of IC_{50} for many kinase inhibitors approximates to, or at least correlates with, K_d. This situation arises when inhibition is noncompetitive with respect to the phosphoacceptor and competitive with ATP if the concentration of ATP is either insufficient to give significant competition, or at least is a constant multiple of K_m.

If the mechanism with respect to ATP is unknown, or varies across the compounds under consideration, values of K_i' measured at $[ATP] = K_m$ can be used to guide structure-based design, because such values are only a 1- to 2-fold multiple of K_i regardless of mechanism of inhibition (Figure 4.6) and so relate to the binding energy for the molecular interaction. However, values from inhibition of catalysis assays are measured on activated kinases, whereas crystal structures normally are determined for non-activated states (because it can be difficult to obtain preparations of phosphorylated kinases which are sufficiently homogeneous to allow crystallization, although some studies mimic phosphorylation by replacing the acceptor Ser or Thr residues with anionic Asp or Glu). Inhibitor binding affinities can be quite different for activated compared with non-activated states, although in many cases binding is not largely affected.[13,37,50] In general, SAR against activated kinase is consistent with crystal structures bound to non-activated kinase, although caution may be advisable, for example because imatinib has quite different binding modes for non-activated ABL and activated SYK.[51]

4.4.1.5   Slow Binding, Tight Binding and Covalent Inhibition

Many IC_{50} values are misleading, because inhibitors give slow or tight binding, perhaps linked with covalent bond formation. Kinetic evaluation of such compounds requires analysis of time-course data and perhaps studies where the concentration of enzyme is varied (See Sections 4.3.2 and 4.3.3.). Assays to estimate potency for slow binding, reversible inhibitors also can be designed carefully, so that enzyme and inhibitor are preincubated to allow attainment of equilibrium and then the reaction is started by addition of a small volume of substrate at a concentration at or below K_m. The choice of volume and concentration of substrate should be insufficient to perturb significantly the established equilibrium.

Differences in rates of onset of inhibition between compounds can suggest differences in molecular mechanism. Compounds that bind to rare forms of the kinase have slower on rates than those that bind abundant forms. This is seen with both DFG-out (p38 and BRAF)[13] and C-helix-out (EGFR) conformations.[15] Compounds with slow on rates must have slow off rates relative to equipotent fast binders. These slow off rates give extended drug-target residence times, which can have clinical implications (see Section 4.4.2.4).

Several parameters are important in the characterisation of slow binding inhibitors (see Section 4.3.2). The magnitude of K_i helps to understand the SAR for formation of any initial complex and that of K_i^* gives insight into the overall steady state affinity. The rate of regain of activity (k_{off}) often approximates to k_4 or k_6 and can be used to estimate the drug target residence time. The rate of onset of inhibition relative to free drug concentration, $k_{obs}/[I]$, also can be important. If it is too slow, then the compound is cleared from the body before exhibiting its full therapeutic effect. The magnitude of k_3 or k_5 is a measure of the inactivation rate constant. It should be sufficiently fast to give a benefit compared to a rapidly reversible inhibitor, but very fast rates for covalent inhibitors are linked with excessively reactive compounds, which react with non-target molecules, leading to adverse effects and depletion of active compound compromising exposure for the true target. It can be useful to optimise the inactivation rate constant relative to such nonspecific reactivity, for example rate of reaction with hydroxide or glutathione.

4.4.2   Clinical Implications

4.4.2.1   Efficacy for Rapidly Reversible Inhibitors

For closely related, rapidly reversible inhibitors following the same mechanism against the same kinase, it may be anticipated that potency *in vitro* may correlate with efficacy *in vivo*. As described in Section 4.3.1, the value of K_m ATP has a large influence on the potency and selectivity of many kinase inhibitors. Most kinases have a K_m for ATP in the range 1–200 µM. The physiological concentrations of ATP and ADP (which also competes) are around 1–5 mM. The potency of compounds, therefore, is typically approximately 10 to 100-fold lower inside cells. The magnitude of this "drop-off" is larger for kinases with lower K_m values. The fall in efficacy (concentration required for a specific biological effect) may be different to the shift in potency (concentration required for inhibition). Prevention of activation often is less susceptible than inhibition of catalysis to competition by ATP, because the non-activated state has a lower affinity for the phosphate donor. Consider a compound, which gives ATP-competitive inhibition of several kinases, each with different values of K_m and K_{is}. It is quite possible that the compound is selective for the kinase with the lowest K_m at the low ATP concentrations typically used during *in vitro* assays, and that selectivity may reverse at physiological ATP, so that the compound is selective for the enzyme with the highest K_m (Figure 4.7). K_m effects also have ramifications in the clinic. An EGFR-TK inhibitor, gefitinib, has modest efficacy in most non-small cell lung cancer patients, perhaps linked with competition by intracellular ATP. However, a subset of patients respond well, especially those whose tumours carry EGFR-TK L858R and ΔE746-A750 mutations.[52,53] Responsiveness may be linked with the kinetic properties of the target protein. L858R increases the affinity for gefitinib, whereas G719S increases the K_m for ATP, reducing its ability to displace the inhibitor from the target enzyme.[54] However, such

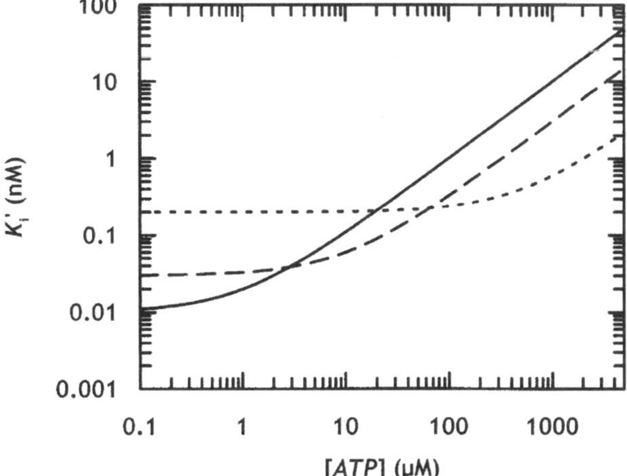

Figure 4.7 Inhibitor selectivity depends on the K_m value for competing ATP. In this theoretical example, the values of K_m and the inhibition constant are such that the compound is selective for one kinase (solid line) in assays at 1 µM ATP and a different kinase (short-dashed line) at physiological [*ATP*] (over 1 mM).

patients tend to become non-responsive, an observation often linked with expression of a gatekeeper mutant of EGFR-TK, commonly T790M.[53] This change has little effect on the affinity of L858R for gefitinib, but does decrease the K_m for ATP, increasing competition for the inhibitor.[55]

4.4.2.2   Inhibitor Selectivity

Most kinase inhibitors use a conserved binding mode in a site utilised by ATP. This gives potential concerns that the compounds may have insufficient selectivity to avoid adverse clinical effects arising from inhibition of non-target kinases. Further issues could come from inhibition of other enzymes that bind ATP or other adenine containing biomolecules such as oxidoreductases (bind NADH and NADPH), methyltransferases (bind *S*-adenosyl methionine) and nucleotide binding enzymes. Structural similarities between inhibitors and adenine also could generate adverse effects through binding to adenosine receptors. During early kinase drug discovery, there were concerns about the prospects for ATP-competitive inhibitors obtaining sufficient selectivity to allow therapeutic activity in the absence of issues with tolerability. Now, clinically approved kinase inhibitors have been shown to exhibit very different degrees of selectivity.[36] Lapatinib seems to effectively inhibit only EGFR-TK and the closely related ERBB2. Conversely, dasatinib and sorafenib each bound around 10% of 317 tested kinases with K_d values within 10-fold of that for their primary targets. Sunitinib bound 57% of 290 tested kinases with a K_d < 3 µM. Selectivity for inhibitors of some kinases in cells has been determined using various methods, which involve monitoring phosphorylation status for proximal downstream substrates.[49]

It is difficult to deconvolute how kinase inhibitor binding mode has implications in the clinic. In general, selectivity tends to increase in the order (i) purine site, (ii) selectivity pocket, and (iii) allosteric site. The ability to overcome resistance mutations in cancer therapy tends to be inversely related to this selectivity pattern. Accordingly, allosteric kinase inhibition may become a preferred mechanism for clinical indications other than cancer.

Kinase signaling networks are subject to homeostasis, which tends to offset the effect of inhibition at any particular point, so that it is perhaps not surprising that inhibition of multiple kinases either may be required for efficacy, or may not severely compromise clinical tolerability. Studies have revealed how inhibition of one kinase can lead to recruitment of alternative kinases to restore signaling. For example during cancer therapy with gefitinib, there can be genetic amplification of a receptor tyrosine kinase, cMET, which effectively replaces the function of EGFR-TK in its signaling pathway.[53] These observations have led to clinical studies of combinations of cMET and EGFR-TK inhibitors. Inhibition of multiple kinases may be achieved by combining different inhibitors, or by using single compounds to inhibit multiple enzymes, sometimes called "pankinase" or "multikinase" inhibitors. There are examples where a single compound with the ability to inhibit several different kinases has been linked with important clinical observations.[53] Both sorafenib and imatinib have proved useful by inhibiting kinases other than their original target. Imatinib was designed to act on ABL in chronic myelogenous leukemia and it also inhibits both cKIT in gastrointestinal stromal tumours, and PDGFR in dermatofibroma sarcoma protruberans. Similarly, sorafenib was aimed to target RAF in lung and pancreas tumours where it has disappointing efficacy, but it has utility in renal cell and hepatocellular carcinomas, which is thought to involve inhibition of respectively VEGF2 and PDGFR. There are examples of clinical compounds, which are thought to act by inhibiting multiple kinases in the same tumour, such as lapatinib (see above) and vandetanib, which is believed to inhibit EGFR, ERBB2 and VEFR2 in non-small cell lung cancer and medullary thyroid cancer.

4.4.2.3 Combination Therapy

Treatment of cancers often has involved co-administration or sequential administration of multiple therapeutic agents, reflecting the complex molecular mechanisms underlying this collection of diseases. Kinase inhibitors can display good clinical efficacy in cancer, but only in a minority of patients and they often go on to develop resistance (Section 4.4.2.1). These observations seem to be linked with the genetic instability of cancer cells and the selective pressure applied by therapy. Combination therapy is a potential approach to dealing with these issues, perhaps by reducing the selection pressure at each individual target and by requiring mutation at two loci. A combined therapy may target multiple kinases (see the EGFR and cMET example in Section 4.4.2.2), or different sites on the same protein (*e.g.* lapatinib and a biologic, trastuzumab, binding respectively to the catalytic and growth factor binding domains of both EGFR and ERBB2 in the treatment of breast cancer.). Combinations of small molecules binding to discrete sites on the same kinase also exhibit promising properties. When GNF-5 is combined with imatinib or nilotinib, which bind at distinct sites on ABL, there is suppression of the emergence of resistance mutations *in vitro* and enhanced activity in biochemical and cellular assays against the resistant gatekeeper mutant T315I.[18] There is positive co-operativity in terms of potency, for example the IC_{50} of nilitonib is reduced from 0.29 to 0.03 μM in the presence of 1 μM GNF-5. The combination of GNF-5 and nilotinib shows efficacy *in vivo*, using a murine bone transplantation model.

4.4.2.4 Slowly Reversible and Irreversible Inhibitors

A slow on rate may compromise activity *in vivo*, because some of the administered drug could be cleared from the body before binding to the target protein. The association rate *in vivo* is likely to be slower than that *in vitro* due to competition by ATP and ADP, and depletion of free drug by nonspecific binding to plasma proteins.

Around 30% of the enzyme inhibitors approved by the FDA between 2001 and 2004 have non-equilibrium kinetics, which involve slow off rates, irreversible binding or inactivation of the target protein.[31] A slow off rate causes the drug to remain bound to the target protein after systemic levels of compound fall between doses, so that pharmacodynamic effect may extend beyond the duration that might be expected from the pharmacokinetic properties of the drug.[56] This effect may be dependent upon the relationship between the drug-target residence time and the rate of clearance from the body. An extended residence time potentially has more influence on pharmacodynamics if clearance is rapid. Generation of new active protein, by mechanisms such as biosynthesis, activation, or release from storage, can complicate the relationship between residence time and pharmacodynamics. There are examples where efficacy *in vivo*, or in the clinic, correlates better with k_{off} than with K_d.[31,57,58] Such observations probably involve compounds having different on rates, otherwise one would expect a correlation between k_{off} and K_d, because $K_d = k_{off}/k_{on}$. Diverse rates of association are linked with different mechanisms of inhibition, so that knowledge of mechanism can give insight into clinical activity. Extended residence times may be linked with increased efficacy, larger interval between doses, higher risk of mechanism-related (on-target) toxicity and decreased off-target toxicity.[57] Accordingly, measurement of dissociation rate constants can give insight into various aspects of clinical and biological activity.

4.4.2.5 *Action on Non-activated Kinases*

Many kinases bind ATP more weakly prior to activation, so that high intracellular concentrations of ATP and ADP are likely to compete less for prevention of activation than inhibition of catalysis. Prevention of activation also may be linked with improved selectivity, because there are diverse mechanisms to retain kinases in an inactive state, whereas activated kinases catalyse similar reactions. In several kinases, prevention of activation results from binding to inactive conformations, which are linked with slow kinetics and so extended drug-target residence times. Slow kinetics are seen for DFG-out binding to p38 and BRAF,[13] resulting in prevention of activation. Some kinase inhibitors have higher affinity for the non-activated state, for example binding to ABL,[50,59] or MEK.[30]

4.4.2.6 *Covalent Inhibitors*

It has been estimated that around 30% of the enzyme inhibitors used as drugs have a mechanism that includes covalent bond formation.[60] This approach has been used in kinase drug discovery in attempts to overcome the competition by ATP and ADP *in vivo*. However, clinical compounds such as CI-1033 and HKI-272 were dose-limited by diarrhoea and skin rash linked with inhibition of the target protein, EGFR-TK, outside the tumour.[61] Recent work has attempted to overcome this difficulty by having inhibitors, which are selective in favour of the mutant kinase in the tumour rather than the unchanged kinase present in the healthy tissue of the same patient. This approach exploits the occurrence of somatic mutations, where the kinase in the tumour has a different sequence to that in the rest of the body.[17,53,61]

4.4.2.7 Inhibition of Mutant Kinases

As described in Section 4.4.2.1, mutation of the gatekeeper residue is commonly linked with resistance to kinase inhibitors in cancer therapy. Around 50% of resistant non-small cell lung cancer patients have tumours carrying the EGFR-TK T790M mutation. These changes often arise by somatic mutation in the tumour, with the original sequence remaining in healthy tissues from the same patient. Somatic gatekeeper mutations are linked with oncogenesis and resistance in animal models.[53] It has been proposed that first generation anilinoquinazoline inhibitors clash with the enlarged gatekeeper, prompting the screening of smaller pyrimidine inhibitors, which contain an acrylamide moiety designed to react with Cys797.[17] This work led to WZ4002 (Figure 4.4), which is bound in a crystal structure of EGFR-TK T790M, well tolerated and effective in a lung cancer model driven by T790M.

Various other approaches have been suggested to overcome resistance mutations. In order to give resistance, a mutation must interfere with inhibition more than catalysis. Accordingly, inhibitors designed to overcome resistance mutations may interact with essential or highly conserved regions of the kinase structure. Compounds using only the purine site, therefore, could be less sensitive to resistance mutations than are compounds that bind in allosteric sites. Addition of hydrophobic groups could help to retain binding upon mutation, because their interactions are less sensitive than hydrogen bonding to distance and angle. The thermodynamic consequences would be a shift from enthalpy-driven towards entropy-driven binding. However, low lipophilicity and enthalpy-driven binding are linked with good bioavailability and an improvement from first in class to best in class.[62,63] Perhaps more effective ways to overcome resistance are to design compounds to have high affinity for common mutants and to form polar interactions with groups on the kinase, which are important for function and so are highly conserved.

4.4.2.8 Patient Selection in Cancer Therapy

Observation of a mutation in a particular kinase does not guarantee response to an inhibitor of the kinase itself or its pathway.[64] For example, K-RAS mutations do not give sensitivity to inhibitors of downstream kinases such as RAF or MEK. Emerging clinical data is helping to identify tumours with "oncogene addiction".[53] Here, the functioning of one gene product or related pathway is required for survival or proliferation of the cancer cells. The same addictions may be found in tumours derived from different anatomical origins, raising the possibility of enrolling patients into clinical trials on the basis of the identity of mutations, rather than tumour origin. Additionally, there may be value in the inclusion of diverse patients in early clinical development and then performing further studies if inhibitors give unexpected effects on particular kinases or tissues. Such approaches revealed how sorafenib could be used for treatment of renal cell and hepatocellular carcinomas (Section 4.4.2.2).

This clinical inhibition of unexpected kinases probably reflects the limitations in our understanding of the functioning of signaling networks—how a particular kinase may be bypassed, replaced or present in excess. Systems biology could be a useful component in target identification. Although valuable for the identification, evaluation and optimisation of inhibitors, *in vitro* assays only have limited relevance for predicting activity *in vivo* due to their use of isolated proteins, model substrates, truncated enzyme constructs and non-physiological substrate concentrations. Assays using cultured transformed cells overcome some of these limitations, but not all, due to the absence of additional factors which influence tumour survival, and because of the diversity of cancer cells in different patients together with the tendency of these cells to change in response to therapy.

4.5 Conclusions

Most kinase inhibitors function by reversible ATP-competitive binding and use similar interactions in the purine site. Despite this lack of diversity, medicinal chemistry has been successful in generating compounds with a range of selectivity profiles. Inhibition of multiple kinases may be required for clinical efficacy and can be reasonably well-tolerated.[64] Small numbers of compounds with more diverse mechanisms have emerged, usually from more complex assay formats. Cell-based assays identified compounds using the myristoyl site of ABL,[65] and the ATP-noncompetitive MEK inhibitor, U-0126,[66] whereas full-length physiological substrates were required for the detection of ATP-non-competitive p38 inhibitors,[21] and favour the detection of prevention of activation compounds.[13,30,50] In some cases, non-activated states or mutant kinases may be more tractable targets than activated wild-type kinases.[37,53]

Assays for the identification and evaluation of kinase inhibitors need to be tailored to address specific questions. Structure-based design usually is best supported by K_d values or potencies measured at $[ATP] = K_m$, whereas understanding activity *in vivo* may require physiological substrate concentrations and perhaps measurement of drug-target residence times.

We are only beginning to understand how potency and selectivity *in vitro* are related to efficacy and tolerability *in vivo*. Over the next few years, I believe that major advances in therapy with kinase inhibitors will be made pragmatically, by studying the causative factors underlying clinical observations and then applying the derived knowledge in judicious selection of patients and treatment regimes. Compounds with improved efficacy and tolerability may be derived from hits identified in assays favouring forms of the kinase, which are common during the disease state.

Acknowledgements

I am very grateful to many people who have made essential contributions to our work on protein kinases, in particular: Andy Barker, Alun Bermingham, Alex Breeze, Doug Campbell, Stefan Gerhardt, Neil Hales, Karl Hård, Geoff Holdgate, Phil Jewsbury, Christine Lambert, Richard Pauptit, Phil Poyser, Gwyn Richards, Pirthipal Singh, Jane Sullivan, Dave Timms and Wendy VanScyoc. I thank Ian Collins, Christine Lambert and Gwyn Richards for comments on a draft manuscript, together with Richard Ward and Fred Goldberg for expert editorial assistance. I am grateful to AstraZeneca for funding much of my work.

References

1. G. Manning, D. B. Whyte, R. Martinez, T. Hunter and S. Sudarsanam, *Science*, 1988, **298**, 1912.
2. L. N. Johnson, M. E. M. Noble and D. J. Owen, *Cell*, 1996, **85**, 149.
3. M. Cherry and D. H. Williams, *Curr. Med. Chem.*, 2004, **11**, 663.
4. J. Stamos, M. X. Sliwkowski and C. Eigenbrot, *J. Biol. Chem.*, 2002, **277**, 46265.
5. C. -H. Yun, T. J. Boggon, Y. Li, M. S. Woo, H. Greulich, M. Meyerson and M. J. Eck, *Cancer Cell*, 2007, **11**, 217.
6. K. S. Gajiwala, J. C. Wu, J. Christensen, G. D. Deshmukh, W. Diehl, J. P. DiNitto, J. M. English, M. J. Greig, Y. A. He, S. L. Jacques, E. A. Lunney, M. McTigue, D. Molina, T. Quenzer, P. A. Wells, X. Yu, Y. Zhang, A. Zou, M. R. Emmett, A. G. Marshall, H. M. Zhang and G. D. Demetri, *Proc. Natl. Acad. Sci. U. S. A.*, 2009, **106**, 1542.
7. R. M. Bukowski, U. Yasothan and P. Kirkpatrick, *Nat. Rev. Drug Discovery*, 2010, **9**, 17.
8. B. Nagar, W. G. Bornmann, P. Pellicena, T. Schindler, D. R. Veach, W. T. Miller, B. Clarkson and J. Kuriyan, *Cancer Res.*, 2002, **62**, 4236.
9. E. Weisberg, P. W. Manley, W. Breitenstein, J. Brüggen, S. W. Cowan-Jacob, A. Ray, B. Huntly, D. Fabbro, G. Fendrich, E. Hall-Meyers, A. L. Kung, J. Mestan, G. Q. Daley, L. Callahan, L. Catley, C. Cavazza, M. Azam, A. Mohammed, D. Neuberg, R. D. Wright, D. G. Gilliland and J. D. Griffin, *Cancer Cell*, 2005, **7**, 129.
10. J. S. Tokarski, J. A. Newitt, C. Y. Chang, J. D. Cheng, M. Wittekind, S. E. Kiefer, K. Kish, F. Y. Lee, R. Borzillerri, L. J. Lombardo, D. Xie, Y. Zhang and H. E. Klei, *Cancer Res.*, 2006, **66**, 5790.

11. P. T. Wan, M. J. Garnett, S. M. Roe, S. Lee, D. Niculescu-Duvaz, V. M. Good, Cancer Genome Project, C. M. Jones, C. J. Marshall, C. J. Springer, D. Barford and R. Marais, *Cell*, 2004, **116**, 855.

12. C. Pargellis, L. Tong, L. Churchill, P. F. Cirillo, T. Gilmore, A. G. Graham, P. M. Grob, E. R. Hickey, N. Moss, S. Pav and J. Regan, *Nat. Struct. Biol.*, 2002, **9**, 268.

13. J. E. Sullivan, G. A. Holdgate, D. Campbell, D. Timms, S. Gerhardt, J. Breed, A. L. Breeze, A. Bermingham, R. A. Pauptit, R. A. Norman, K. J. Embrey, J. Read, W. S. VanScyoc and W. H. J. Ward, *Biochemistry*, 2005, **44**, 16475.

14. J. F. Ohren, H. Chen, A. Pavlovsky, C. Whitehead, E. Zhang, P. Kuffa, C. Yan, P. McConnell, C. Spessard, C. Banotai, W. T. Mueller, A. Delaney, C. Omer, J. Sebolt-Leopold, D. T. Dudley, I. K. Leung, C. Flamme, J. Warmus, M. Kaufman, S. Barrett, H. Tecle and C. A. Hasemann, *Nat. Struct. Mol. Biol.*, 2004, **11**, 1192.

15. E. R. Wood, A. T. Truesdale, O. B. McDonald, D. Yuan, A. Hassell, S. H. Dickerson, B. Ellis, C. Pennisi, E. Horne, K. Lackey, K. J. Alligood, D. W. Rusnak, T. M. Gilmer and L. Shewchuk, *Cancer Res.*, 2004, **64**, 6652.

16. G. Bollag, P. Hirth, J. Tsai, J. Zhang, P. N. Ibrahim, H. Cho, W. Spevak, C. Zhang, Y. Zhang, G. Habets, E. A. Burton, B. Wong, G. Tsang, B. L. West, B. Powell, R. Shellooe, A. Marimuthu, H. Nguyen, K. Y. Zhang, D. R. Artis, J. Schlessinger, F. Su, B. Higgins, R. Iyer, K. D'Andrea, A. Koehler, M. Stumm, P. S. Lin, R. J. Lee, J. Grippo, I. Puzanov, K. B. Kim, A. Ribas, G. A. McArthur, J. A. Sosman, P. B. Chapman, K. T. Flaherty, X. Xu, K. L. Nathanson and K. Nolop, *Nature*, 2010, **467**, 596.

17. W. Zhou, D. Ercan, L. Chen, C. H. Yun, D. Li, M. Capelletti, A. B. Cortot, L. Chirieac, R. E. Iacob, R. Padera, J. R. Engen, K. K. Wong, M. J. Eck, N. S. Gray and P. A. Jänne, *Nature*, 2009, **462**, 1070.

18. J. Zhang, F. J. Adrián, W. Jahnke, S. W. Cowan-Jacob, A. G. Li, R. E. Iacob, T. Sim, J. Powers, C. Dierks, F. Sun, G. R. Guo, Q. Ding, B. Okram, Y. Choi, A. Wojciechowski, X. Deng, G. Liu, G. Fendrich, A. Strauss, N. Vajpai, S. Grzesiek, T. Tuntland, Y. Liu, B. Bursulaya, M. Azam, P. W. Manley, J. R. Engen, G. Q. Daley, M. Warmuth and N. S. Gray, *Nature*, 2010, 463, 501.

19. S. F. Barnett, D. Defeo-Jones, S. Fu, P. J. Hancock, K. M. Haskell, R. E. Jones, J. A. Kahana, A. M. Kral, K. Leander, L. L. Lee, J. Malinowski, E. M. McAvoy, D. D. Nahas, R. G. Robinson and H. E. Huber, *Biochem. J.*, 2005, **385**, 399.

20. W. I. Wu, W. C. Voegtli, H. L. Sturgis, F. P. Dizon, G. P. Vigers and B. J. Brandhuber, *PLoS One*, 2010, **5**, e12913.

21. W. Davidson, L. Frego, G. W. Peet, R. R. Kroe, M. E. Labadia, S. M. Lukas, R. J. Snow, S. Jakes, C. A. Grygon, C. Pargellis and B. G. Werneburg, *Biochemistry*, 2004, **43**, 11658.

22. J. L. Stebbins, S. K. De, T. Machleidt, B. Becattini, J. Vazquez, C. Kuntzen, L. H. Chen, J. F. Cellitti, M. Riel-Mehan, A. Emdadi,

G. Solinas, M. Karin and M. Pellecchia, *Proc. Natl. Acad. Sci. U. S. A.*, 2008, **105**, 16809.

23. K. M. Comess, C. Sun, C. Abad-Zapatero, E. R. Goedken, R. J. Gum, D. W. Borhani, M. Argiriadi, D. R. Groebe, Y. Jia, J. E. Clampit, D. L. Haasch, H. T. Smith, S. Wang, D. Song, M. L. Coen, T. E. Cloutier, H. Tang, X. Cheng, C. Quinn, B. Liu, Z. Xin, G. Liu, E. H. Fry, V. Stoll, T. I. Ng, D. Banach, D. Marcotte, D. J. Burns, D. J. Calderwood and P. J. Hajduk, *ACS Chem. Biol.*, 2011, **6**, 234.

24. A. Converso, T. Hartingh, R. M. Garbaccio, E. Tasber, K. Rickert, M. E. Fraley, Y. Yan, C. Kreatsoulas, S. Stirdivant, B. Drakas, E. S. Walsh, K. Hamilton, C. A. Buser, X. Mao, M. T. Abrams, S. C. Beck, W. Tao, R. Lobell, L. Sepp-Lorenzino, J. Zugay-Murphy, V. Sardana, S. K. Munshi, S. M. Jezequel-Sur, P. D. Zuck and G. D. Hartman, *Bioorg. Med. Chem. Lett.*, 2009, **19**, 1240.

25. D. Vanderpool, T. O. Johnson, C. Ping, S. Bergqvist, G. Alton, S. Phonephaly, E. Rui, C. Luo, Y. L. Deng, S. Grant, T. Quenzer, S. Margosiak, J. Register, E. Brown and J. Ermolieff, *Biochemistry*, 2009, **48**, 9823.

26. Y. Tal-Gan, N. S. Freeman, S. Klein, A. Levitzki and C. Gilon, *Bioorg. Med. Chem.*, 2010, **18**, 2976.

27. D. B. Lipka, L. S. Hoffmann, F. Heidel, B. Markova, M. C. Blum, F. Breitenbuecher, S. Kasper, T. Kindler, R. L. Levine, C. Huber and T. Fischer, *Mol. Cancer Ther.*, 2008, **7**, 1176.

28. Y. -C. Cheng and W. H. Prusoff, *Biochem. Pharmacol.*, 1973, **22**, 3099.

29. D. A. Roberts and W. H. J. Ward, in *Medicinal Chemistry Principles and Practice*, ed. F. D. King, Royal Society of Chemistry, Cambridge, 2nd edn, 2002, chapter 4, p 64.

30. W. S. VanScyoc, G. A. Holdgate, J. E. Sullivan and W. H. J. Ward, *Biochemistry*, 2008, **47**, 5017.

31. D. C. Swinney, *Curr. Top. Med. Chem.*, 2006, **6**, 461.

32. J. F. Morrison and C. T. Walsh, *Adv. Enzymol. Relat. Areas Mol. Biol.*, 1988; **61**, 201.

33. J. W. Williams and J. F. Morrison, *Methods Enzymol.*, 1979, **63**, 437.

34. P. Singh and W. H. J. Ward, *Expert Opin. Drug Discovery*, 2008, **3**, 819.

35. M. A. Fabian, W. H. Biggs, D. K. Treiber, C. E. Atteridge, M. D. Azimioara, M. G. Benedetti, T. A. Carter, P. Ciceri, P. T. Edeen, M. Floyd, J. M. Ford, M. Galvin, J. L. Gerlach, R. M. Grotzfeld, S. Herrgard, D. E. Insko, M. A. Insko, A. G. Lai, J. M. Lélias, S. A. Mehta, Z. V. Milanov, A. M. Velasco, L. M. Wodicka, H. K. Patel, P. P. Zarrinkar and D. J. Lockhart DJ, *Nat. Biotechnol.*, 2005; **23**, 329.

36. M. W. Karaman, S. M. I. Davis, P. T. Edeen, R. Faraoni, M. Floyd, J. P. Hunt, D. J. Lockhart, Z. V. Milanov, M. J. Morrison, G. Pallares, H. K. Patel, S. Pritchard, L. M. Wodicka and P. P. Zarrinkar, *Nat. Biotechnol.*, 2008, **26**, 127.

37. L. M. Wodicka, P. Ciceri, M. I. Davis, J. P. Hunt, M. Floyd, S. Salerno, X. H. Hua, J. M. Ford, R. C. Armstrong, P. P. Zarrinkar and D. K. Treiber, *Chem. Biol.*, 2010, **17**, 1241.
38. H. Nordin, M. Jungnelius, R. Karlsson and O. P. Karlsson, *Anal. Biochem.*, 2005, **340**, 359.
39. G. A. Holdgate and W. H. J. Ward, *Drug Discovery Today*, 2005, **10**, 1543.
40. R. R. Kroe, J. Regan, A. Proto, G. W. Peet, T. Roy, L. D. Landro, N. G. Fuschetto, C. A. Pargellis and R. H. Ingraham, *J. Med. Chem.*, 2003, **46**, 4669.
41. J. R. Simard, C. Grütter, V. Pawar, B. Aust, A. Wolf, M. Rabiller, S. Wulfert, A. Robubi, S. Klüter, C. Ottmann and D. Rauh D, *J. Am. Chem. Soc.*, 2009, **131**, 18478.
42. M. Pellecchia, I. Bertini, D. Cowburn, C. Dalvit, E. Giralt, W. Jahnke, T. L. James, S. W. Homans, H. Kessler, C. Luchinat, B. Meyer, H. Oschkinat, J. Peng, H. Schwalbe and G. Siegal, *Nat. Rev. Drug Discovery*, 2008, **7**, 738.
43. N. Thorne, D. S. Auld and J. Inglese, *Curr. Opin. Chem. Biol.*, 2010, **14**, 315.
44. J. S. Albert, N. Blomberg, A. L. Breeze, A. J. Brown, J. N. Burrows, P. D. Edwards, R. H. Folmer, S. Geschwindner, E. J. Griffen, P. W. Kenny, T. Nowak, L. L. Olsson, H. Sanganee and A. B. Shapiro, *Curr. Top. Med. Chem.*, 2007, **7**, 1600.
45. J. Yang, R. A. Copeland and Z. Lai, *J. Biomol. Screening*, 2009, **14**, 111.
46. B. Y. Feng, A. Shelat, T. N. Doman, R. K. Guy and B. K. Shoichet, *Nat. Chem. Biol.*, 2005, **1**, 146.
47. M. R. Grace, C. T. Walsh and P. A. Cole, *Biochemistry*, 1997, **36**, 1874.
48. Z. A. Knight and K. M. Shokat, *Chem. Biol.*, 2005, **12**, 621.
49. L. A. Smyth and I. Collins, *J. Chem. Biol.*, 2009, **2**, 131.
50. B. Okram, A. Nagle, F. J. Adrián, C. Lee, P. Ren, X. Wang, T. Sim, Y. Xie, X. Wang, G. Xia, G. Spraggon, W. Warmuth, Y. Liu and N. S. Gray, *Chem. Biol.*, 2006, 13, 779.
51. S. Atwell, J. M. Adams, J. Badger, M. D. Buchanan, I. K. Feil, K. J. Froning, X. Gao, J. Hendle, K. Keegan, B. C. Leon, H. J. Müller-Dieckmann, V. L. Nienaber, B. W. Noland, K. Post, K. R. Rajashankar, A. Ramos, M. Russell, S. K. Burley and S. G. Buchanan, *J. Biol. Chem.*, 2004, **279**, 55827.
52. T. J. Lynch, D. W. Bell, R. Sordella, S. Gurubhagavatula, R. A. Okimoto, B. W. Brannigan, P. L. Harris, S. M. Haserlat, J. G. Supko, F. G. Haluska, D. N. Louis, D. C. Christiani, J. Settleman and D. A. Haber, *N. Engl. J. Med.*, 2004, 350, 2129.
53. P. A. Jänne, N. Gray and J. Settleman, *Nat. Rev. Drug. Discovery*, 2009, **8**, 709.
54. C. H. Yun, T. J. Boggon, Y. Li, M. S. Woo, H. Greulich, M. Meyerson and M. J. Eck, *Cancer Cell*, 2007, **11**, 217.

55. C. H. Yun, K. E. Mengwasser, A. V. Toms, M. S. Woo, H. Greulich, K. K. Wong, M. Meyerson and M. J. Eck, *Proc. Natl. Acad. Sci. U. S. A.*, 2008, **105**, 2070.
56. P. J. Tummino and R. A. Copeland, *Biochemistry*, 2008, **47**, 5481.
57. D. C. Swinney, *Curr. Opin. Drug Discovery Dev.*, 2009, **12**, 31.
58. H. Lu and P. J. Tonge, *Curr. Opin. Chem. Biol.*, 2010, **14**, 467.
59. T. Schindler, W. Bornmann, P. Pellicena, W. T. Miller, B. Clarkson and J. Kuriyan, *Science,* 2000, **289**, 1938.
60. J. G. Robertson, *Biochemistry*, 2005, **44**, 5561.
61. J. Singh, R. C. Petter and A. F. Kluge, *Curr. Opin. Chem. Biol.*, 2010, **14**, 475.
62. E. Freire, *Drug Discovery Today*, 2008, **13**, 869.
63. F. E. Torres, M. I. Recht, J. E. Coyle, R. H. Bruce and G. Williams, *Curr. Opin. Struct. Biol.*, 2010, **20**, 598.
64. Z. A. Knight, H. Lin and K. M. Shokat, *Nat. Rev. Cancer*, 2010, **10**, 130.
65. F. J. Adrián, Q. Ding, T. Sim, A. Velentza, C. Sloan, Y. Liu, G. Zhang, W. Hur, S. Ding, P. Manley, J. Mestan, D. Fabbro and N. S. Gray, *Nat. Chem. Biol.*, 2006, **2**, 95.
66. M. F. Favata, K. Y. Horiuchi, E. J. Manos, A. J. Daulerio, D. A. Stradley, W. S. Feeser, D. E. Van Dyk, W. J. Pitts, R. A. Earl, F. Hobbs, R. A. Copeland, R. L. Magolda, P. A. Scherle and J. M. Trzaskos, *J. Biol. Chem.*, 1998, **273**, 18623.

Kinase Mutations and Resistance in Cancer

JACK ANDREW BIKKER

World Wide Medicinal Chemistry, Pfizer Global Research and Development, 445 Eastern Point Road, Groton CT 06340, United States of America

5.1 Introduction

The story of kinase inhibitors as drugs is largely also the story of the interplay between kinase mutations and drug discovery for cancer.[1] In 2001, the small molecule kinase inhibitor, imatinib mesylate (Gleevec, **5.1**) was first approved for use in various leukemias in the United States.[2] This drug represented a breakthrough for several reasons. For patients with several types of leukemia, it represented additional years of life and progress to a cure. For researchers, its path to the clinic has been, in retrospect, instructive. The program started from a scaffold that inhibited cyclin-dependent kinases and protein kinase C, and grew from a program targeted to Src kinase, which had been identified as a viral oncogene caused by an introduced activating mutation. It eventually proved its clinical success in chronic myeloid leukemia (CML), which is driven by a fusion protein of BCR and the Abl kinase. It is the latter protein that is inhibited by imatinib. To emphasize the importance of research to this effort, it is worth noting that both the eventual clinically-relevant kinase target and the inactive-form binding mode of imatinib were completely unanticipated when the research program began. However, this narrative has since become instructive for a third reason; after a clinically effective period, drug resistance emerged in patients, which has been traced to several mutants in the kinase domain of Abl. These have been shown to directly affect the binding of

RSC Drug Discovery Series No. 19
Kinase Drug Discovery
Edited by Richard A. Ward and Frederick Goldberg
© Royal Society of Chemistry 2012
Published by the Royal Society of Chemistry, www.rsc.org

imatinib while preserving the binding of ATP, thereby leading to drug resistance. Mutations have therefore been essential to identifying and justifying the research effort to kinases, and have more recently become a significant driver for continued development of new drugs to augment and replace clinically useful kinase inhibitors. At the moment, approximately 12 kinase inhibitors are approved for various indications in the United States. Imatinib **5.1**, dasatinib **5.2** and nilotinib **5.3** are indicated for various leukemias and gastrointestinal stromal tumors, due to their inhibition of Abl kinase and Kit kinase. Sunitinib **5.4** and sorafenib **5.5** inhibit a number of tyrosine kinases; the latter notably inhibits b-Raf and c-Raf. Gefitinib **5.6** and erlotinib **5.7** are EGFR inhibitors that have been shown to be effective in some non-small cell lung cancers (NSCLCs). Lapatinib **5.8** is a recently approved Her2 inhibitor that is indicated in some breast cancer treatments. Pirfenidone **5.9** is a p38 kinase inhibitor that has been approved in Europe (but not the United States as of June 2011) for iodiopathic pulmonary fibrosis. Sirolimus **5.10**, also known as rapamycin, and its analogues zotarolimus, temsirolimus **5.11**, zotarolimus **5.12**, and everolimus **5.13** bind allosterically to mTOR kinase and have uses in immunology. Temsirolimus **5.11** has also been approved to treat some renal cancers. Over the past decade, several tactics have emerged to overcome resistance mutations. However, early indications suggest that the continuing clinical use of kinase inhibitors in cancer may involve cycles of new drug introductions followed by the emergence of new clinical resistance.

5.1 imatinib

5.2 nilotinib

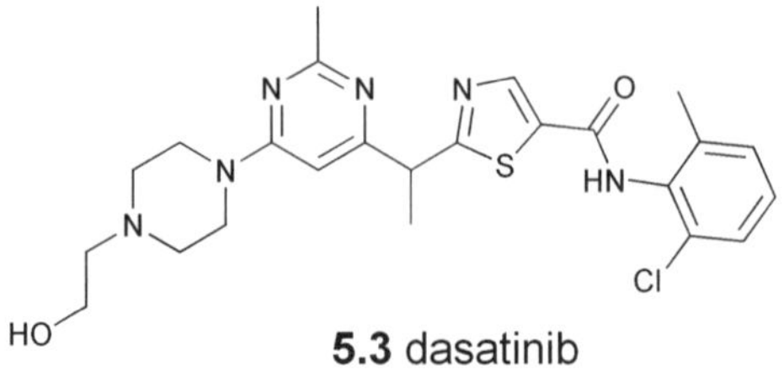

5.3 dasatinib

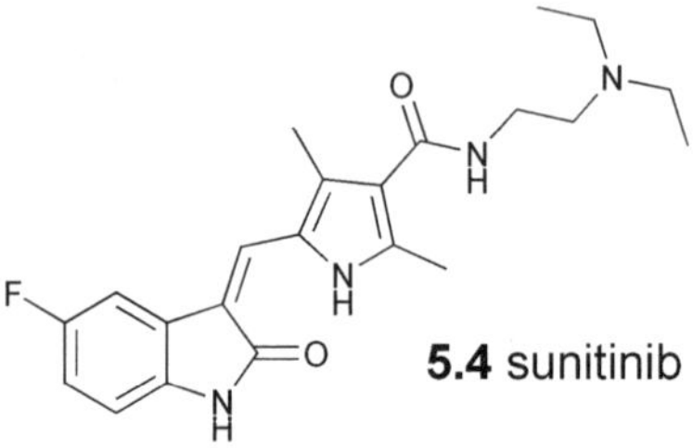

5.4 sunitinib

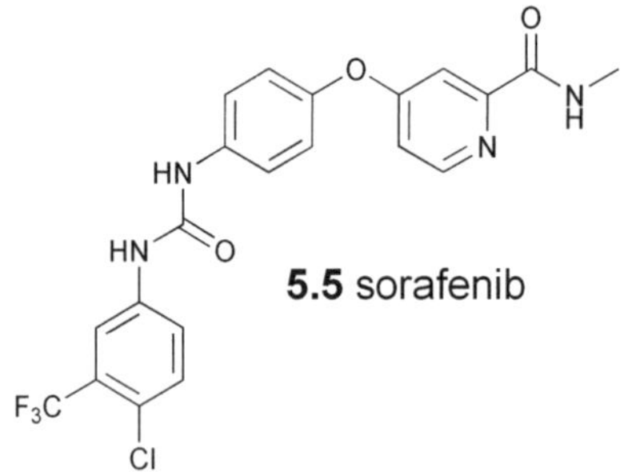

5.5 sorafenib

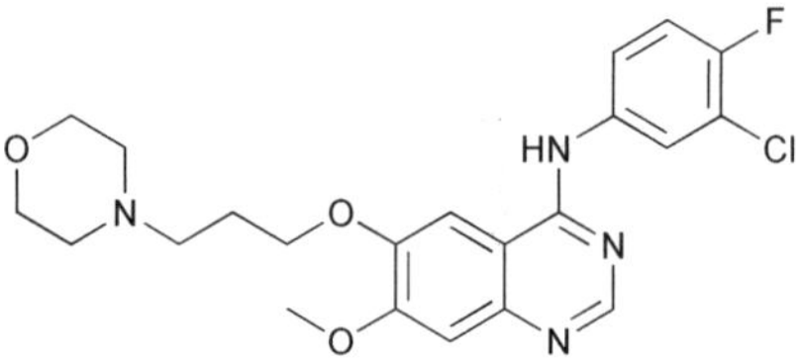

5.6 gefitinib

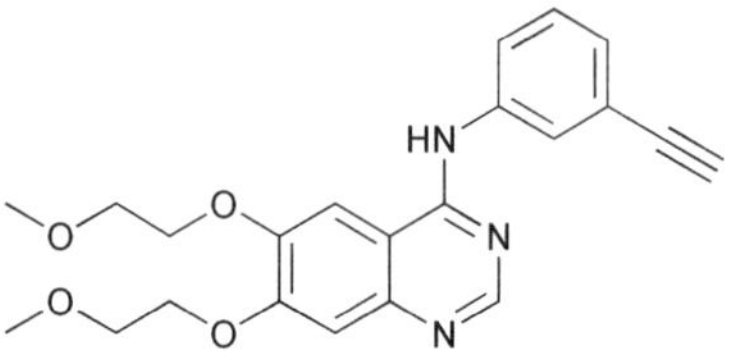

5.7 erlotinib

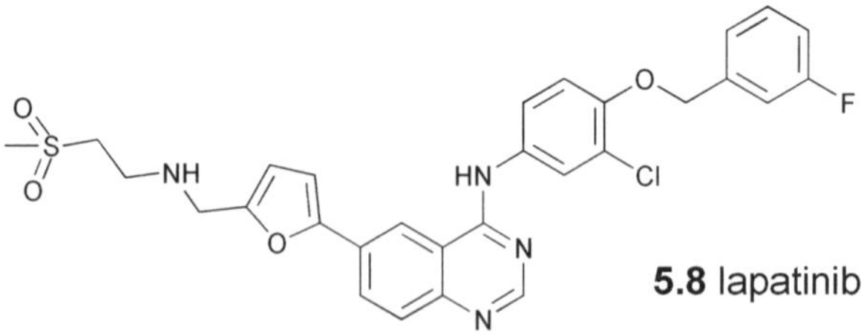

5.8 lapatinib

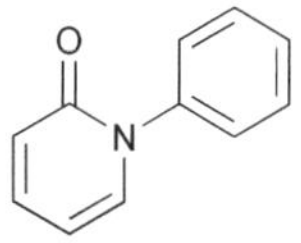

5.9 pirfenidone

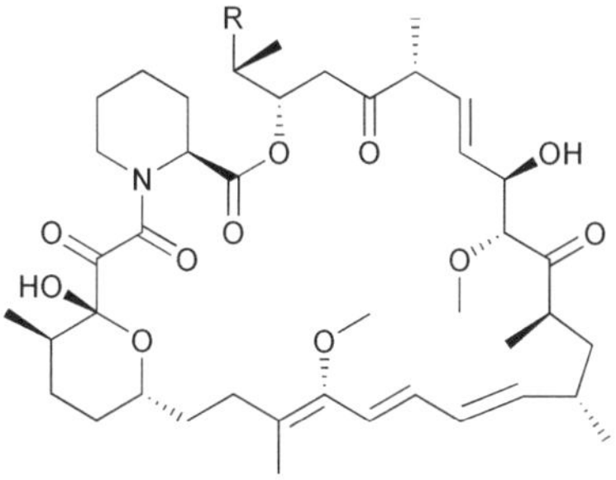

R

5.10 sirolimus

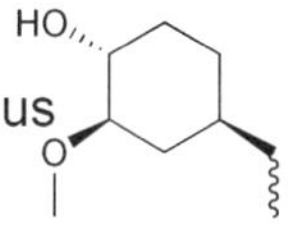

5.11 temsirolimus

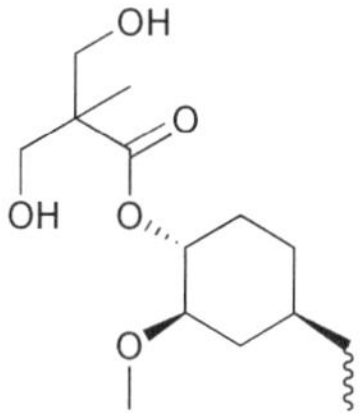

5.12 zotarolimus

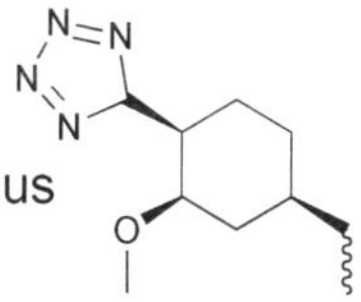

5.13 everolimus

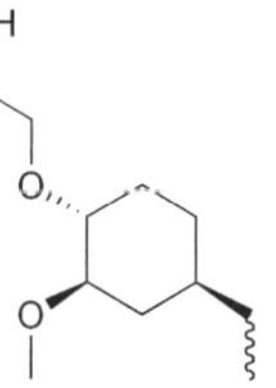

5.1.1 Kinase activation

An understanding of structure and function of kinases provides necessary context to understand the effect of activating and resistance mutations. Kinases are essential signaling molecules in many pathways,[3] and act by phosphorylation of many endogenous substrates. These can include specific serine, threonine, or tyrosine sidechains on a range of proteins, as well as precursors to secondary messengers such as phosphatidylinositol-3,4,5-triphosphate (PIP3). This phosphorylation event, especially when it serves to activate other kinases, can cause amplification of an intracellular signal or recruitment of adjacent pathways. The activation of a kinase generally involves the shift to an ATP-bound form of the kinase that can selectively bind the relevant protein and transfer the gamma phosphate group of ATP onto the relevant hydroxyl group of the substrate. Endogenous regulatory mechanisms control the proportion of the kinase in the active state.

Many crystal structures are available of the active binding conformation of kinases, both with ATP analogues and with bound inhibitors (For example, dasatinib, shown in Figure 5.1a.). The ATP molecule binds in a cleft between an N-terminal domain consisting of mostly beta sheet structure, and a C-terminal domain consisting of alpha helices. In addition, there are a number of important loop regions that affect the ATP binding and protein function. The phosphate-binding loop (or P-loop) binds the phosphate groups of ATP and plays a key role in ATP accessibility to the binding site. The activation loop, near the protein binding domain, shifts from blocking the ATP site in many inactive states to opening the ATP binding cleft in the active form of the kinase. In many kinases, phosphorylation of residues on this loop is critical in activating the kinase and shifting the equilibrium to its active form. Three residues, generally Asp-Phe-Gly (DFG), immediately before the activation loop, have been used as conformational indicators of the state of the kinase. If the Asp and Phe residues are pointed into the ATP site, the kinase is characterized as in a "DFG-in" conformation.

Several inactive states of the kinase have been identified and been characterized by protein crystallography and NMR studies (See Table 5.1 for a listing of crystal structures of marketed kinase inhibitors). One important class of inactive conformation is the "DFG-out", in which the activation loop partly occludes the ATP site, with the Asp and Phe residues pointed away from the ATP site.[4] This has been extremely important to drug discovery efforts as a

(a)   **(b)**

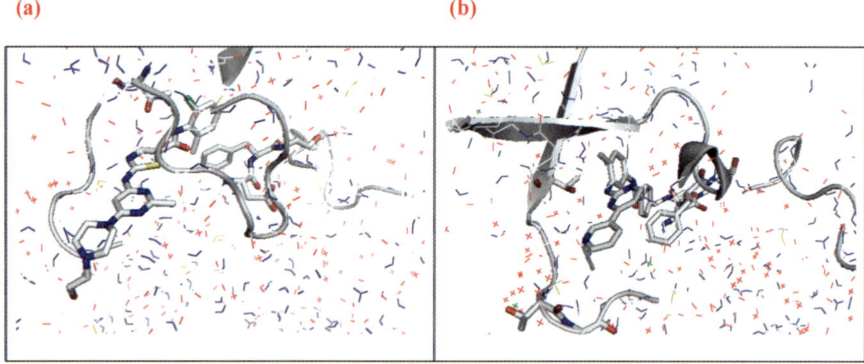

Figure 5.1   (a, left): The kinase domain of BCR-ABL bound to dasatinib is shown in an active conformation of the kinase. The hinge region (to the left of the inhibitor), the P-loop (above and to the right of the inhibitor) and the activation loop (to the right, in the background) are shown. The phenyl ring of the DFG motif is pointed into the ligand binding site. (b) On the right, the inactive form of the kinase bound to imatinib is shown. The phenyl ring of the DFG motif makes van der Waals contacts with the inhibitor. The inhibitor extends well past the gatekeeper threonine into the selectivity pocket.

selectivity pocket deep in the ATP site opens up in this state (Figure 1b). Clinically relevant inhibitors selective for this state of the kinase have been developed, and include imatinib,[5] nilotinib,[6] sorafenib,[7] and sunitinib,[8] (Table 5.1). A somewhat different inactive state has been observed in other kinases such as EGFR[9,10] and CDK2.[11] In this form, the one helix in the N-terminal domain, known as the C-helix, shifts out and leads to an alternative inactive form of the kinase. For these kinases, phosphorylation of the activation loop is often less important to activation.[12] Kinase domain dimerization (EGFR) or binding an additional protein (*e.g.* cyclin for CDK2) serve to shift the equilibrium to the active form for kinases dependent on this inactive state.

5.1.2 Type I, Type II and Type III inhibitors

Drug discovery directed to inactive states of kinases serves to overcome two issues in kinase drug discovery. The first is the issue of selectivity. Kinases remain under evolutionary pressure to ensure that they continue to bind ATP. This imparts a general shape similarity to all ATP binding clefts, which also opens the probability that a compound binding at any one kinase will bind at multiple additional kinases. By designing to an inactive form, the expectation is that the chances of promiscuous binding will be reduced, as the shape of the sites can be expected to differ. The second challenge these address is that of the endogenous concentration of ATP, which exists at millimolar levels. Any ATP inhibitor needs to out-compete this very high concentration of the endogenous ligand. An inactive-form binder stabilizes an alternative conformation of the kinase, thereby adding a possible kinetic component to an equilibrium process. Once bound, the inhibitor may release very slowly.

Table 5.1 Selected crystal structures of kinase inhibitors bound to their clinically relevant targets.

Inhibitor	PDB ID	Protein	State
Imatinib	1IEP	C-ABL[50]	DFG-out inactive
Imatinib	1T46	c-KIT[124]	DFG-out inactive
Nilotinib	3CS9	BCR-ABL[6]	DFG-out inactive
Dasatinib	2GQG	ABL[47]	DFG-in active
Sunitinib	3GOE	KIT[8]	DFG-out inactive
Sunitinib	3GOF	(D816H)-Kit[8]	DFG-in active
Sorafenib	1UWH	b-Raf[7]	DFG-out inactive
Sorafenib	1UWJ	(V599E) b-Raf[7]	DFG-out inactive
Gefitinib	2ITY	EGFR[68]	Active
Erlotinib	1M17	EGFR[69]	Active
Lapatinib	1XKK	EGFR[70]	C-helix-out inactive
Lapatinib	3BBT	Her4[125]	C-helix-out inactive
Sirolimus (Rapamycin)	1FAP	mTOR[126]	Allosteric

Inhibitors of the active form of the kinase are classified as Type I inhibitors, and inhibitors of inactive forms of the kinases are Type II inhibitors. Recently, there has been considerable interest in identifying additional binding sites remote from the active site but which can modulate kinase activity. Inhibitors binding to such allosteric or substrate sites would be considered type III inhibitors.

5.1.3   Chapter overview

This chapter will review the clinically important kinase inhibitors, as well as several kinase inhibitors emerging in late stage clinical trials. The importance of understanding the impact of mutations on the disease state itself, the effect of drug action, and clinical drug resistance will be reviewed. Several medicinal chemistry approaches to designing drugs to overcome or mitigate drug resistance will then be reviewed. Moreover, efforts to predict clinical resistance mutations preclinically will be reviewed. Several pertinent resistance mechanisms not driven by kinase domain mutations will also be highlighted. This should provide perspective on the relevance of kinase domain mutations to the underlying cancer and emerging drug resistance, as well as providing prospective means of mitigating the effect of future mutations on emerging drugs.

5.2   Kinases inhibited by imatinib

Imatinib was initially approved for use in chronic myeloid leukemia, which is driven by the BCR-Abl (Philadelphia chromosome) mutation.[13–15] Driven partly by the emergence of drug resistance, considerable effort has gone into the development of additional inhibitors to BCR-Abl. There are presently three ABL kinase inhibitors approved for use in the United States. These are imatinib **5.1**, nilotinib **5.2**, and dasatinib **5.3**. In addition to Abl kinase itself, these inhibitors also potently inhibit additional kinases such as c-Kit and PDGFR. This has broadened their indications to provide clinically important treatment for additional types of cancer, and have also provided additional targets in which to observe emerging clinical resistance mutants. Imatinib was found to inhibit c-Kit kinase, which is overexpressed in 90% of gastrointestinal stromal tumors (GISTs).[16–19] Furthermore, imatinib binding to PDGFR has been exploited to other diseases in which PDGFR overexpression has been linked to the disease. This includes chronic myelomonocytic leukemia (CMML),[20,21] AML, chronic myeloproliferative disorders (CMPD) and dermatofibrosarcoma proterbans (DFSP).[22,23] In this way, lack of selectivity of an otherwise very selective kinase inhibitor led to the broadening of its clinical use.

Nilotinib and dasatinib both emerged into clinical use after the experience of imatinib resistant clinical mutations had been identified and characterized.

This second generation of Abl kinase inhibitors therefore exemplifies two strategies to overcome clinical resistance.

In the following sections, the cancers relevant to kinases inhibited by imatinib will be reviewed. Clinical resistance mutations will be described, most of which were identified based on the clinical experience with imatinib. The development of the two second-generation inhibitors will be described, and subsequent clinical experience will be reviewed. Finally, additional kinase inhibitors to this group of kinases will be reviewed.

5.2.1 Resistance mutations to imatinib and successor compounds (BCR-Abl)

Imatinib was first approved as a treatment for chronic myeloid leukemia (CML).[24] CML is caused by a mutation in which the breakpoint cluster region (BCR) gene is aberrantly spliced into Abelson's kinase (Abl), creating a BCR-Abl hybrid. This splice is estimated to account for 15–20% of all cases of adult leukemia.[25] This splice causes constitutive activation of the hybrid BCR-Abl kinase, and subsequent uncontrolled proliferation of myeloid cells in the bone marrow.[25] Imatinib acts by competitively inhibiting the binding of ATP to the Abl kinase domain, which attenuates the kinase activity. Imatinib is notably effective at arresting the chronic phase of CML, with typically 0.5–3% of patients annually advancing to an accelerating stage of CML.[26] When this does occur, approximately 40–60% of patients exhibit tumors that now include resistance mutations in the kinase binding domain.

Many mutations of residues in the binding site of cAbl have been observed in these resistant tumors. The most common mutations have included T315I, E255K and M351T. Together, these account for approximately 60% of all cases.[27–29] The locations of these mutations are shown in Figure 5.2.

Of the resistance mutations, these are believed to act by two mechanisms. Some of these mutations occur at inward-facing residues of the ATP binding site, which also binds the inhibitor. In this case, a significant decrease in inhibitor binding potency occurs, but the binding of ATP tolerates the mutated residue. The second effect of resistance mutations is to shift the equilibrium of the kinase to its active form. Since imatinib binds to the inactive form of the kinase, this will reduce the availability of its binding state.

A number of mutations in the hinge region (T315I–D–N, F317L, and G231W) have a probable direct effect on inhibitor binding. Threonine 315 is the gatekeeper residue in BCR-Abl. Its hydroxyl (OH) group donates a hydrogen bond to imatinib as seen in multiple crystal structures (See examples listed in Table 5.1.). The mutations remove the hydrogen bond donor, thereby depriving imatinib of a key interaction with the kinase binding pocket. Furthermore, these residues are larger than threonine and cause additional steric clashes with the inhibitor.

In addition to a direct effect on the shape and character of the binding pocket, recent mutagenesis suggests that gatekeeper mutations in several

P-Loop	ABL	242	ITMKHKLGGGQYGEVYVG
Hinge /Gatekeeper	ABL	310	PFYIITEFMTYGNLLDY
Hydrophobic pocket	ABL	350	AMEYLEKKNFIHRDLAARN
Activation Loop	ABL	381	DFGLSRLMTGDTYTAHAG

Figure 5.2 Sequences of important regions of Abl kinase, highlighting sites of resistance mutations identified from clinical tumor samples. Several mutations including Y253F–H, E255 K–V, and T315I have a > 10-fold impact on imatinib potency when tested in a Ba–F3 cellular proliferation assay.[31] Only T315I retains such a significant resistance effect on the second generation inhibitors dasatinib and nilotinib.

kinases, including Abl, have increased activity. When transfected, these kinases can make BaF3 cells independent of IL-3 for survival. This is attributed to a strengthening of a "hydrophobic spine" in the active site that favors the active form of the kinase.[30]

The F317L mutation is located in the hinge region two residues away from the gatekeeper. This residue may shield the important hinge hydrogen bonds from solvent by creating a hydrophobic environment. Furthermore, the aromatic character may also participate in *pi* interactions with both ATP and inhibitors with an aromatic character such as imatinib. Mutation to leucine both changes the character of the interactions and affects the shape of the pocket.

The final hinge motif interaction is G321W, which replaces a small, highly mobile residue with the largest aromatic residue. The resulting steric clash and change in kinase dynamics may combine to reduce the potency of imatinib to this mutated kinase.

A number of P-loop mutations have been observed in BCR-Abl, which may have several ways of affecting imatinib binding. These include steric clashes, especially in the DFG-out form to which imatinib binds. Leu 248 has clinically important mutations to valine and arginine. In the former case, an important van der Waals interaction to imatinib is lost; in the latter, the larger residue may cause a steric clash. Mutations of Gly 250 and Gln 252 may also have a steric effect, as well as changing the flexibility and allowed motions of the P-loop. The Y253F mutation may act by removing a key hydrogen bond. Tyr 253 may form hydrogen bonds to a backbone NH and to N254. Mutation to phenylalanine will remove this OH group and destabilize these interactions, possibly changing the active state-inactive state equilibrium of the kinase. The final residue just at the edge of the P-loop is Glu255. In various crystal structures, this residue forms hydrogen bonds to Tyr 257 and Lys 247. Mutation to the observed lysine or valine will destabilize these interactions, and leads to a 50-fold change in binding.

A recent perspective article has analyzed the incidence of clinical resistance mutations observed with imatinib, dasatinib, and nilotinib.[31] The authors note several studies that suggest the frequency of mutations increases in accelerated phase or blast crisis disease progression states compared to chronic phase progression.[15,32] BCR-ABL appears to be able to induce reactive oxygen species, which may lead to increased genetic instability.[33,34] The authors cite several factors that may contribute to mutation incidence, including the effect of the mutation on inhibitor binding, the underlying incidence of nucleotide exchange, and whether the mutation is gain-of-function *versus* loss-of-function. Overall, mutations that result in significant loss of inhibitor binding are favored, especially T315I. Nucleotide exchange suggests that C to T transitions alone account for 40% of observed mutations.[35]

Additional mechanisms of imatinib resistance have been identified that do not involve kinase domain mutations.[36] These affect a number of other proteins that generally affect imatinib transport, metabolism, and efflux. In treatment-naive patients, the predominant resistance gene identified was that coding for PTGS1 prostaglandin-endoperoxide synthase, which is implicated in imatinib metabolism. Secondary resistance mutations were far more diverse, and up to 15 emerged from multivariate analyses from gene expression profiles. These include the ABCB1 (PGP) and ABCG2 transport proteins, solute carrier proteins, and the STAT5A and RUNX3 transcription factors. Overall, no strong single effect was identified that correlated with secondary imatinib resistance. Another study probed the effect of mutations in the regulatory domains (SH2, SH3) of the kinase.[37] This study identified one mutation, T212R, which was associated with relapse, and was linked to increased kinase activity *in vitro*. This suggests that the mutation may stabilize the active form of the kinase. Notably, however, many mutations in this region predicted *in vitro,*[27] were not observed clinically, nor were 6 of 7 mutations in this region observed in patients associated with imatinib resistance and relapse. For many of these resistance mechanisms, the hope and expectation is that these will be chemotype-specific, and that inhibitors with significantly different chemical structures may overcome these resistance mechanisms in subsequent treatment.

5.2.2 Resistance mutations to imatinib and successor compounds (KIT kinase)

Inhibitor resistance mutants have also been identified in tumors from patients with gastrointestinal stromal tumors (GISTs).[38] These tumors are driven by activating mutations of KIT and PDGFRa receptor tyrosine kinases.[39] Both imatinib and sunitinib are indicated to treat surgically inoperable GISTs. However, complete responses are rare and many patients go on to develop secondary resistance.[40]

In one study, a total of 53 tumor samples were obtained from 14 patients that progressed on imatinib or sunitinib. These were subject to D-HPLC and mutation-specific PCR. KIT oncogenic mutations were found in 11 patients, and

Hinge /Gatekeeper	ABL	310	PFYIITEFMTYGNLLDY
	KIT	651	PTLVITEYCCYGDLLNF
Activation Loop	ABL	381	DFGLSRLMTGDTYTAHAG
	KIT	810	DFGLARDIKNDSNYVVKGN

Figure 5.3 Sequences of important regions of c-Kit kinase, highlighting sites of resistance mutations identified from clinical tumor samples. Important mutations include V654A, D816 H–V, D820G, N822K, and Y823D. The sequence of Abl kinase from Figure 5.2 is included for reference.

secondary mutations were identified in nine of these patients. These secondary mutations included V654A in the ATP binding pocket and D816H, D820G, N822K, and Y823D in the activation loop. Notably, a number of patients displayed tumors with heterogeneous resistance mutants. Furthermore, patients treated with either drug displayed these mutants, suggesting that a common, inactive, state of the kinase binds both ligands in KIT. Crystallographic studies of sunitinib bound to the wild type enzyme as well as a D816H mutated enzyme has demonstrated that sunitinib binds to the DFG-out form of the kinase in the wild type enzyme and an active form of the kinase in the mutated form (Table 5.1).[8] The effect of the V654A mutation was evaluated both alone and in combination with the juxtamembrane V560G activating mutation.[41] This study demonstrated that V654A alone conferred resistance to imatinib, which had an especially pronounced effect in the presence of the V560G activating mutation. Valine 654 is a residue that lies adjacent to the gatekeeper threonine residue. Replacement of this residue with the smaller alanine might remove important van der Waals contacts to the inhibitors that align them correctly in the binding pocket. Interestingly, nilotinib, an inhibitor that is somewhat more optimized for the selectivity pocket, was less affected by this mutation. It may obtain less of its binding energy from the hinge region. Additionally, this study also noted that the D816V resistance mutation, which shifts the equilibrium to favor the active state of the kinase, conferred significant resistance to both imatinib and nilotinib.

5.2.3 Design strategies to overcome resistance mutations

A number of design strategies have been conceived and applied to circumvent the emergent drug resistance. Several recent reviews provide considerable depth into the details of each strategy as applied by several organizations.[42,43] Broadly, the approaches that have met with some success include (a) increasing inhibitor potency to both wild-type and mutant kinases, (b) designing inhibitors for different states of the kinase, and (c) designing inhibitors that do not interact with the gatekeeper threonine.

The observation that various resistance mutations lead to a drop in inhibitor potency opens the possibility that a simple increase in potency might overcome

Table 5.2   Design strategies to overcome resistance mutations in Abl and Kit.

Strategy	*Examples*
Increasing potency	Nilotinib
Designing to the active state of the kinase	Dasatinib
Avoiding interaction with the gatekeeper	VX-680, danusertib

emerging resistance. This appears to have occurred with nilotinib **5.2**, which is a Type II inhibitor that is approximately 10–30 fold more potent than imatinib **5.1** *in vitro*.[44, 45] This increase in potency appears to have made it less susceptible to the effect of activating mutations; it maintains sufficient inhibitory activity to several resistant mutants for which imatinib inhibition is lost.[44] Unfortunately, it is still highly dependent on forming a hydrogen bond with Thr 315, and is therefore still susceptible to the T315I mutation. A more recent, preclinical example is that of GNF-7 (**5.14**), which has been shown to be more potent than imatinib *in vitro* and much less susceptible to various mutants, including T315I.[46] In this case, however, docking also suggests that the molecule may bind without forming a hydrogen bond to Thr 315.

An alternative approach to overcome mutations that cause kinase activation is to target the active state of the kinase. This is exemplified by dasatinib **5.3**, which is an active state inhibitor of BCR-Abl.[45] In this case, the attendant concerns about achieving acceptable selectivity against other kinases and competing with physiological concentrations of ATP appear to have been overcome by this potent inhibitor. Because it binds to the active state, this inhibitor is not affected by most activating mutations that affect imatinib. However, it contains a hydrogen bond to Thr 315,[47] and is also still susceptible to the T315I mutation.

More recently, several compounds have been disclosed that are specifically designed to avoid making a hydrogen bond to Thr 315(reviewed by Liao,[42]). Several of these were designed to be inhibitors of the Aurora kinases.[48] These kinases have binding sites that are remarkably similar to Abl kinase, but have leucine rather than threonine as their gatekeeper residue. This chemical similarity of the binding sites increases if the Aurora kinases are compared to the T315I mutant in Bcr-Abl. One example of an compound designed as an Aurora kinases inhibitor but effective in Bcr-Abl is VX-680 **5.15**,[49] which has been shown both *in vitro*,[50] and clinically,[51] to retain potency against the T315I mutation. Notably, in a Phase I trial, patients with accelerated or blast-phase CML responded to VX-680 (now named MK-0457), based on hematological and cytogenetic responses.[52] A second example is danusertib **5.16**, which was a pan-Aurora kinase inhibitor developed around a tetra-hydropyrrolopyrazole core. It is currently undergoing clinical trials as a therapy for imatinib-resistant Bcr-Abl positive leukemia.[53]

An example of a specifically designed inhibitor is shown with the design of AP-23846 **5.17** based on its parent compound AP-23464, **5.18**. In this case, the key hydrogen bond to Thr 315 was avoided, which lost some potency *in vitro*, but retained activity against the resistant mutant.[54] Another example is provided by SGX-70393, whose structure is not disclosed but is presumed to be similar to PPY-A (**5.19**). SGX-393 showed *in vitro* potency against known imatinib-resistant cell lines.[55] A crystal structure of PPY-A in both the wild type (pdb code 2q0h) and T315I mutant kinase (pdb code 2z60) showed no hydrogen bond to Thr 315.[56]

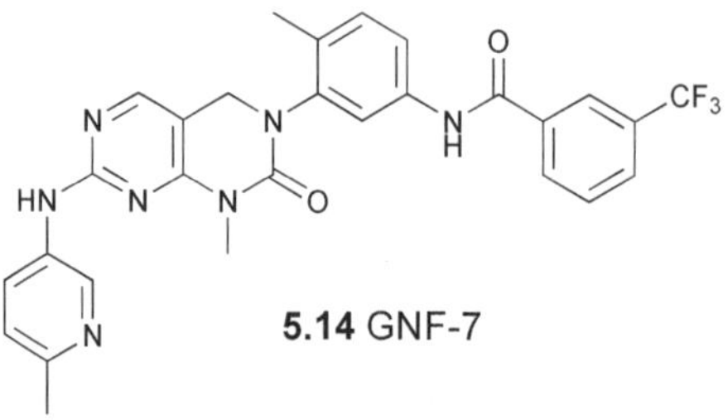

5.14 GNF-7

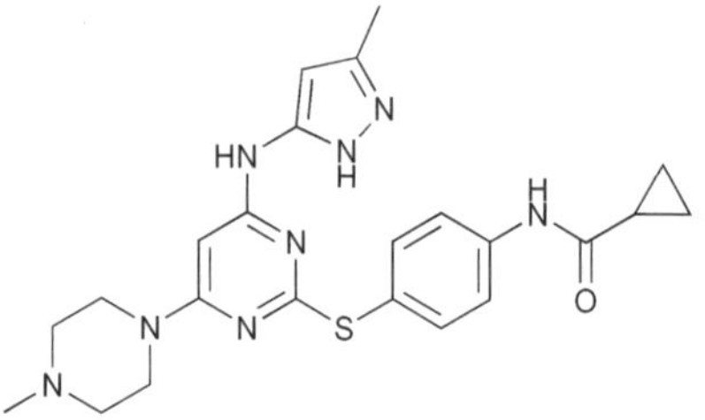

5.15 MK-0457 (VX-680)

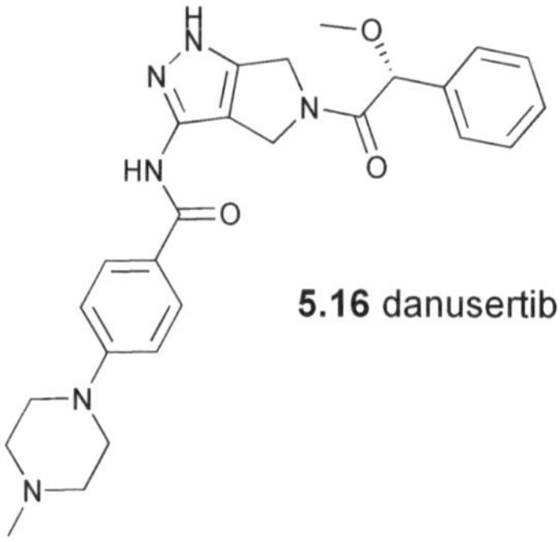

5.16 danusertib

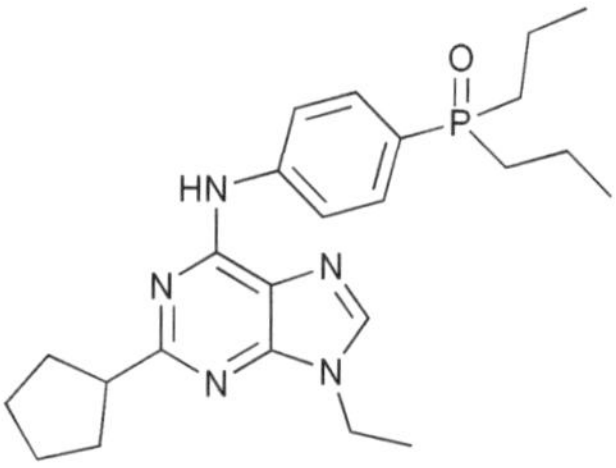

5.17 AP23846

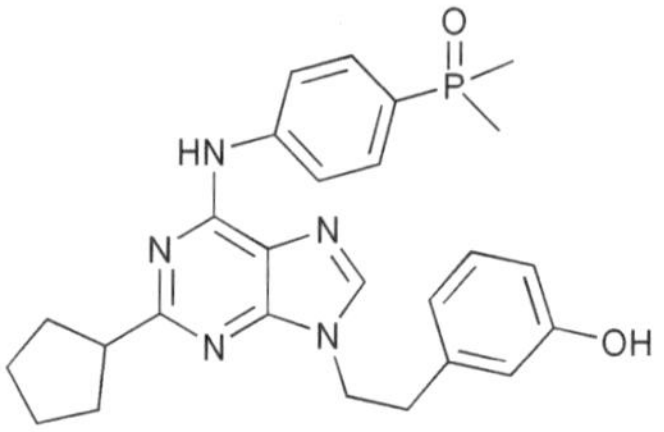

5.18 AP23464

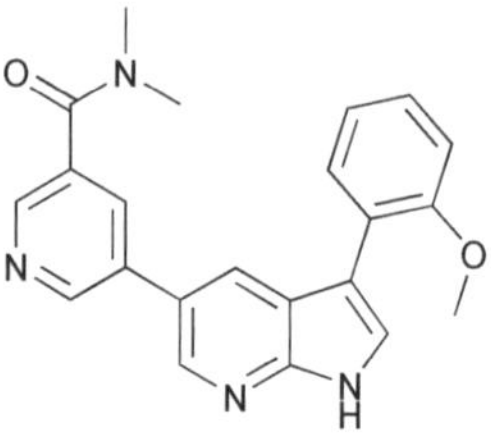

5.19 PPY-A

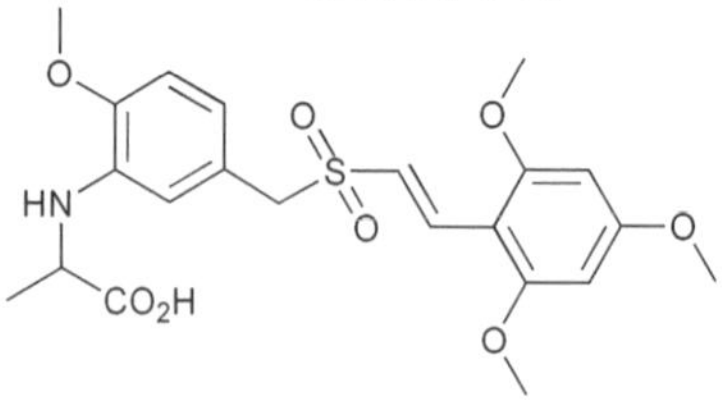

5.20 ON012380

A novel alternative to attempting to develop a potent and selective ATP-competitive inhibitor is to target the peptide site of the kinase. This strategy has been applied with the design of ONO-12380 **5.20**, which is non-competitive to ATP and showed no loss of potency to 16 mutant strains of BCR-Abl.[57]

5.3 EGFR Kinases

The epidermal growth factor (EGF) pathway is essential to cell proliferation and differentiation. Four kinases recognize EGF, and are, respectively, EGFR, HER-2, HER-3 and HER-4. HER-2 lacks an endogenous ligand, and HER-3 lacks kinase activity.[58] Since the early days of cancer research, various cancers have been recognized to be associated with over-expression or constitutive activation of these kinases, leading to breast, lung, or prostate cancers.[59] Two inhibitors that inhibit EGFR preferentially (gefitinib **5.6** and erlotinib **5.7**) have been approved for non-small cell lung cancers, and lapatinib **5.8**, a HER-2 inhibitor, has been approved for breast cancer treatment in combination with various antibody approaches. Both activating and resistance mutations are important to the clinical effectiveness of these therapies.[60] This section will review activating mutations of EGF kinases and the effect of resistance mutants.

5.3.1 Activating mutations leading to drug susceptibility

Only a subset of patients with non-small cell lung cancer (NSCLC) responds to treatment with erlotinib and gefitinib.[61–65] This unexpected finding was traced to a number of mutations in the active site of EGFR kinase that appeared to shift the kinase equilibrium to its active state. These mutations include G719S in the P-loop, L858R in the activation loop, and a number of deletion mutants prior to the C-helix of which del746-749 is the most abundant. A recent review,[66] that surveyed many NSCLC genotyping studies,[67] has concluded that approximately 90% of cancers driven by EGFR mutations are related to L758R mutation or the del-19 mutations. A recent paper investigating the effect of these mutations demonstrated that the G719S and L858R mutants led to catalytically competent enzymes that were 20-fold and 50-fold more active than the wild-type kinase.[68] This was interpreted as resulting from the mutation shifting the equilibrium toward the active state of the kinase. At the same time, the Michaelis–Menten constant (K_m) for the L858R mutant decreased by 28 fold compared to wild type, suggesting that the mutant enzyme had a decreased affinity for ATP. Taken together, this argued that the susceptibility of these tumors to erlotinib and gefitinib was driven by the decreased affinity of the mutant enzymes by ATP, and therefore a greater ability for the inhibitors to outcompete ATP at physiological concentrations of ATP.

Crystallography of known inhibitors to EGFR and mutant EGFR are consistent with these observations. Both erlotinib,[69] and gefitinib,[68] have been crystallized to EGFR, and show the active form of the kinase (Table 5.1). Both

P-Loop	ABL	248	**L**G**G**G**QY**G**E**VYVGVWKKYS----LTVAVKTLKED**T**--
	EGFR	718	L**G**SGAFGTVYKGLWIPEGEKVKIPVAIK**ELREATS**P
Hinge /Gatekeeper	ABL	310	P**F**YII**TEF**MTY**G**NLLDY
	EGFR	785	TVQLI**T**QLMPFG**C**LLDY
Activation Loop	ABL	381	DFGLSR**L**MTGDTYTA**H**AG
	EGFR	842	DFG**L**AK**L**LGAEEKEYHAE

Figure 5.4 Sequences of important regions of EGFR kinase, showing activating and resistance mutations. G719S, L858R, and the deletion mutants in the flexible region between the P-loop and C-helix serve as activating mutations. T790M, the gatekeeper mutation, is the most significant resistant mutation. The sequence of Abl kinase is shown as a reference.

inhibitors accept a key positioning hydrogen bond from the backbone NH of MET 793 in the hinge region. In addition, the inhibitor accepts a hydrogen bond from THR 790 through a bridging water molecule. The second aromatic ring binds adjacent to the catalytic lysine, but makes primary van der Waals contact with the protein.

In contrast, the EGFR–Her-2 inhibitor lapatinib binds to a C-helix-out inactive form of the kinase.[70] While maintaining the anchoring hydrogen bond to the hinge region, the benzyloxyphenyl group extends into a hydrophobic selectivity pocket past the gatekeeper residue that is further opened by the shift outward of the C-helix. In a concerted action, the P-loop also closes down on the inhibitor. This conformation appears to be common to the C-helix-out inactive form, as a second crystal structure of an ATP analogue binding to the inactive form is very consistent with the lapatinib bound form.[12]

The information arising from crystallography of both active and inactive forms of the kinases can help interpret the effect of mutant forms of the kinases. The G719S mutant has been solved in the presence of gefitinib.[68] This P-loop mutation does not appear to have a direct effect on the binding of the inhibitor itself. However, the conformation of this glycine in the C-helix-out inactive form of the kinase is one that is unique to glycine and would be highly strained when replaced by serine.[71]

The L858R mutation also is best interpreted in the context of the inactive state of the kinase. In the inactive conformation, this leucine is tightly packed against Leu 747, Met 766, Leu 777, Leu 788, and Leu 862. Replacing this leucine with the larger and charged arginine would destabilize this conformation in favor of the active conformation, in which Arg 858 points out to solvent and is removed from the binding site.[68]

The activating effect of the various deletion mutants prior to the C-helix is also interpretable by a review of the crystal structures of the inactive form of the kinase. In this C-helix-out form, the helix itself is unwound a turn compared to the active form, and the amino acid chain connecting the P-loop to the helix is extended.[70] Removing a portion of this sequence would result in further shortening the C-helix and changing the composition of the connecting fragment. Together, these would presumably further destabilize the inactive form, while still allowing a viable ATP-binding site in its active form.

5.3.2 EGFR resistance mutations

With prolonged treatment of either erlotinib or gefitinib, resistant tumors eventually emerged and were subsequently characterized to identify resistant mutants.[65,72] These were traced to the appearance of a gatekeeper mutation T790M,[66,72] either alone or combined with either L858R or a deletion mutant.[72,73] When characterized, the K_m value for ATP in the double mutant T790M–L858R is 8.4 µM, which is significantly lower than the value for the L858R mutant alone (184 µM) and much closer to the wild-type value (5.2 µM). This suggests that the effect of the double mutation is to restore the ATP affinity that was lost with the single activating mutation. This is supported by binding studies of gefitinib to the L858R mutant ($K_d = 2.4$ nM), the T790M–L858R double mutant ($K_d = 10.9$ nM) and the wild type enzyme ($K_d = 35$ nM).[74] Together, these indicate that the binding for the double mutation is within three-fold of the native enzyme, suggesting that it is the restoration of ATP affinity rather than a loss of inhibitor binding affinity that drives the resistance mutation.[74]

These observations are consistent with clinical observation. The T790M has been detected in 50% of patients resistant to gefitinib or erlotinib.[75,76] This resistance, as measured by circulating tumor cells harboring T790M, has been shown to increase over time,[77] and to be present in untreated patients.[78] In the latter study, detection of T790M prior to treatment reduced progression free survival times from 16.5 months to 7.7 months.[77]

A crystal structure of the mutant T790M kinase is available that shows the new gatekeeper methionine able to move to accommodate a bound inhibitor. This is consistent with the observed binding of inhibitors to mutant and double mutant kinases containing this mutation.[74]

This has significant implications to the design of inhibitors that circumvent the effect of the resistance mutations. Unlike Bcr-Abl driven mutations, for which designing inhibitors that avoid the interaction with the T315I mutation is a potential strategy, this would suggest that the mutation acts primarily to restore affinity for ATP. Therefore, designing inhibitors that avoid interactions with Thr 790, although feasible, is unlikely alone to lead to improved clinical effectiveness. Because the inhibitor of the doubly-mutated kinase must outcompete restored physiological ATP concentrations, follow-on treatments must necessarily focus on supra-potent inhibitors, which may include covalent inhibitors.[76,79]

5.3.3 Inhibitors and design strategies

A number of design strategies have been applied to EGFR inhibitors and have made it into the clinic and, in the case of lapatinib, onto the market. Due to research timelines, many of the compounds present in the clinic were designed prior to the emergence of resistance information in the 2002–2006 timeframe. However, subsequent clinical experience has identified that the design of irreversible inhibitors may prove to be a viable strategy to overcome resistance.[80]

EGFR kinase is one of 10 kinases that has a cysteine residue in a homologous location in the binding site located in such a way that it is potentially accessible to inhibitors containing chemistry that would react to form covalent bonds.[80] These open the possibility of increasing selectivity, ATP-competition, and long duration binding by an appropriate inhibitor. The concern with any covalent bond is that of non-specific reaction leading to safety issues. This concern is clearly manageable, as demonstrated by several important clinical drugs.[81] In addition, but changing the reversible nature of the binding event itself, they may open the possibility of overcoming the effect of resistance mutations.[73] Researchers at both Wyeth and Parke Davis (both now Pfizer) have introduced Michael acceptors onto reversible kinase inhibitors to create molecules that reactive to form covalent bonds with this binding site cysteine.[79,82-84] The Parke Davis inhibitor (canertinib, **5.21**) was discontinued due to its potency and safety profile,[85] but the Wyeth inhibitor (neratinib, **5.22**) continues in clinical trials.[86] The latter is distinguished by both binding to the C-helix-out inactive form of the kinase, as well as forming a covalent bond *in situ*.[74] Two additional irreversible inhibitors, BIBW-2992 (**5.22**),[87] and PF-00299804,[88] (a follow-on to **5.21**, structure not disclosed), have both advanced to Phase III trials.[80]

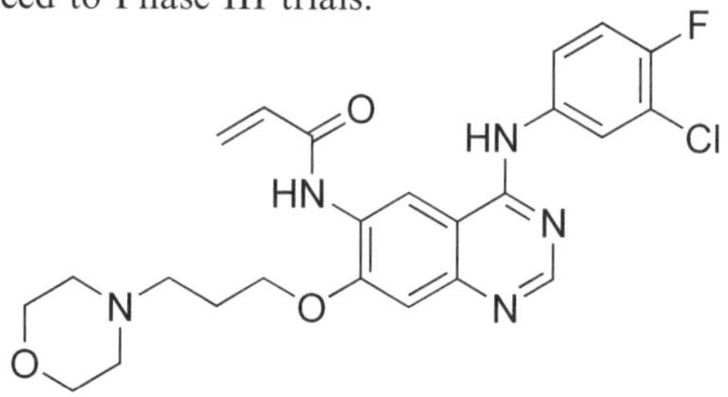

5.21 canertinib (CI-1033)

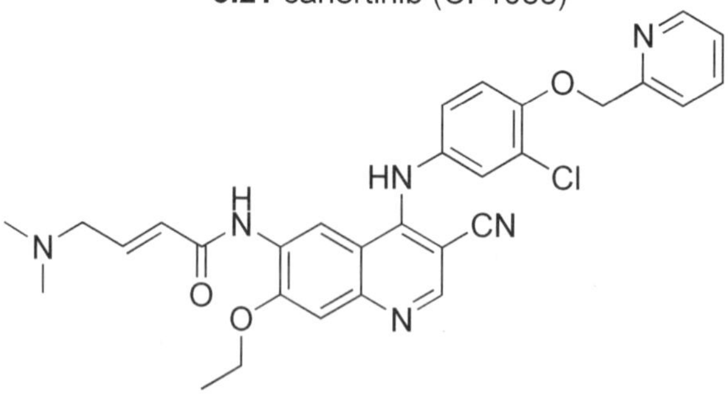

5.22 neratinib (HKI-272)

5.23 BIBW-2992

5.24 EKI-785

5.25

Irreversible inhibitors of EGFR have shown an ability to bind to mutant forms of the EGFR receptor *in vitro*.[89] The prototypical irreversible inhibitor EKI-785 (**5.24**) maintained good inhibitory activity against cells engineered with both the activating del747-752 mutation and the resistant del747-752–L790M double mutation.[72] In another study, irreversible inhibitors continued to inhibit an engineered cell line with the resistant double mutant L858R–T790M, while gefitinib and erlotinib were not effective.[64,90] An extensive *in vitro* chemical biology investigation into several matched irreversible and reversible EGFR inhibitors concluded that the T790M mutation did affect the binding of the irreversible inhibitors significantly.[91] For at least one inhibitor, the doubling of covalent bond formation time was observed, allowing a mechanistic explanation for this observation.

Recent reports outline the development of a novel irreversible inhibitor **5.25** specifically targeted to the EGFR containing the T790M mutation.[76,92] The researchers used growth-resistant PC9 cells containing both the (activating) 746–750 deletion mutant and the (resistant) T790M mutant as the basis to drive their discovery program. Using knowledge obtained from the crystal structures, an irreversible inhibitor was developed that had two-digit nanomolar affinity in cells to both the activated and resistant PC9 cells. The inhibitor was based on a pyrimidine core, which had no direct interactions with the gatekeeper residues (Thr or Met). Furthermore, the molecule included an acrylamide group to interact with Cys 797 in kinase binding site, whereas the methoxyl group *meta* to the solubilizing piperazine group appears to improve selectivity against other kinases that contain a homologous cysteine in the active site. However, this effort remains in the discovery phase; no PK or *in vivo* data has been reported to date.

Another possible strategy is to inhibit multiple targets in the pathway, to compensate for the resistance that may emerge in a single pathway. For example, an *in vitro* combination of an irreversible EGFR inhibitor and an mTOR inhibitor was effective at inducing apoptosis in a T790M cell line, albeit without statistical significance.[91]

5.4 Preclinical prediction of kinase resistance mutations

Given the important clinical implications of the emergence of kinase mutations in the clinic, several efforts have been pursued to attempt to identify possible emergent mutations preclinically and integrate this information as part of the design effort. This has been pursued against several kinase targets, including BCR-ABL,[27,93] PI3 kinase,[94] and Aurora kinase.[95] Although the details are generally target-specific, the general strategy is to incubate a susceptible and preferably hyper-mutagenic cell line with an inhibitor of interest, and incubate daughter cell lines that show increasing degrees of resistance. These can then be genotyped to identify the relevant mutations.

An example of this is provided by the work of Girdler *et al.* on Aurora kinases.[95] In this case, the candidate Aurora kinase inhibitor ZM-447439 (**5.26**) was incubated with HCT-116 cells, starting with a low concentration of the

Table 5.3 Design strategies to overcome resistance mutations in EGFR and Her2.

Strategy	*Examples*
Increasing potency	Erlotinib *vs.* gefitinib
Designing covalent inhibitors	Neratinib, CI-1033
Targeting the inactive form of the kinase	Lapatinib

inhibitor and ramping up. This process resulted in seven resistant cell lines, of which two maintained cell division and histone H3 methylation, which are markers of active Aurora kinase. When sequenced, five mutations were identified in the binding site region, including Y156H, G160E, G160V, H250Y and L308P. The most resistant mutation was shown to be G160V in Aurora B, followed by Y156H and H250Y. These mutants did not have compromised catalytic activity, but conferred resistance appeared to cross over to other inhibitors. This led to a chemical biology demonstration that VX-680, which inhibits both Aurora A and Aurora B kinases, predominantly exerts its cytotoxic effect through Aurora B. VX-680 inhibited both kinases and caused apoptosis in wild-type cells, but was not cytotoxic to cell lines with resistance mutations in Aurora B.

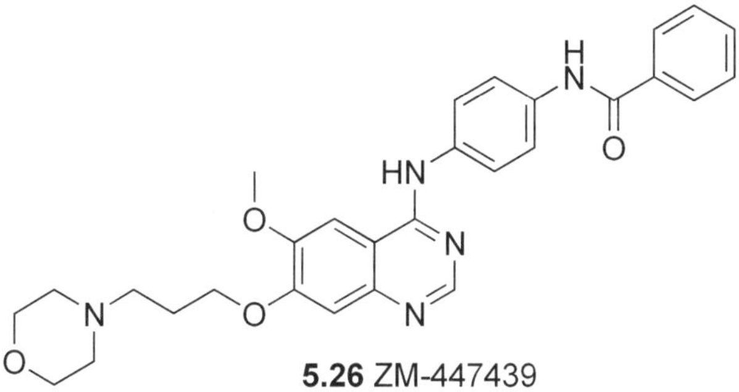

5.26 ZM-447439

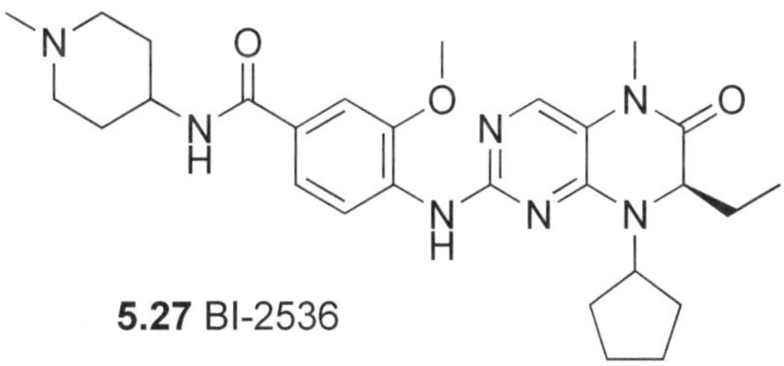

5.27 BI-2536

Another study has been performed to exploit the potential of resistance mutants to map biological pathways *in vitro*. This study used resistance mutation to Aurora A, Aurora B, and PLK1 to probe the pharmacology of VX-680 in the Aurora kinases.[96] Several mutations were engineered in Aurora A kinase, notably G216L which was resistant to VX-680. VX-680 resistant G160L was also engineered into Aurora B, and T136G, resistant to the PLK1 inhibitor BI 2536 (**5.27**), was engineered into PLK1. Using mutant cell lines that were generated incorporating the mutants above, the effect of inhibiting Aurora B signaling was associated with the antiproliferative effect of VX-680. The investigators also observed that PLK1 mediated the effect of Aurora A autophosphorylation, although the exact mechanism by which this happened was unclear. This study demonstrates the utility of resistance mutations in probing cellular function. This use does not depend on whether the resistance mutation engineered or selected ultimately is shown to have clinical relevance.

5.28 AZD-6244

5.29 CI-1040

Another example is afforded by the attempt to identify mutations that might confer resistance to MEK1 inhibitors.[97] In this study, random mutagenesis was used to generate a library of mutations in A375 melanoma cells. These cells include V600E BRAF and are consequently highly MEK-dependent. The cells were cultured for 4 weeks with either 1.5 µM of AZD-6244(**5.28**) or 2 µM of CI-1040 (**5.29**), both MEK1 inhibitors. Two sets of resistant mutants were identified after sequencing the *ca.* 1100 clones that resulted. The primary set of resistance mutations was associated with the allosteric pocket adjacent to the C-helix. This is also the region in which the two inhibitors bound the kinase. Since the MEK inhibitors shift the helix to an inactive form, these mutations can serve either to directly interfere with ligand binding, or to shift the conformation of helix-C to an active state. A second set of mutants were identified near the N-terminal helix-A, which is known to exert a negative regulatory effect on the kinase. These mutants appear to up-regulate the kinase. Interestingly, sequencing of MEK1 obtained from tumors of AZD-6244-resistant patients found one mutant (P124L) in this region. This mutant corresponded to two mutations at the same location as the P124Q and P124S mutations identified by this *in vitro* screening. *Ex vivo* testing of this mutant kinase confirmed that the mutation affected AZD6244 binding. Overall, the *in vitro* study appears to predict possible mutations, but it is uncertain (due to low patient sample numbers) whether the incidence suggested by the *in vitro* study bears any correlation with the *in vivo* incidence.

A study of predicted EGFR mutations in response to treatment by erlotinib, lapatinib, and CI-1033 has been reported.[98] Mutations were generated using error-prone PCR and introduced into Ba–F3 cells transfected with EGFR or both EGFR and ErbB2, after which they were incubated in the presence of one of the three listed inhibitors. In all cases, the clinically observed T790M emerged as a candidate resistance mutation. However, the overall predictive power of this approach was somewhat limited. Of 19 known clinical mutations associated with erlotinib use, this method identified only four mutations, at the expense of predicting five additional mutations not observed clinically. The addition of ErbB2 to the cell line identified six of the clinical mutations, but also identified seven mutations not observed clinically. A similar trend in predictive power was observed with lapatinib, although here additional mutations in the selectivity pocket were predicted, which were observed clinically with erlotinib. Finally, treatment with CI-1033 identified the C797S mutation as abolishing binding. This is the site of covalent linkage to the kinase. Interestingly, two mutations at C796 (to R and C) were also predicted, which have been observed clinically with erlotinib treatment. Overall, this approach appears to be able to predict some, but not all clinical mutations. However, even these predictions are complicated by predicting resistance mutations not observed clinically.

Several mutagenesis studies were performed *in vitro* to predict important clinical resistance mutations to nilotinib and dasatinib.[99–101] As reviewed by O'Hare *et al.*,[31] there was limited overlap among the studies with nilotinib, with mutations of Y253, E 255, and T315 the most common across all three studies. Studies with dasatinib also identified mutations of T315, but now also included additional residues such as F317 and V299.[102] Of these mutations, only T315I showed unambiguous clinical relevance during clinical trials.

5.5 Resistance mechanisms not involving kinase domain mutations

Kinase domain mutations are not the only mechanism that may lead to drug resistance. As noted in Section 5.2, several additional mechanisms have been identified by which cells can escape or mitigate the effect of dosing with imatinib. In several other cells, resistance has been traced to one or several factors associated with changes in signaling or metabolic pathways. Several examples are highlighted in this section, in which resistance has been traced to one or several factors that do not involve kinase domain mutations.

Mutations of the BRAF gene were identified from a number of cancer cell lines and tumors.[103] B-RAF occurs in the signaling pathway downstream of several growth factor receptors, including EGFR (see Figure 5.5). Approximately 89% of the mutations identified in the kinase domain occurred in the activation loop, with V600E being the predominant activating mutation. The remaining activating mutations were associated with the glycine-rich loop. The mutations of the activation loop generally led to RAS-independent signaling through the normal endogenous MEK and MAPK pathways, whereas mutations in the glycine-rich loop generally remained RAS-dependent. Mutations of RAF appear to result from up-regulation of α-MSH dependent cAMP signaling which occurs as a consequence of UVB radiation.

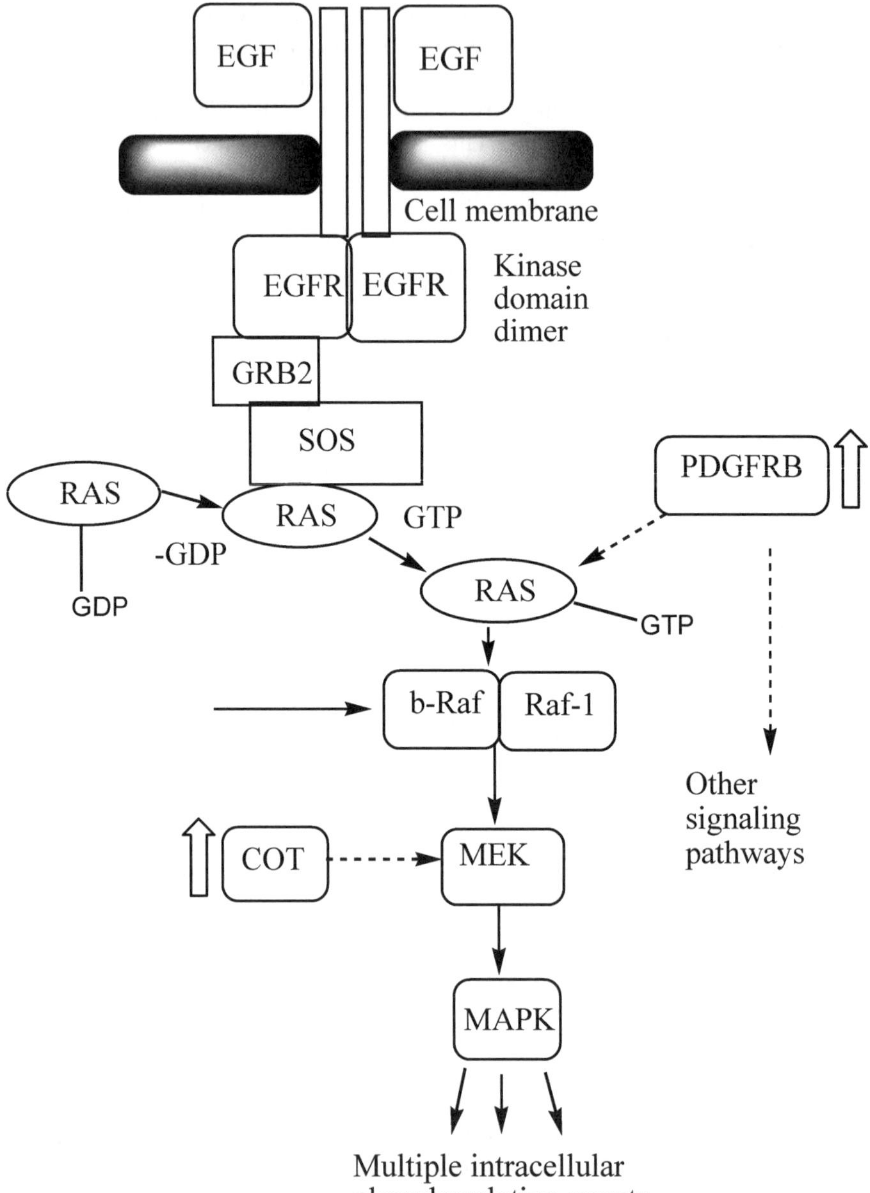

Figure 5.5 The RAS–RAF–MEK pathway, showing mechanisms of b-Raf resistance that have been proposed based on tumor sequencing studies. These include pathway reactivation due to over expression of Cot kinase, mutations that affect RAS (especially the N-RAS isoform) and cause reactivation, possibly through Raf-1 homodimers, and overexpression of PDGFRB that might affect MEK signaling or may proceed *via* alternative pathways.[123]

5.30 PLX-4032

Recently, clinical trials have been initiated to evaluate inhibitors designed specifically to block B-Raf, including the oncogenic mutant variants such as V600E. In a recent trial of PLX-4032 (**5.30**), initial response was followed by relapse in several patients.[104] The tumors were sequenced, and resistance was shown to be connected to reactivation of the signaling pathway, rather than by secondary mutations in the kinase domain itself.[105] The nature of the relapse patients was traced to re-activation of the MAPK pathway. Several potential mechanisms may be involved, including mutation-dependent N-RAS activation and up-regulation of PDGFRB signaling.[105] The MAPK reactivation was also traced to over-expression of COT (MAP3K8) kinase, which was shown to activate the pathway in a BRAF-independent manner. Indeed, *de novo* overexpression of COT in cancer cell lines conferred resistance to BRAF or BRAF(V600E) inhibitors.[106] This effect appeared to be MEK1 dependent, and did not require RAF signaling. Interestingly, a separate paper suggested that mutations in MEK, including the clinically observed P124L mutation, also conferred BRAF resistance.[97] Since MEK is downstream of BRAF, this would suggest that the mutant shifts signaling away from the RAS–RAF, MEK, and ERK pathway to another signaling cascade.

Resistance mechanisms for trastuzumab and for rapamycin have been shown to involve the PI3K signaling pathway. PI3K signaling is an important survival pathway for cells, including cancer cells.[107] The most important member of the PI3K family for proliferation is PI3Ka, which consists of a P110alpha catalytic subunit and a P85 regulatory subunit (see Figure 5.6). Growth factor binding to growth factor receptors can lead to the growth factor receptors phosphorylating the P85 regulatory subunit, which leads to activation of the P110alpha kinase. This converts phosphatidylinositol 2-phosphate to phosphatidylinositol-3 phosphate, an important intracellular messenger. Kinases that contain a pleckstrin homology domain are recruited to the cell surface. These include PDK1 and AKT. At the surface, PDK1 phosphorylates AKT, which can also be phosphorylated at a different serine by the mTOR TORC2 kinase. Activation of AKT leads to a signal cascade that ultimately results in cell survival, metabolism, protein translation, cell cycle progression, and RNA transcription. PTEN acts to dephosphorylate PIP3 to PIP2, and thereby acts to suppress the activation of AKT. PTEN can

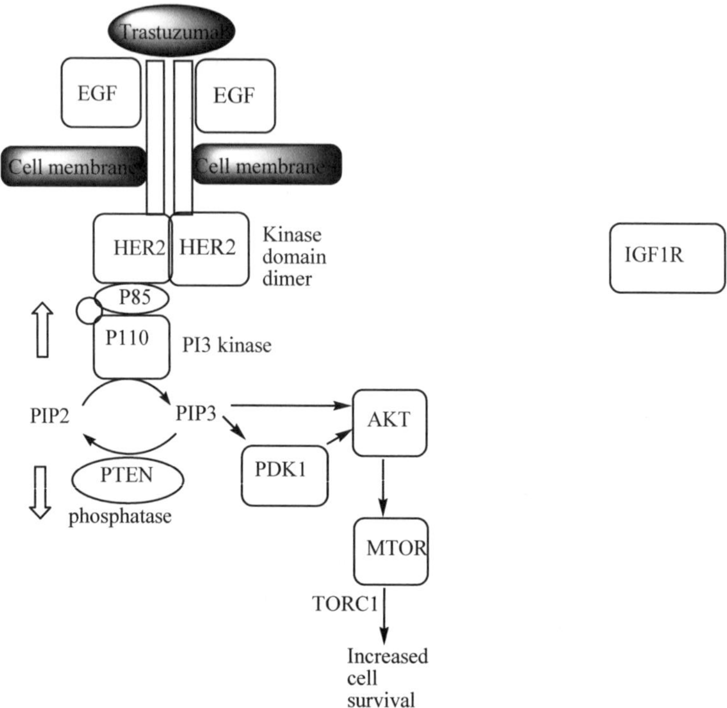

Figure 5.6 The Her2-dependent survival pathway is activated in many EGF-dependent breast cancers. Suppression of Her-2 by the binding of trastuzumab normally decreases PI3K kinase activation of AKT, and therefore reduces the downstream cell survival signal. Activating mutations of PI3K kinase (*e.g.* by reducing the binding of the P85 regulatory unit) or suppression of PTEN expression both increase AKT signaling and therefore cell survival. Both are implicated in trastuzumab resistance in the clinic.

therefore act as a tumor repressor; proliferating cancer cells often contain mutations of PTEN.

Several activating mutations have been identified in sequenced tumors. The PIK3CA gene encodes for the PI3Ka protein, which includes the P85 binding domain, RAS binding domain, a C2 domain, a helical segment, and the 110alpha kinase domain. The most frequent mutants include those of E542 and E545 in the helical domain, and H1047 in the kinase domain. The latter occurs on the P110–P85 interface; presumably, it disrupts the protein-protein association, thereby leading to constitutive activation of the kinase. Mutations of PIK3CA varied by cancer type, with 40% incidence in breast cancers, 18% in colorectal cancers, and 6% in primary epithelial ovarian cancers.[108] In the latter, PI3KCA cancers were only observed in cancers without gene amplification. Depending on cancer type, PIK3CA mutations and PTEN mutations can coexist (*e.g.* in endometrial carcinoma cancers,[109]) or appear mutually exclusive (*e.g.* in several types of brain cancers,[110] or breast cancers,[111]).

Trastuzumab resistance has been traced to changes in two proteins in the PI3K pathway (See Figure 5.6). Trastuzumab is an antibody used clinically in the treatment of Her2-dependent tumors. During recent clinical studies, upregulation of the AKT–PI3K cell survival pathway,[112] either by low expression of the PTEN phosphatase that acts as an endogenous inhibitor,[113,114] or by activating mutations of PI3K,[114] (especially between the P85 regulatory and P110 kinase subunits) has been associated with resistance to trastuzumab. Interestingly, the effect of the upregulation of the PI3K–AKT pathway predicted both for trastuzumab resistance and for lapatinib success.[114] This suggests that the two inhibitors have differing effects on pathways downstream of EGFR.

Another example of compensatory pathway upregulation is provided by a feedback mechanism identified in rapamycin and its clinical cousins temsirolimus and everolimus.[115] The only target of rapamycin is mTOR (TORC1), which is in the PI3K–AKT pathway downstream of AKT. This kinase phosphorylates two downstream kinases, S6K and 4E-BP1, leading to cell proliferation.[116] However, inhibition of mTOR TORC1 paradoxically *increases* AKT phosphorylation and activity, leading to rapamycin-resistant cell survival.[117] Two feedback mechanisms have emerged to explain this observation. S6K (see Figure 5.7) is implicated in a feedback loop that reduces IGF1R signaling through PI3K. Normally, insulin acts at IGFR1, which has two substrates, IRS-1 and IRS-2 that lead to PI3K activation and signaling through AKT. S6K in turn phosphorylates IRS-1 and IRS-2 leading to the destabilization of these substrates and decoupling of PI3K from insulin signaling. Paradoxically, inhibition of mTOR (TORC1) reduces S6K kinase activity, and increases the amount of viable IRS-1 and IRS-2 available, ultimately leading to a net increase in PI3K–AKT signaling. There is evidence that activation of PI3K–AKT can lead to activation of an alternative mTOR-independent pathway that leads to cell proliferation. A second isoform of mTOR, TORC2, is also activated by PI3K–AKT and is associated with downstream effects on cell adhesion and morphology. It is not inhibited by rapamycin and its analogues, but may exacerbate the resistance mechanisms because it phosphorylates AKT in a positive reinforcement loop.[115] A recent study that combined rapamycin with the AKT inhibitor perifosine has shown that the combination results in the induction of apoptosis and autophagy in multiple myeloma cells *in vitro*.[118]

5.6 Outlook

Kinase mutations have emerged as an important consideration in cancer drug discovery, both as a means of understanding potential targets, and as a challenge to clinical use as resistance emerges with continued use. The industry as a whole has embarked on a prolonged campaign to develop drugs to each of the key resistant strains, with the hope that eventually most resistant strains can also be inhibited.[65,119] What infuses this effort with some hope is that the

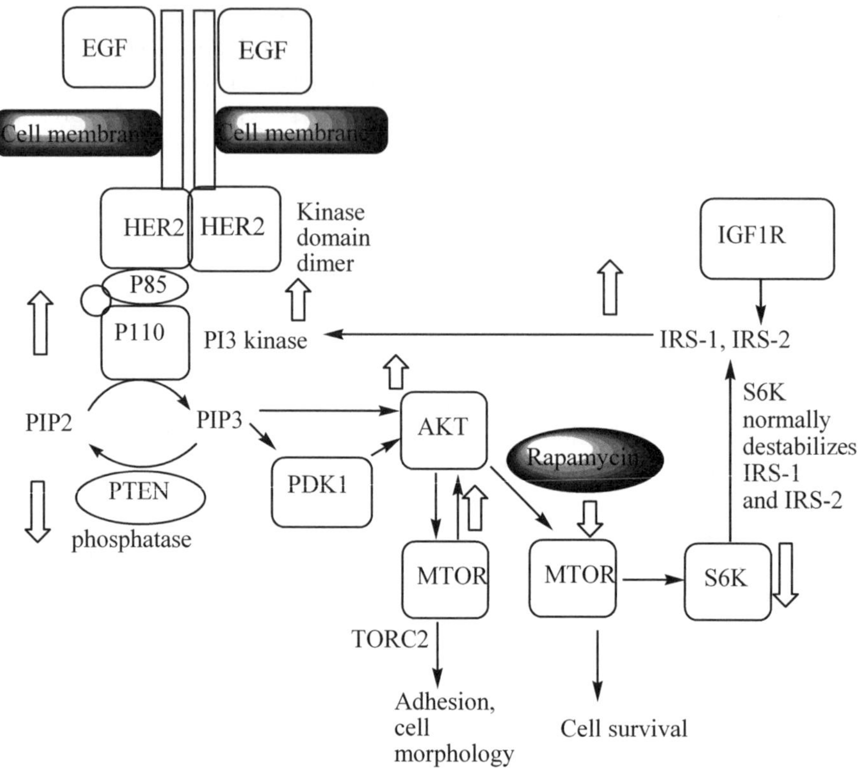

Figure 5.7 A more detailed view of the PI3K–AKT–mTOR pathway, showing the mechanism by which S6K modulates IGF1R activation of PI3K. By reducing TORC1 signaling, the downstream effect of S6K is also reduced, which leads to a net increase in IRS-mediated IGF1R signaling through PI3K. This has a net effect of increasing AKT activation, which can lead to cell proliferation by alternative pathways. In addition, TORC2, which is not affected by rapamycin, can directly phosphorylate AKT and also leads to increased AKT signaling.

number of emergent resistance mutations, while large, is hopefully somewhat finite, as the mutation needs to both reduce the binding of the synthetic inhibitor while continuing to support normal, ATP-driven kinase activity. With new inhibitors capable of overcoming families of resistant mutations, the challenge looks large but ultimately sustainable.

The development of drugs to specific mutant kinases will require methods to routinely genotype cancer tissues diagnostically. The kinase mutation literature has provided a considerable baseline of polymerase chain reaction (PCR) priming sequences and the DNA sequences of the mutant kinases.[15,40] Several detection methods are available, of which the most robust is direct sequencing.[14,15,28,35,78] However, if specific sequences are being probed, techniques such as allele-specific PCR,[120] pyrosequencing,[121] or denaturing HPLC,[122] may be more amenable to routine diagnosis. This will be a key complementary technology to the kinase inhibitors themselves.

Kinases are also under investigation for indications other than oncology. Will resistance emerge? Much of the evolutionary pressure to develop resistance arises from the inhibition of key, life-sustaining processes by kinases inhibited for oncology indications. With other uses, the kinase inhibitor will be used to attenuate an over-expressing or over-active kinase, while leaving normal kinase function uninhibited. This may reduce the evolutionary pressure to develop resistant strains substantially.

In summary, the key emergent resistance mutants have been reviewed for the clinically relevant inhibitors that are targeted at either the BCR-ABL, cKit, and PDGFR kinases or the EGFR-Her2 kinases. Where some structural rationale is available, this has been highlighted. A number of strategies to overcome resistant mutations have been developed, and these have been shown with examples. Finally, a brief overview of progress toward pre-clinically predicting clinically relevant mutations has been described.

References

1. J. Zhang, P. L. Yang and N. S. Gray, *Nat. Rev. Cancer*, 2009, **9**, 28–39.
2. P. W. Manley, *Eur. J. Cancer*, 2002, **38** S19–S27.
3. G. Manning, D. B. Whyte, R. Martinez, T. Hunter and S. Sudarsanam, *Science*, 2002, **298**, 1912–1934.
4. Y. Liu and N. S. Gray, *Nat. Chem. Biol.*, 2006, **2**, 358–364.
5. T. Schindler, W. G. Bornmann, P. Pellicena, W. T. Miller, B. Clarkson and J. Kuriyan, *Science*, 2002, **289**, 1938–1942.
6. E. Weisberg, P. W. Manley, W. Breitenstein, J. Brueggen, S. W. Cowan-Jacob, A. Ray, B. Huntly, D. Fabbro, G. Fendrich, E. Hall-Meyers, A. L. Kung, J. Mestan, G. Q. Daley, L. Callahan, L. Catley, C. Cavazza, M. Axam, D. Neuberg, R. D. Wright, D. G. Gilliland and J. D. Griffin, *Cancer Cell*, 2005, **7**, 129–141.
7. P. T. C. Wan, M. J. Garnett, S. M. Roe, S. Lee, D. Niculescu-Duvaz, V. M. Good, C. G. Project, C. M. Jones, C. J. Marshall, C. J. Springer, D. Barford and R. Marais, *Cell*, 2004, **116**, 855–867.
8. K. S. Gajiwala, J. C. Wu, J. Christensen, G. D. Deshmukh, W. Diehl, J. P. DiNitto, J. M. English, Y. A. He, S. L. Jacques, E. A. Lunney, M. McTigue, D. Molina, T. Quenzer, P. A. Wells, X. Yu, Y. Zhang, A. Zou, M. R. Emmett, A. G. Marshall, H. M. Zhang and G. D. Demetri, *Proc. Natl. Acad. Sci. U. S. A.*, 2009, **106**, 1542–1547.
9. X. Zhang, K. A. Pickin, R. Bose, N. Jura, P. A. Cole and J. Kuriyan, *Nature*, 2007, **450**, 741–745.
10. X. Zhang, J. Gereasko, K. Shen, P. A. Cole and J. Kuriyan, *Cell Cycle*, 2006, **125**, 1137–1149.
11. P. D. Jeffrey, A. A. Russo, K. Polyak, E. Gibbs, E. Hurwitz, J. J. Massague and N. P. Pavletich, *Nature*, 1995, **376**, 313–320.

12. X. Zhang, J. Gureasko, H. Shen, P. A. Cole and J. Kuriyan, *Cell*, 2006, **125**, 1137–1149.

13. D. H. Boschelli, *Top. Med. Chem.*, 2007, **1**, 407–444.

14. M. E. Gorre, M. Mohammed, K. Ellwood, N. Hsu, R. L. Paquette, P. N. Rao and C. L. Sawyers, *Science*, 2001, **293**, 876–880.

15. S. Branford, Z. Rudzki, S. Walsh, I. Parkinson, A. Grigg, J. Szer, K. Taylor, R. Herrmann, J. F. Seymour, C. Arthur, D. Joske, K. Lynch and T. Hughes, *Blood*, 2003, **102**, 276–283.

16. M. Miettinen, M. Sarlomo-Rikala and J. Lasota, *Hum. Pathol.*, 1999, **30**, 1213–1220.

17. S. Hirota, K. Isozaki, Y. Moriyama, K. Hashimoto, T. Nishida, S. Ishiguro, K. Kawano, M. Hanada, A. Kurata, T. Muhammad, Y. Matsuzawa, Y. Kanakura, Y. Shinomura and Y. Kitamura, *Science*, **279**, 577–580.

18. J. Verweij, P. G. Casali, J. Zalcberg, A. LeCesne, P. Reichardt, J. -Y. Blay, R. Issels, A. Van Oosterom, P. C. W. Hegendoorn, M. Van Glabbeke, R. Bertulli and I. Judson, *Lancet*, **364**, 1127–1134.

19. H. Joensuu, P. J. Roberts, M. Sarloma-Rikala, L. C. Andersson, P. Tervahartiala, D. Tuveson, S. L. Silberman, R. Capdeville, S. Dimitrijevic, B. Druker and G. D. Demetri, *N. Engl. J. Med.*, 2001, **344**, 1052–1056.

20. T. Sjolblom, A. Boureux, L. Ronnstarnd, C. H. Heldin, J. Ghysdael and A. Ostman, *Oncogene*, 1999, **18**, 7055–7062.

21. C. Jousset, C. B. Carron, A. Boureux, C. T. Quang, C. Oury, I. Dusanter-Fourt, M. Charon, J. Levin, O. Bernard and J. Ghysdael, *EMBO J.*, 1997, **16**, 69–82.

22. A. Shimizu, K. P. O'Brien, T. Sjolblom, K. Pietras, E. Buchdunger, V. P. Collins, C.-H. Heldin, J. P. Dumanski and A. Ostman, *Cancer Res.*, 1999, **59**, 3719–3723.

23. M. P. Simon, F. Pedeutour, N. Sirevent, J. Grosgeorge, F. Minoletti, J. M. Coindre, M. J. Terier-Lacombe, N. Mandahl, R. D. Craver, N. Clin, G. Sozzi, C. Turc-Carel, K. P. O'Brien, D. Kedra, I. Fransson, C. Guildbaud and J. P. Dumanski, *Nat. Genet.*, 1997, **15**, 95–98.

24. B. J. Druker, S. Tamaru, E. Buchdunger, S. Ohno, G. M. Segal, J. Zimmermann and N. B. Lydon, *Nat. Med.*, 1996, **2**, 561–566.

25. A. Quintas-Cardama, H. Kantarijian and J. Cortes, *Nat. Rev. Drug Discovery*, 2007, **6**, 834–848.

26. B. Druker, *N. Engl. J. Med.*, 2006, **355**, 2408–2417.

27. M. Azam, R. R. Latek and G. Q. Daley, *Cell*, 2003, **112**, 831–843.

28. A. Hochhaus, S. Kreil, A. S. Corbin, P. La Rosee, M. C. Muller, T. Lahaye, B. Hanfstein, C. Schoch, N. C. Cross, U. Berger, H. Gschaidmeier, B. J. Druker and R. Hehlmann, *Leukemia*, 2002, **16**, 2190–2196.

29. N. P. Shah, J. M. Nicoll, B. Nagar, M. E. Gorre, R. L. Paquette, J. Kuriyan and C. L. Sawyers, *Cancer Cell*, 2002, **2**, 117–125.

30. M. Azam, M. A. Seeliger, N. S. Gray, J. Kuryan and G. Q. Daley, *Nat. Struct. Mol. Biol.*, 2008, **15**, 1109–1118.
31. T. O'Hare, C. A. Eide and M. W. N. Deininger, *Blood*, 2007, **110**, 2242–2249.
32. S. Soverini, S. Colaraossi, A. Gnani, G. Rosti, F. Castagnetti, A. Poerio, I. Iacobucci, M. Amabile, E. Abruzzese, E. Orlandi, F. Radaelli, F. Ciccone, M. Tiribelli, R. di Lorenzo, C. Caracciolo, B. Izzo, F. Pane, G. Saglio, M. Maccarani and G. Martinelli, *Clin. Cancer Res.*, 2006, **12**, 7374–7379.
33. T. Skorski, *Oncogene*, 2002, **21**, 8591–8604.
34. M. Koptyra, R. Falinski, M. O. Nowicki, T. Stoklosa, I. Majsterek, M. Nieborowska-Skorska, J. Blasiak and T. Skorski, *Blood*, 2006, **108**, 319–317.
35. T. Hughes, M. W. N. Deininger, A. Hochhaus, S. Branford, J. Radich, J. Kaeda, M. Baccarani, J. Cortes, N. C. P. Cross, B. Druker, J. Gabert, D. Grimwade, R. Hehlmann, S. Kamel-Reid, J. H. Lipton, J. Longtine, G. Martinelli, G. Saglio, S. Soverini, W. Stock and J. M. Goldman, *Blood*, 2006, **108**, 28–37.
36. W. W. Zhang, J. E. Cortes, H. Yao, L. Zhang, N. G. Reddy, E. Jabbour, H. M. Kantarjian and D. Jones, *J. Clin. Oncol.*, 2009, **27**, 3642–3649.
37. D. W. Sherbenou, O. Hantschel, I. Kaupe, S. G. Willis, T. Bumm, L. P. Turaga, T. Lange, K.-H. Dao, R. D. Press, B. J. Druker, G. Spuperti-Furga and M. W. N. Deininger, *Blood*, 2010, **116**, 3278–3285.
38. B. Liegl, I. Kepten, C. Le, M. Zhu, G. D. Demetri, M. C. Heinrich, C. D. M. Fletcher, C. L. Corless and J. A. Fletcher, *J. Pathol.*, 2008, **216**, 64–74.
39. M. C. Heinrich, C. L. Corless, G. D. Demetri, C. D. Blanke, M. Von Mehren, H. Joensuu, L. S. McGreevey, C.-J. Chen, A. D. Van der Abbeele, B. Druker, B. Kiese, B. L. Eisenberg, P. J. Roberts, S. Singer, C. D. M. Fletcher, S. L. Silberman, S. Dimitrijevic and J. A. Fletcher, *J. Clin. Oncol.*, 2003, **21**, 4342–4349.
40. M. C. Heinrich, C. L. Corless, C. D. Blanke, G. D. Demetri, H. Joensuu, P. J. Roberts, B. L. Eisenberg, M. von Mehren, C. D. M. Fletcher, K. Sandau, K. McDougall, W.-B. Ou, C.-J. Chen and J. A. Fletcher, *J. Clin. Oncol.*, 2006, **24**, 4764–4774.
41. K. G. Roberts, A. F. Odell, E. M. Byrnes, R. M. Baleato, R. Griffith, A. B. Lyons and L. K. Ashman, *Mol. Cancer Ther.*, 2007, **6**, 1159–1166.
42. J. J.-L. Liao, *J. Med. Chem.*, 2007, **50**, 409–424.
43. J. J.-L. Liao and R. C. Andrews, *Curr. Top. Med. Chem.*, 2007, **7**, 1394–1407.
44. P. le Coutre, O. G. Ottmann, F. J. Giles, D.-W. Kim, J. Cortes, N. Gattermann, J. F. Apperley, R. A. Larson, E. Abruzzese, S. G. O'Brien, K. Kuliczkowski, A. Hochhaus, F.-X. Majon, G. Saglio, M. Gobbi, Y.-L. Kwong, M. Baccarani, T. Hughes, G. Martinelli, J. Radich, M. Zheng, Y. Shou and H. Kantarijian, *Blood*, 2008, **111**, 1834–1839.

45. T. O'Hare, D. K. Walters, E. P. Stofftregen, T. Jia, P. W. Manley, J. Mestan, S. W. Cowan-Jacob, F. Y. Lee, M. C. Heinrich, M. W. N. Deininger and B. J. Druker, *Cancer Res.*, 2005, **65**, 4500–4505.

46. H. G. Choi, P. Ren, F. Adrian, F. Sun, H. S. Lee, X. Wan, Q. Ding, G. Zhang, Y. Xie, J. Zhang, Y. Liu, T. Tuntland, M. Warmuth, P. W. Manley, J. Mestan, N. S. Gray and T. Sim, *J. Med. Chem.*, 2010, **53**, 5439–5448.

47. J. S. Tokarski, J. Newitt, C. Y. J. Chang, J. D. Cheng, M. Wittekind, S. E. Kiefer, K. Kish, F. Y. F. Lee, R. Borzilerri, L. J. Lombardo, D. Xie, Y. Zhang and H. E. Klei, *Cancer Res.*, 2006, **66**, 5790–5797.

48. A. Gontarewicz and T. H. Burmmendorf, *Recent Results Cancer Res.*, 2010, 199–214.

49. E. A. Harrington, D. Bebbington, J. Moore, R. K. Rasmussen, A. O. Ajose-Adeogun, T. Nakayama, J. A. Graham, C. Demur, T. Hercend, A. Diu-Hercend, M. Su, J. M. C. Golec and K. M. Miller, *Nat. Med.*, 2004, **10**, 262–267.

50. B. Nagar, W. G. Bornmann, P. Pellicena, T. Schindler, D. R. Veach, W. T. Miller, B. Clarkson and J. Kuriyan, *Cancer Res.*, 2002, **62**, 4236–4243.

51. N. P. Shah, B. J. Skaggs, S. Brankford, T. P. Hughes, J. M. Nioll, R. L. Paquette and C. L. Sawyers, *J. Clin. Invest.*, 2007, **117**, 2562–2569.

52. T. Hampton, *JAMA*, 2007, **297**, 457–458.

53. A. Gontarewicz and T. H. Brummendorf, *Recent Results Cancer Res.*, 2010, **184**, 199–214.

54. M. Azam, V. Nardi, W. C. Shakespeare, C. A. I. Metcalf, R. S. Bohacek, Y. Wang, R. Sundaramoorthi, P. Sliz, D. R. Veach, W. G. Bornmann, B. Clarkson, D. C. Dalfarno, T. K. Sawyer and G. Q. Daley, *Proc. Natl. Acad. Sci., U. S. A.*, 2006, **103**, 9244–9249.

55. T. O'Hare, C. A. Eide, J. W. Tyner, A. S. Corbin, M. J. Wong, S. Buchanan, K. Holme, K. A. Jessen, C. Tang, H. A. Lewis, R. D. Romero, S. K. Burley and M. W. N. Deininger, *Proc. Natl. Acad. Sci. U. S. A.*, 2008, **105**, 5507–5512.

56. T. Zhou, L. Parillon, F. Li, Y. Wang, J. Keats, S. Lamore, Q. Xu, W. C. Shakespeare, D. Dalgarno and X. Zhu, *Chem. Biol. Drug Des.*, 2007, **70**, 171–181.

57. K. Gumireddy, S. J. Baker, S. C. Cosenza, P. John, A. D. Kang, K. A. Robell, M. V. R. Reddy and E. P. Reddy, *Proc. Natl. Acad. Sci. U. S. A.*, 2005, **102**, 1992–1997.

58. N. E. Hynes and H. A. Lane, *Nat. Rev. Cancer*, 2005, **5**, 341–354.

59. A. F. Gazdar, H. Shigematsu, J. Herz and J. D. Minna, *Trends Mol. Med.*, 2004, **10**, 481–486.

60. A. F. Gazdar, *Oncogene*, 2009, **28**, S24–S31.

61. T. J. Lynch, D. W. Bell, R. Sordella, S. Gurubhagavatual, R. A. Okimoto, B. W. Brannigan, P. L. Harris, S. M. Haserlat, J. G. Supko,

F. G. Haluska, D. N. Louis, D. C. Christiani, J. Settleman and D. A. Haber, *N. Engl. J. Med.*, 2004, **350**, 2129–2139.

62. J. G. Paez, P. A. Janne, J. C. Lee, S. Tracy, H. Greulich, S. Gabriel, P. Herman, F. J. Kaye, N. Lindeman, T. J. Boggon, K. Naoki, H. Sasaki, Y. Fujii, M. J. Eck, W. R. Sellers, B. E. Johnson and M. Meyerson, *Science*, 2004, **304**, 1497–1500.

63. W. Pao, V. A. Miller, K. A. Politi, G. I. Reily, R. Somwar and M. F. Zakowski, *Proc. Natl. Acad. Sci. U. S. A.*, 2004, **101**, 13306–13311.

64. H. Ji, X. Zhao, Y. Yuza, T. Shimamura, D. Li, A. Protopopov, B. L. Jung, K. McNamara, H. Xia, K. A. Glatt, R. K. Thomas, H. Sasaki, J. W. Horner, M. J. Eck, C. M. Discafani, E. Maher, G. I. Shapiro, M. Meyerson and K.-K. Wong, *Proc. Natl. Acad. Sci. U. S. A.*, 2006, **103**, 7817–7822.

65. J. E. Dowell and J. D. Minna, *N. Engl. J. Med.*, 2005, **352**, 830–832.

66. R. Rosell, T. Moran, E. Carcereny, V. Quiroga, M. A. Molina, C. Costa, S. Benlloch and M. Taron, *Clin. Transl. Oncol.*, 2010, **12**, 75–80.

67. D. M. Jackman, V. A. Miller, L. A. Cioffredi, B. Y. Yeap, P. A. Jaenne, G. J. Riely, M. G. Ruiz, G. Giaccone, L. V. Sequist and B. E. Johnson, *Clin. Cancer Res.*, 2009, **15**, 5267–5273.

68. C.-H. Yun, T. J. Boggon, Y. Li, M. S. Woo, H. Greulich, M. Meyerson and M. J. Eck, *Cancer Cell*, 2007, **11**, 217–227.

69. J. Stamos, M. X. Sliwkowski and C. Eigenbrot, *J. Biol. Chem.* , 2002, **277**, 46265–46272.

70. E. R. Wood, A. T. Tuesdale, O. B. McDonald, D. Yan, A. Hassell, S. H. Dickerson, B. Ellis, C. Pennisi, E. Horne, K. Lakcey, K. J. Alligood, D. W. Rusnak, T. M. Gilbmer and L. Shewchuck, *Cancer Res.*, 2004, **64**, 66452–66659.

71. A. Kumar, E. T. Petri, B. Halmos and T. J. Boggon, *J. Clin. Oncol.*, 2008, **26**, 1742–1751.

72. S. Kobayashi, T. J. Boggon, T. Dayaram, P. Janne, O. Kocher, M. Meyerson, B. E. Johnson, M. J. Eck, D. G. Tenen and B. Halmos, *N. Engl. J. Med.*, 2005, **352**, 786–792.

73. E. L. Kwak, R. Sordella, D. W. Bell, N. Godin-Heymann, R. A. Okimoto, B. W. Brannigan, P. L. Harris, D. R. Driscoll, P. Fidias, T. J. Lynch, S. K. Rabindran, J. P. McGinnis, A. Wissner, S. V. Sharma, K. J. Isselbacher, J. Settleman and D. A. Haber, *Proc. Natl. Acad. Sci. U. S. A.*, 2005, **102**, 7665–7670.

74. C.-H. Yun, K. E. Mengwasser, A. V. Toms, M. S. Woo, H. Greulich, K.-K. Wong, M. Meyerson and M. J. Eck, *Proc. Natl. Acad. Sci. U. S. A.*, 2008, **105**, 2070–2075.

75. K. S. Nguyen, S. Kobayashi and D. B. Costa, *Clin. Lung Cancer*, 2009, **10**, 281–289.

76. W. Zhou, D. Ercan, L. Chen, C.-H. Yun, D. Li, M. Capelletti, A. B. Cortot, L. Chirieeac, R. E. Iacob, R. Padera, J. R. Engen, K.-K. Wong, M. J. Eck, N. S. Gray and P. A. Janne, *Nature*, 2009, **462**, 1070–1074.

77. S. Maheswaran, L. V. Sequist, S. Nagrath, L. Ulkus, B. Brannigan, C. V. Collura, E. Inserra, S. Diederichs, A. J. Iafrete and D. W. Belle, *N. Engl. J. Med.*, 2008, **359**, 366–377.

78. M. Inukai, S. Toyooka, S. Ito, H. Asano, S. Ichihara, J. Soh, H. Suihisa, M. Ouchida, K. Aoe, M. Aoe, K. Kiura, N. Shimizu and H. Date, *Cancer Res.*, 2006, **66**, 7854–7858.

79. A. Wissner and T. S. Mansour, *Arch. Pharm. Chem. Life Sci.*, 2008, **341**, 465–477.

80. J. Singh, R. C. Petter and A. F. Kluge, *Curr. Opin. Chem. Biol.*, 2010, **14**, 475–480.

81. M. H. Potashman and M. E. Duggan, *J. Med. Chem.*, 2009, **52**, 1231–1236.

82. W. A. Denny, *Pharmacol. Ther.*, 2002, **93**, 253–261.

83. A. Michalczyk, S. Kluter, H. B. Rode, J. R. Simard, C. Grutter, M. Baviller and D. Rauh, *Bioorg. Med. Chem.*, 2008, **16**, 3482–3488.

84. A. J. Bridges, *Curr. Top. Med. Chem.*, 1999, **6**, 825–843.

85. P. A. Janne, J. von Pawel, R. B. Cohen, L. Crino, C. A. Butts, S. S. Olson, I. A. Eiseman, A. A. Chiappori, B. Y. Yeap, P. F. Lenehan, K. Dasse, M. Sheeran and P. D. Bonomi, *J. Clin. Oncol.*, 2007, **25**, 3936–3944.

86. K.-K. Wong, *Clin. Cancer Res.*, 2007, **13**, 4593–4596.

87. N. Minkovsky and A. Berezov, *Curr. Opin. Invest. Drugs*, 2008, **9**, 1336–1346.

88. A. J. Gonzales, K. E. Hook, I. W. Althaus, P. A. Ellis, E. Trachet, A. M. Delaney, P. J. Harvey, T. A. Ellis, D. M. Amato, J. M. Nelson, D. W. Fry, T. Zhu, C.-M. Loi, S. A. Fakhoury, K. M. Schlosser, K. E. Sexton, R. T. Winters, J. E. Reed, A. J. Bridges, D. J. Lettiere, D. A. Baker, J. Yang, H. T. Lee, H. Tecle and P. W. Vincent, *Mol. Cancer Ther.*, 2008, **7**, 1880–1889.

89. S. K. Rabindran, *Cancer Res.*, 2004, **64**, 3958–3965.

90. H. Ji, D. Li, L. Chen, T. Shimamura, S. Kobayahshi, K. McNamara, U. Mahmood, A. Mitchell, Y. Sun, R. Al-Hashem, L. R. Chirieac, R. Padera, R. T. Bronson, W. Kim, P. Janne, G. I. Shapiro, D. G. Tenen, B. E. Johnson, R. Weissleder, N. E. Sharpless and K.-K. Wong, *Cancer Cell*, 2006, **9**, 485–495.

91. M. L. Sos, H. B. Rode, S. Heynick, M. Peifer, F. Fischer, S. Kluter, V. G. Pawar, C. Reuter, J. M. Heuckmann, J. Weiss, L. Ruddigkeit, M. Rabiller, M. Koker, J. R. Simard, M. Getlik, Y. Yuza, T.-H. Chen, H. Greulich, R. K. Thomas and D. Rauh, *Cancer Res.*, 2010, **70**, 868–874.

92. W. Zhou, D. Ercan, P. A. Janne and N. S. Gray, *Bioorg. Med. Chem. Lett.*, 2011, **21**, 638–643.

93. N. von Bubnoff, D. R. Veach, H. van der Kuip, W. E. Aulitzky, J. Sander, P. Seipel, W. G. Bornmann, C. Peschel, B. Clarkson and J. Duyster, *Blood*, 2005, **105**, 1652–1659.

94. E. R. Zunder, Z. A. Knight, B. T. Houseman, B. Apsel and K. M. Shokat, *Cancer Cell*, 2008, **14**, 180–192.

95. F. Girdler, F. Sessa, S. Patercoli, F. Villa, A. Musacchoi and S. Taylor, *Chem. Biol.*, 2008, **15**, 552–562.

96. P. J. Scutt, M. L. Chu, D. A. SLoane, M. Cherry, C. R. Bignell, D. H. Williams and P. A. Eyers, *J. Biol. Chem.*, 2009, **284**, 15880–15893.

97. C. M. Emery, K. G. Vijayendran, M. C. Zipser, A. M. Sawyer, L. Niu, J.vJ. Kim, C. Hatton, R. Chopra, P. A. Overholzer, M. B. Karpova, L. E. MacConaill, J. Zhang, N. S. Gray, W. R. Sellers, R. Dummer and L. A. Garraway, *Proc. Natl. Acad. Sci. U. S. A.*, 2009, **106**, 20411–20416.

98. E. Avizienyte, R. A. Ward and A. P. Garner, *Biochem. J.*, 2008, **415**, 197–206.

99. H. A. Bradeen, C. A. Eide, T. O'Hare, K. J. Johnson, S. G. Willis, F. Y. Lee, B. Druker and M. W. N. Deininger, *Blood*, 2006, **108**, 2332–2338.

100. A. Ray, S. W. Cowan-Jacob, P. W. Manley, J. Mestan and J. D. Griffin, *Blood*, 2007, **109**, 5011–5015.

101. N. von Bubnoff, P. W. Manley, J. Mestan, J. Sanger, C. Peschel and J. Duyster, *Blood*, 2006, **108**, 1328–1333.

102. M. R. Burgess, B. J. Skaggs, N. P. Shah, F. Y. Lee and C. L. Sawyers, *Proc. Natl. Acad. Sci. U. S. A.*, 2005, **102**, 3395–3400.

103. H. Davies, G. R. Bignell, C. Cox, P. Stephens, S. Edkins, S. Clegg, J. Teague, H. Woffendin, M. J. Garnett, W. Bottomley, N. Davis, E. Dicks, R. Ewing, Y. Floyd, K. Gray, S. Hall, R. Hawes, J. Hughes, V. Kosmidou, A. Menzies, C. Mould, A. Parker, C. Stevens, S. Watt, S. Hooper, R. Wilson, H. Jayatilake, B. A. Gusterson, C. Cooper, J. Shipley, D. Hargrave, K. Pritchard-Jones, N. Mailtland, G. Chenevix-Trench, G. J. Riggins, D. D. Bigner, G. Palmieri, A. Cossu, A. Flanagan, A. Nicholson, J. W. C. Ho, S. Y. Leung, S. T. Yuen, B. L. Weber, H. F. Seigler, T. L. Darrow, H. Paterson, R. Marais, C. J. Marshall, R. Wooster, M. R. Stratton and P. A. Futreal, *Nature*, 2002, **417**, 949–954.

104. K. T. Flaherty, I. Puzanov, M. B. Kim, A. Ribas, G. A. McArthurs, J. A. Sosman, P. J. O'Dwyer, R. J. Lee, J. F. Grippo, K. Nolop and P. B. Chapman, *N. Engl. J. Med.*, 2010, **363**, 809–819.

105. R. Nazarian, H. Shi, Q. Wang, X. Kong, R. C. Koya, H. Lee, Z. Chen, M.-K. Lee, N. Attar, H. Sazegar, T. Chodon, S. F. Nelson, G. A. McArthur, J. A. Sosman, A. Ribas and R. S. Lo, *Nature*, 2010, **468**, 973–977.

106. C. M. Johannessen, J. S. Boehm, S. Y. Kim, S. R. Thomas, L. Wardwell, L. A. Johnson, C. M. Emery, N. Stransky, A. P. Cogdill, J. Barretina, G. Caponigro, H. Hieronymus, R. R. Murray, K. Salehi-Ashtiani, D. E. Hill, M. Vidal, J. J. Zhao, X. Yang, O. Alkan, S. Kim, J. L. Harris, C. J. Wilson, V. E. Myer, P. M. Finan, D. E. Root, T. M. Roberts, T. Golub, K. T. Flaherty, R. Dummer, B. L. Weber, W. R. Sellers, R. Schlegel, J. A. Wargo, W. C. Hahn and L. A. Garraway, *Nature*, 2010, **468**, 968–972.

107. G. Ligresti, L. Militello, L. S. Steelman, A. Cavallaro, F. Basile, F. Nicoletti, F. Stivala, J. A. McCubrey and M. Libra, *Cell Cycle*, 2009, **8**, 1352–1358.
108. I. G. Campbell, S. E. Russell, D. Y. Choong, K. G. Montgomery, M. L. Ciavarella, C. S. Hooi, B. E. Cristiano, R. B. Pearson and W. A. Phillips, *Cancer Res.*, 2004, **64**, 7678–7681.
109. K. Oda, D. Stokoe, Y. Taketani and F. McCormick, *Cancer Res.*, 2005, **65**, 10669–10673.
110. D. K. Broderick, C. Di, T. J. Parrett, Y. R. Samuels, J. M. Cummins and R. E. CMcLendon, *Cancer Res.*, 2004, **64**, 5048–5050.
111. L. H. Saal, K. Holm, M. Maurer, L. Memeo, T. Su and X. Wang, *Cancer Res.*, 2005, **65**, 2554–2559.
112. K. Berns, H. M. Horlings and B. T. Hennessy, *Cancer Cell*, 2007, **12**, 395–402.
113. Y. Nagata, K. H. Lan and X. Zhou, *Cancer Cell*, 2004, **6**, 117–127.
114. B. Dave, I. Migliaccio, M. C. Gutierrez, M.-F. Wu, G. C. Chamnes, H. Wong, A. Narasanna, A. Chakrabarty, S. G. Hilsenbeck, J. Huang, M. Rimawi, R. Schiff, C. Arteaga, C. K. Osborne and J. C. Chang, *J. Clin. Oncol.*, 2011, **29**, 166–173.
115. G. R. Hudes, *Cancer Cell*, 2009, **May 15**, 2313–2320.
116. K. Inoki and K.-L. Guan, *Trends Cell Biol.*, 2006, **16**, 206–212.
117. S.-Y. Sun, L. M. Rosenberg, X. Wang, Z. Zhou, P. Yue, H. Fu and F. R. Khuri, *Cancer Res.*, 2005, **65**, 7052–7058.
118. D. Cirstea, T. Hideshima, S. Rodig, L. Santo, S. Pioszzi, S. Vallet, H. Ikeda, G. Perrone, G. Gorgun, K. Patel, N. Desai, P. Sportelli, S. Kapoor, S. Vali, S. Mukherjee, N. C. Munshi, K. C. Anderson and N. Raje, *Mol. Cancer Ther.*, 2010, **9**, 963–975.
119. A. Quintas-Cardama and J. Cortes, *Clin. Cancer Res.*, 2008, **14**, 4392–4399.
120. S. G. Willis, T. Lange, S. Demehri, S. Otto, S. Crossman, D. Niederwieser, E. P. Stoffregen, S. McWeeney, I. Kovacs, B. Park, B. J. Druker and M. W. N. Deininger, *Blood*, 2005, **106**, 2128–2137.
121. C. Roche-Lestienne, J. L. Lai, S. Darre, T. Facon and C. Preudhomme, *N. Engl. J. Med.*, 2003, **348**, 2265–2266.
122. M. W. N. Deininger, L. S. McGreevey, S. G. Willis, T. M. Bainbridge, B. J. Druker and M. C. Heinrich, *Leukemia*, 2004, **18**, 864–871.
123. D. Solit and C. L. Sawyers, *Nature*, 2010, **468**, 902–903.

Non-Protein Kinases as Therapeutic Targets

JEROEN C. VERHEIJEN*[a], DAVID J. RICHARD[b] AND ARIE ZASK[c]

[a] AstraZeneca Pharmaceuticals LP, 35 Gatehouse Drive, Waltham, MA 02451, USA; [b] Pfizer Global Research, 200 Cambridgepark Drive, Cambridge, MA 02140, USA; [c] Columbia University, Department of Chemistry, 3000 Broadway, New York, NY 10027, USA

6.1 Introduction

As illustrated in the preceding chapters, the last decade has seen substantial progress in the development of protein kinase inhibitors as drugs. Significantly fewer drug development projects aimed at inhibiting non-protein kinases have been described, despite the essential role these kinases play in modulating cellular pathways by phosphorylation of sugars and lipids as well as nucleosides and nucleotides. In this chapter, we provide an overview of some of the efforts aimed at the inhibition of non-protein kinases for drug development.

We will discuss small molecule inhibitors of three main classes of nonprotein kinases: sugar kinases, nucleoside kinases and lipid kinases. This chapter will be limited to discussion of the inhibition of human kinases. Several non-human non-protein kinases have been the focus of drug development efforts as well, including choline kinase for inhibition of *Plasmodium*, uridinecytidine kinase as an anti-parasitic target and thymidine kinase inhibitors as an anti-mycobacterial treatment or for treatment of Herpes simplex viral infections. These applications will not be discussed in this chapter. Peptide

RSC Drug Discovery Series No. 19
Kinase Drug Discovery
Edited by Richard A. Ward and Frederick Goldberg
© Royal Society of Chemistry 2012
Published by the Royal Society of Chemistry, www.rsc.org

inhibitors and other large molecule inhibitors of non-protein kinases also fall outside the scope of this chapter.

Due to the large number of non-protein kinases as potential drug development targets, this chapter does not provide an exhaustive review of all efforts in this field, but rather provides the reader with a sampling of representative examples and leading references, with a focus on some of the more recent drug development efforts.

6.2 Sugar Kinases

According to the Warburg hypothesis,[1] tumor cell survival and proliferation critically depend on aerobic glycolysis rather than mitochondrial glucose oxidative metabolism as an energy source. This makes the inhibition of glycolytic enzymes, such as the sugar kinases, attractive for the development of anticancer drugs.

6.2.1 Hexokinases (HK)

There are four human hexokinases (hexokinases I–III and hexokinase IV, which is also known as glucokinase). Hexokinases phosphorylate glucose, mannose and fructose to the corresponding hexose-6-phosphates, a starting point for their breakdown into pyruvate by glycolysis. Research aimed at the identification of hexokinase inhibitors dates back several decades.[2–4] These early efforts involved mainly sugar derivatives with poor drug-like properties, although 2-deoxyglucose and 2-fluorodeoxy-glucose have shown some promise as therapeutic agents.[5] In addition to the studies mentioned above, Coats *et al.* used glucosamine derived inhibitors to explore the hexokinase glucose binding site using molecular modeling.[6]

In addition to sugar analogs, several other compounds have been identified as hexokinase inhibitors. Lonidamine ((**1**), Figure 6.1), a derivative of indazole-3-carboxylic acid that was launched approximately 30 years ago as an anticancer drug, also functions by inhibiting hexokinase. More recently, this drug was investigated for the treatment of benign prostatic hyperplasia by Threshold Pharmaceuticals.[7] Prodrugs of lonidamine have been explored by Giorgioni *et al.*,[8] and Threshold Pharmaceuticals.[9] A patent application by Mallinckrodt describes another class of hexokinase inhibitors, phloretin analogs, as diagnostics and therapeutics.[10]

Although interest in inhibition of glycolytic enzymes as anticancer treatment had waned somewhat in the late 1990s, the description of the promising *in vivo* activity of 3-bromopyruvate (**2**), an inhibitor of hexokinase II, has prompted a renewed interest in this field. Efficacy of 3-bromopyruvate was reported against various cancer cell lines in animal studies at Johns Hopkins.[11] Further investigation has revealed that direct intra-arterial administration to the site of the primary tumor may be especially effective.[12,13] Despite the promising pre-clinical activity of 3-bromopyruvate, no human clinical trials have been reported.

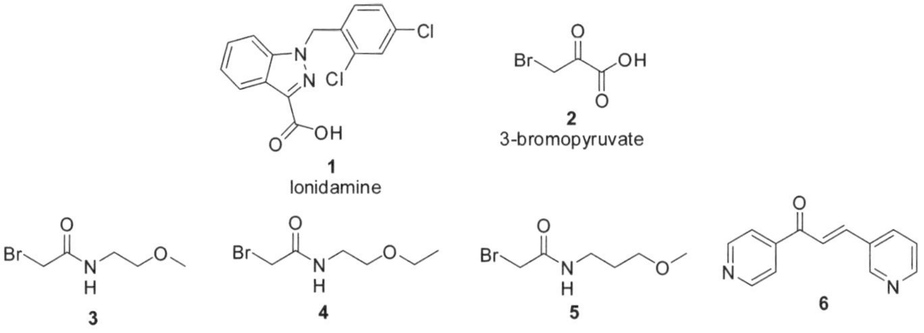

Figure 6.1 Sugar kinase inhibitors.

To date, inhibitors stemming from modern medicinal chemistry approaches combining high throughput screening with rational design and utilizing computational or structural support have not been reported. This may change now that this field has received renewed interest due to the activity of 3-bromopyruvate. In this respect, a recent publication describing a new high throughput screening method for the identification of such compounds is noteworthy.[14]

6.2.2 Ketohexokinase (KHK, fructokinase)

Researchers at the University of Florida have proposed targeting ketohexokinase (fructokinase), one of the main enzymes responsible for fructose metabolism, as a therapy for cardiovascular disease, metabolic syndrome and renal diseases.[15] They propose that the rise in serum uric acid levels following fructose intake induces metabolic syndrome by inhibiting endothelial NO. The authors suggest that metabolic syndrome and cardiovascular disease can be controlled by inhibiting the enzymes responsible for fructose metabolism. No further drug discovery efforts have been described.

6.2.3 6-Phosphofructo-2-kinase–Fructose-2,6-bisphosphatase (PFK2–FBPase2)

6-Phosphofructo-2-kinase (Phosphofructokinase 2, PFK2), also known as fructose biphosphatase 2 (FBPase2), is a bifunctional enzyme. It is a homo-dimer, with each polypeptide chain containing independent kinase and phosphatase domains. The ratio between kinase and phosphatase activity is controlled by phosphorylation, with phosphorylation of Ser-32 leading to a conformational shift favoring FBPase2 activity; otherwise, PFK2 activity is favored. Four isozymes have been identified to date. The PFKFB3 isozyme (previously known as placenta type PFK2) has a very high kinase : phosphatase ratio and is overexpressed in leukemias and various solid tumors. Hence, this isozyme has been the primary focus of efforts aimed at the development of antineoplastic agents that target glycolysis.

Starting from a known irreversible inhibitor, *N*-bromoacetylethanolamine phosphate, novel analogs ((**3–5**), Figure 6.1) were developed that inhibited PFKFB3 isozyme kinase activity *in vitro* and in cells.[16] Modest *in vivo* effects were also observed. Clem *et al.* identified a micromolar inhibitor (**6**) of the PFKFB3 isozyme using virtual screening in a PFKFB3 homology model derived from the X-ray structure of rat testes PFKFB4.[17] Administration of compound (**6**) to mice was shown to reduce glycolysis and tumor growth rate *in vivo*. Interestingly, an opposite approach was advocated by Shaikh *et al.*[18] They proposed the development of inhibitors of the phosphatase domain rather than the kinase domain of this bifunctional enzyme to overcome insulin resistance as a potential treatment for diabetes.

6.2.4 Galactokinase

Although no drug discovery efforts have been reported to date, Wierenga *et al.* proposed inhibition of galactokinase for treatment of classic galactosemia and described an assay to screen potential inhibitors.[19,20]

6.2.5 Conclusion

After the initial interest in sugar kinase inhibitors had worn off in the late 1990s, over the course of the last decade a renewed interest in inhibitors of these kinases has developed. Interest in the development of hexokinase inhibitors as new anticancer agents with potentially reduced side effects has increased in recent years. In addition, several other sugar kinases have been identified as potential therapeutic targets. The majority of inhibitors reported to date have been substrate analogs or serendipitously discovered small molecules. Given their pre-eminent role in sugar metabolism, it is conceivable that future drug discovery efforts incorporating the power of modern medicinal chemistry drug design will yield novel sugar kinase inhibitors for treatment of cancer, metabolic disorders–diabetes and cardiovascular disease.

6.3 Nucleoside Kinases

Nucleoside kinases phosphorylate the 5'-hydroxy group on (desoxy)ribonucleosides and play an essential role in the salvage pathway. The importance of the nucleoside salvage pathway for cancer and inflammation has long been recognized.[21,22]

6.3.1 Uridine-cytidine Kinase

Uridine-cytidine kinase is an important enzyme in pyrimidine nucleotide biosynthesis through the salvage pathway. In addition, this enzyme is also responsible for the 5'-phosphorylation of pyrimidine nucleoside analogs used in chemotherapy, an essential step to convert these chemotherapeutics into the biologically active triphosphates.

Inhibitors of uridine-cytidine kinase were explored in the 1980s, but there has been little or no recent activity in this field.

Cytidine analogs (**7**) were evaluated as inhibitors of uridine-cytidine kinase in leukemia cell extracts, with the best inhibitors showing ~80% inhibition of enzyme activity at 1.5–1.75 mM.[23] More potent inhibitors were obtained when cyclopentenyl pyrimidine nucleosides, analogs of antitumor antibiotic nucleoside neplanocin A ((**8**), Figure 6.2), were prepared.[24] Cytidine analog (**9**) inhibited enzyme activity by 57% at 100 μM. More importantly, cell growth was inhibited 99% at that concentration, whereas 58% inhibition of cell growth was observed when inhibitor (**9**) was present at 30 nM. It is not clear whether uridine-cytidine kinase is still inhibited at 30 nM or whether the observed cytotoxicity is caused by off-target effects.

Although no recent activity has been reported in this field, it remains possible that more potent inhibitors of this enzyme have potential applications as anticancer therapeutics. It is also possible that the combination of salvage uridine-cytidine monophosphate synthesis inhibitors with inhibitors of *de novo* uridine-cytidine monophosphate synthesis (such as PALA or pyrazofurin) may result in more efficacious cancer therapeutics.

6.3.2 Thymidine Kinase

Thymidine kinase (TK) is an essential enzyme for phosphorylation of deoxy-thymidine and DNA synthesis. In addition, it is the enzyme responsible for activation and–or metabolism of a variety of nucleoside analog drugs (for example DNA chain terminators such as AZT). This enzyme has received significant interest for treatment of herpes simplex virus infections (*cf.* acyclovir) or mycobacterium infections. In both cases, therapies have been developed that seek to exploit the differences between viral or bacterial TK and human TK. Two forms of thymidine kinase are found in human cells: TK1 (cytosolic, active in replicating cells) and TK2 (mitochondrial, constitutively active). Some antiviral–anticancer drugs are metabolized by either one of the isozymes. In some cases, this may be associated with either lack of efficacy or

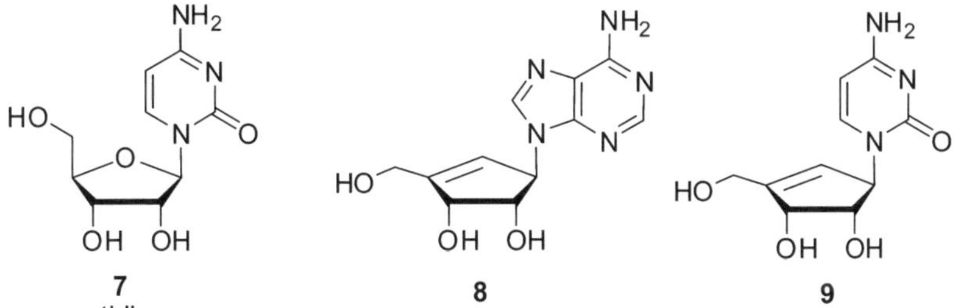

Figure 6.2 Uridine-cytidine kinase substrate and inhibitors.

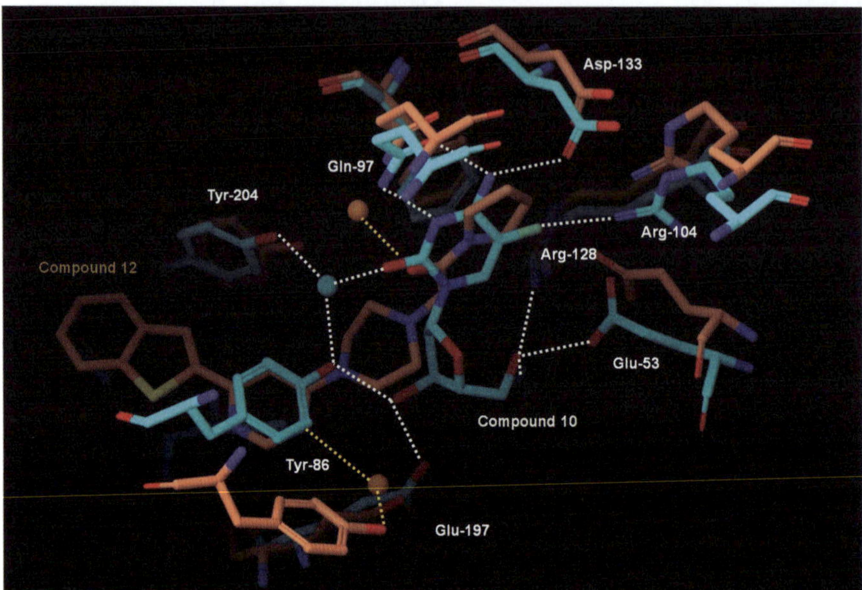

Figure 6.3   Overlay of the co-crystal structures of **10** (cyan) and **12** (brown) with deoxycytidine kinase.[35] The following interactions are indicated. Compound **10**: H-bonds from the 5'-OH to Glu-53, Arg-128 and water; from the 4'-OH to Glu-197 and Tyr-86; from the carbonyl to water to Tyr-204 and Tyr-86; from the pyrimidine N to the Gln-97 backbone carbonyl; from the exocyclic NH_2 to the Gln-97 backbone NH and the Asp-133 side chain; and from F to Arg-104. The binding mode of compound **12** is characterized by a H-bond between the amide carbonyl and a conserved water molecule; and by a water-mediated H-bond between the pyrimidine N and Tyr-86. The overlay of the two structures also clearly shows how a new pocket is created to accommodate the biaryl portion of compound **12** (*cf.* movement of Tyr-86).

toxicity. For example, TK2-mediated phosphorylation of AZT has been suggested to lead to AZT triphosphate accumulation in mitochondria and result in mitochondrial damage. Sub-type selective inhibitors have been designed to study the role of TK1,[25] and TK2,[26–29] in different physiological processes. Further clarification of these roles should provide a clear picture of the potential of human TK as a drug target.

6.3.3   Deoxycytidine Kinase

In addition to possible applications for the development of cancer drugs, deoxycytidine kinase (dCK) inhibitors also hold promise for the treatment of immune disorders. Drug discovery efforts aimed at inhibiting deoxycytidine kinase have been reported by Lexicon Pharmaceuticals. They disclosed an HTS assay for the discovery of dCK inhibitors.[30]

Based on a co-crystal structure of (**10**) bound to deoxycytidine kinase (Figure 6.3), non-substrate inhibitors were designed that explored the extensive hydrogen bond network available to the cytosine base in the enzyme's substrate pocket, leading to compound (**11**) (Figure 6.4).[31] During the course of these efforts, the co-crystal structure of HTS lead (**12**),[30] became available. The bi-aryl portion of compound (**12**) occupied a newly formed pocket of the enzyme, resulting from reorientation of four amino acid side chains (Figure 6.3). Docking studies suggested that compounds (**11**) and (**12**) occupied similar areas of the binding pocket, but derived most of their potency from different (complementary) interactions with the enzyme. Combination of molecular motifs responsible for enzyme binding into a single hybrid molecule led to LP-661438 ((**13**), Figure 6.4), a potent, selective, and orally bioavailable inhibitor of dCK both in primary T cells and *in vivo*.[31–33]

Lexicon also disclosed another class of dCK inhibitors. Compound (**14**) potently inhibited dCK in enzyme and cellular assays, engaged the target in isolated mouse cells and prevented the incorporation of radiolabeled ^{3}H-dC *in vivo*.[34]

6.3.4 Adenosine Kinase

Of all the nucleoside kinases, the development of adenosine kinase inhibitors has received the most attention. This can be ascribed to the unique role that adenosine plays as a secondary messenger, modulating neuronal activity

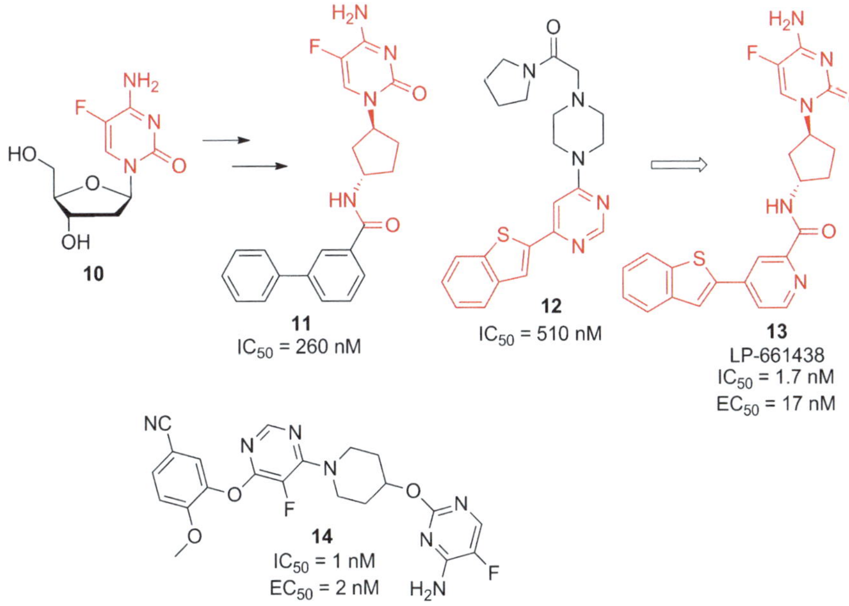

Figure 6.4 Lexicon's deoxycytidine inhibitors.

and inflammation. Whereas direct adenosine agonists have suffered from mechanism-related side effects, it has been postulated that indirect regulation of adenosine activity by inhibiting its metabolism may result in a wider therapeutic window. Since damaged tissues produce elevated levels of adenosine, inhibition of adenosine kinase, which plays a major role in metabolizing and inactivating adenosine, may result in locally increased adenosine signaling without the systemic side effects associated with adenosine agonists. In the following sections we will highlight some adenosine kinase inhibitor drug discovery efforts. Adenosine kinase inhibitors,[36,37] and the therapeutic potential of adenosine kinase inhibition,[38] have been reviewed.

6.3.4.1 Abbott Inhibitors

Like many efforts in this field, initial inhibitors reported by Abbott were nucleoside analogs.[39] Later efforts aimed to improve on the prototypical nucleoside analogs and led to the first non-nucleoside inhibitors of adenosine kinase. Starting with an SAR by NMR study,[40] in addition to traditional SAR studies, two classes of compounds were identified: alkynylpyrimidines and pyridopyrimidines. SAR studies and optimization of the physicochemical properties of the alkynylpyrimidines led to inhibitors such as ((**15**) Figure 6.6).[41] Compound (**15**) had an IC_{50} of 15 nM and displayed cellular activity at 100 nM. *In vivo* activity was shown at 3–10 μmol kg^{-1}. The binding mode of alkynylpyrimidines was shown by X-ray crystallography to be distinct from traditional nucleoside inhibitors (*e.g.* 5-IT, (**20**), Figure 6.7), involving a significant conformational change in the enzyme (Figure 6.5).[42] Molecular modeling studies of these distinct binding modes have also been published.[43]

SAR studies and optimization of physicochemical properties for the second series, the pyridopyrimidines, initially led to ABT-702 ((**16**), Figure 6.6),[44] a potent inhibitor with promising *in vitro*,[45] as well as *in vivo*,[46] activity. ABT-702 inhibited the enzyme at 1.7 nM and was active in cells at 50 nM. It showed *in vivo* activity following doses of 0.7 μmol kg^{-1} po. However, ABT-702 suffered from a short half-life and moderate bioavailability, prompting the quest for additional analogs with improved properties. Thus, compounds such as (**17**),[47] and (**18**),[48] were reported. Compound (**17**) had a sub-nanomolar IC_{50} (0.17 nM). Compound (**18**) maintained potency while having an improved half-life and bioavailability, but the further development of (**18**) was hampered by locomotor side effects. Replacement of the morpholine ring with more polar groups led to (**19**), which despite a loss in enzyme inhibitory potency (IC_{50} 8 nM) remained quite potent in cells (50 nM). Compound (**19**) showed analgesic effects *in vivo* following a 4 μmol kg^{-1} po dose and no locomotor effects after 30–100 μmol kg^{-1} po.[49] The analgesic effect is expected to stem from adenosine kinase inhibition at spinal sites whereas the locomotor side effects are thought to derive from inhibition at supraspinal sites. It was postulated that the increased selectivity for analgesic effects over locomotor side effects was not a reflection of decreased CNS penetration, but rather

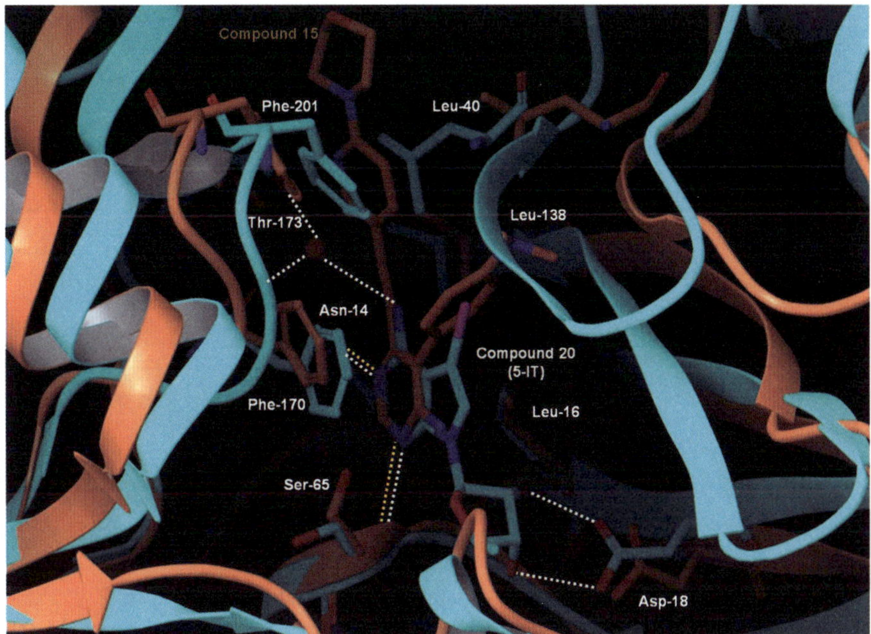

Figure 6.5 Overlay of the co-crystal structures of Abbott's alkynylpyrimidine **15** (brown) and the prototypical nucleoside inhibitor 5-IT (**20**, cyan) bound to adenosine kinase. The binding mode of 5-IT shows H-bonds from the ribose 2'-OH and 3'-OH to Asp-18. The exocyclic amine is engaged in a water-mediated H-bond to the main chain of Phe-170 and the side chain of Ser-173. One of the pyrimidine nitrogens interacts with the nitrogen of the side chain of Asn-14, while the other bonds to the main-chain nitrogen of Ser-65. The Abbott inhibitor (**15**) is engaged in the same interactions with Ser-65 and Asn-14. The morpholinopyridine intersects the space between the Phe-201 and Leu-40 side chains, causing a significant conformational change of the enzyme (compare brown and cyan ribbons).

seemed to be caused by preferential partitioning of the inhibitors in the CNS. Several later studies described additional analogs of both series.[50–53]

6.3.4.2 Gensia Sicor – Metabasis Inhibitors

Gensia Sicor initially developed analogs of adenosine and the natural adenosine kinase inhibitor 5-iodotubercidin (5-IT, (**20**), Figure 6.7, IC_{50} 26 nM),[54,55] such as GP515 ((**21**), IC_{50} 1.5 nM),[56] and GP3269 (**22**),[57] focusing on their antiseizure activity. Although GP3269 went into human clinical trials in 1996, it suffered from poor solubility and CNS side effects. In 1997, Gensia Sicor transferred all its drug discovery and development assets to a new company to create Metabasis.

Further analogs of GP515 and GP3269 were explored at Metabasis. Although very potent inhibitors derived from 5-iodotubercidin were reported

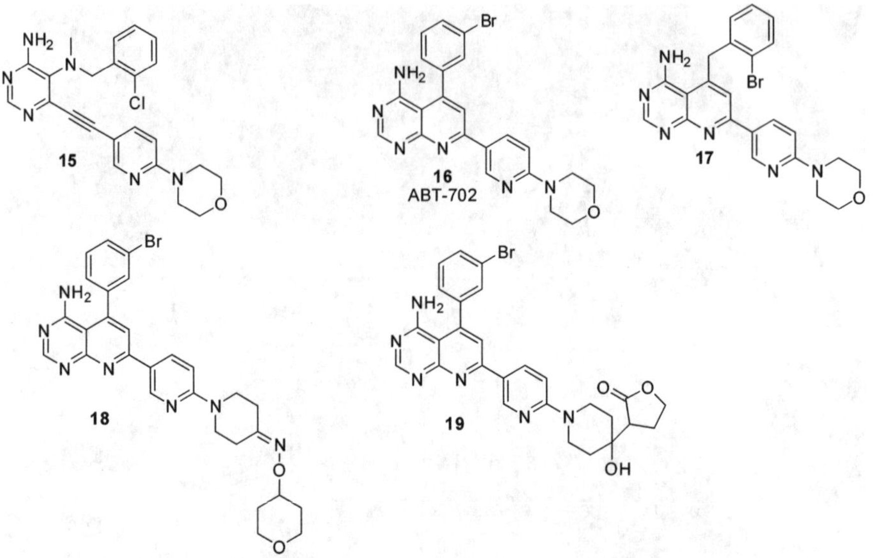

Figure 6.6 Abbott's non-nucleoside adenosine kinase inhibitors.

((**23–25**), IC_{50} values 0.1–0.6 nM), no compounds with a safety, efficacy and side effect profile suitable for further development were identified in this study.[58] Combining features of GP515 and GP3269 into a single molecule resulted in potent tubercidin analogs with improved *in vivo* activity (*e.g.* (**26**), IC_{50} 1 nM).[59] Inversion of the C4' substituent, in an attempt to avoid metabolism by phosphorylation, resulted in the potent xylose derivative GP790 ((**27**), IC_{50} 0.5 nM).[60] Further analog design led to compounds such as (**28**),[61] GP3966 (**29**),[62] and (**30**).[63] Although compound (**30**) possessed good *in vivo* potency and significantly increased solubility (32 µg mL^{-1}) compared to GP3269, chronic dosing of (**30**) led to loss of efficacy and resulted in death. A QSAR model for these types of inhibitors has been developed by Singh *et al.*[64] The model used a training set of 66 previously reported compounds and either a Fujita-Ban or Hansch approach to analyze relative contributions to the IC_{50} of the various substituents. The Fujita-Ban approach proposed a compound (**31**) that was expected to be 1.5 orders of magnitude more potent than any compound in the training set, but this compound has yet to be synthesized.

6.3.4.3   *Other Inhibitors*

Several other groups have explored inhibitors of adenosine kinase. For example, Bauser and co-workers at Bayer reported that compound (**32**) (Figure 6.8, $IC_{50} < 10$ nM) was more potent in their hands than ABT-702 (IC_{50} 10 nM).[65] A group at Universidade Estadual de Campinas (Unicamp, Brazil) disclosed potent quinazoline-based inhibitors such as (**33**) ($EC_{50} =$

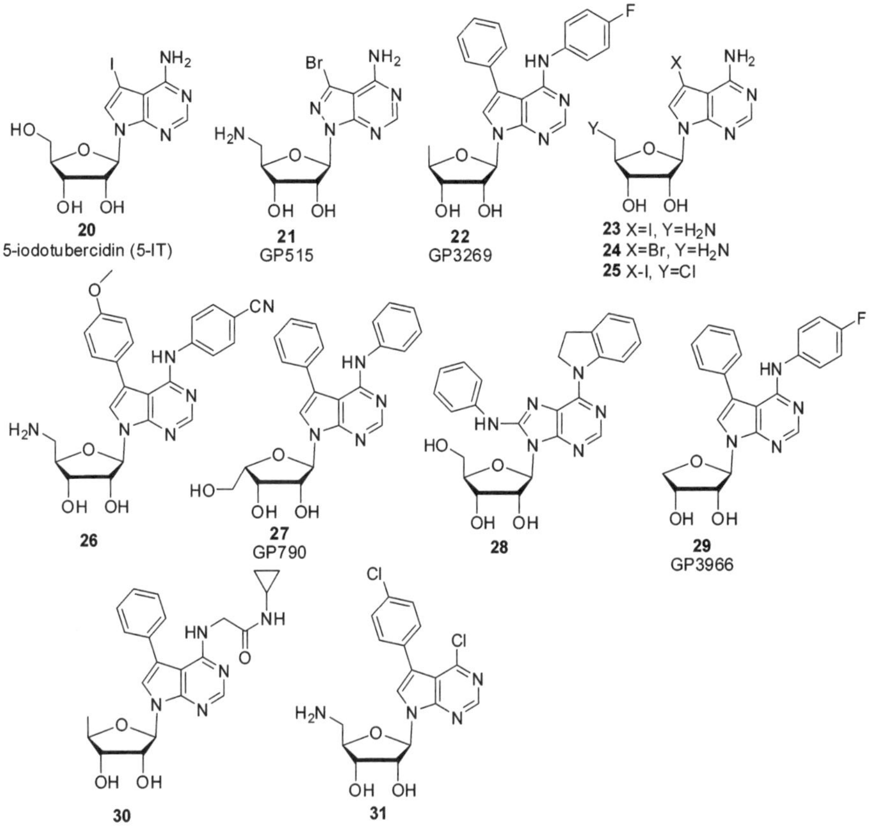

Figure 6.7 Nucleoside derivatives developed as adenosine kinase inhibitors by Gensia Sicor / Metabasis.

6 nM) and (**34**) (EC_{50} = 0.1 nM).[66] Gupta and co-workers at McMaster University identified eight non-nucleoside inhibitors of adenosine kinase from a screen of 1040 compounds.[67] The most potent of these inhibitors, 8008–6198 (**35**), had an IC_{50} of 0.38 μM and was competitive with respect to adenosine but not ATP, similar to the Abbott non-nucleoside inhibitors. These compounds also inhibited uptake of adenosine and its metabolism in cultured mammalian cells.

Non-competitive, allosteric inhibitors of adenosine kinase were reported by Butini *et al.*[68] A biological screen of an in-house database of proprietary compounds identified pyrrolobenzoxazepines, representative of a class of non-nucleoside inhibitors of HIV-1 reverse transcriptase, with a unique "T shape" geometry as micromolar inhibitors of human adenosine kinase (*e.g.* compound (**36**), IC_{50} 80 μM). SAR studies led to compound (**37**) (IC_{50} 1.2 μM) which was selective over adenosine deaminase. Selectivity over HIV-1 reverse transcriptase was not reported. Molecular modelling suggested that compound (**37**) bound to adenosine kinase at a novel, allosteric site and by doing so inhibited the enzyme through interference with the enzyme's conformational plasticity. Modeling also confirmed that (**37**) adopted the unique "T shape" geometry. Compound (**37**) had pro-apoptotic efficacy in cells and was cytostatic to tumor cells while displaying low cytotoxicity.

A pharmacophore model for new inhibitors was designed by Lee *et al.*, although no inhibitory activity for the proposed inhibitors was determined experimentally.[69] Long *et al.* have designed adenosine kinase inhibitors to target *Mycobacterium tuberculosis*.[70,71] Finally, a patent application from McMaster University disclosed phosphonates as millimolar adenosine kinase inhibitors.[72]

6.3.5 Conclusion

Nucleoside kinase inhibitors hold promise for the development of anticancer drugs by inhibition of the salvage pathway. Although previous attempts to develop these types of inhibitors into drugs have failed, the majority of these efforts focused on nucleoside analogs with poor physicochemical properties. Several non-nucleoside inhibitors have been reported for adenosine kinase and deoxycytidine kinase, with promising initial properties. It is possible that future efforts may result in the identification of non-nucleoside inhibitors that possess an appropriate efficacy-side effect profile.

Adenosine kinase inhibitors take a special place in this group, due to the additional roles adenosine plays in biological processes as a second messenger. It remains to be seen whether the central nervous system side effects observed in clinical trials of GP3269 were target-mediated or compound specific.

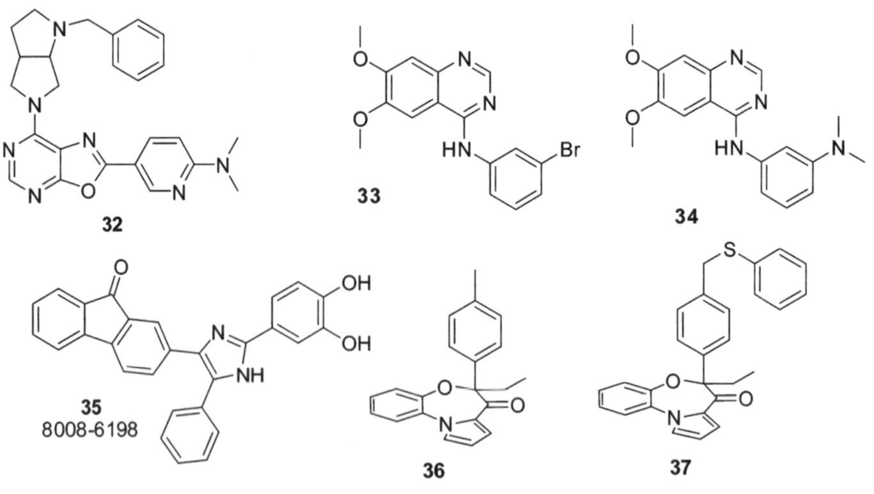

Figure 6.8 Other adenosine kinase inhibitors.

6.4 Lipid Kinases

Of the different classes of non-protein kinases, the development of inhibitors of lipid kinases has received by far the most attention. A majority of this work centers on inhibitors of phosphatidylinositol kinases, which reflects the important role that phosphatidylinositols play as second messengers. In this section, we review inhibitors of PI3K and related kinases, as well as inhibitors of other phosphatidyl inositol kinases. We also discuss sphingosine kinase inhibitors and we end this section with a brief summary of inhibitors of other lipid kinases.

6.4.1 PI3K and Structurally Related Kinases

The phosphatidylinositol 3-kinases (PI3Ks) regulate cellular metabolism and growth by phosphorylation of phosphatidylinositol at the 3-position to create the second messenger phosphatidylinositol triphosphate (PIP3). Upregulation of the PI3K pathway, resulting in signaling amplification, is one of the most frequently encountered abnormalities in human cancers. Therefore, design of inhibitors of PI3K has received significant attention. PI3K-α is the predominant target for anticancer drugs; we will limit our review to the most advanced inhibitors of this isoform. Compounds targeting the other PI3K isoforms have been studied for other therapeutic applications and will be discussed in a separate section. Even though these are not lipid kinases *per se*, we will also discuss efforts toward the design and synthesis of inhibitors of a class of enzymes that are structurally related to PI3K, the phosphatidyl inositol kinase-related kinases (PIKKs), including the mammalian target of rapamycin (mTOR).

6.4.1.1 *PI3K-α Inhibitors, pan-PI3K Inhibitors and Dual mTOR– PI3K-α Inhibitors*

Due to the homology between PI3K and the PIKKs, most notably mTOR, many PI3K inhibitors also inhibit mTOR. Two prototypical PI3K inhibitors have been used extensively as pharmacological tools: the natural product, irreversible inhibitor wortmannin ((**38**), Figure 6.16); and the reversible inhibitor LY294002 (**40**), a derivative of the natural product quercetin.

The binding mode of these two inhibitors was elucidated by co-crystallization with PI3K-γ.[73] As is typically observed for ATP competitive kinase inhibitors, PI3K inhibitors have a hydrogen bond acceptor in a position equivalent to N1 of ATP to form the crucial hinge region interaction with the backbone amide of Val-862 (Figure 6.9). In the case of wortmannin, the 17-keto oxygen forms this interaction (Figure 6.10). A unique feature of the binding mode of wortmannin involves attack of the furan ring by the primary amine of the active site Lys-833 side chain, resulting in irreversible enzyme inhibition.

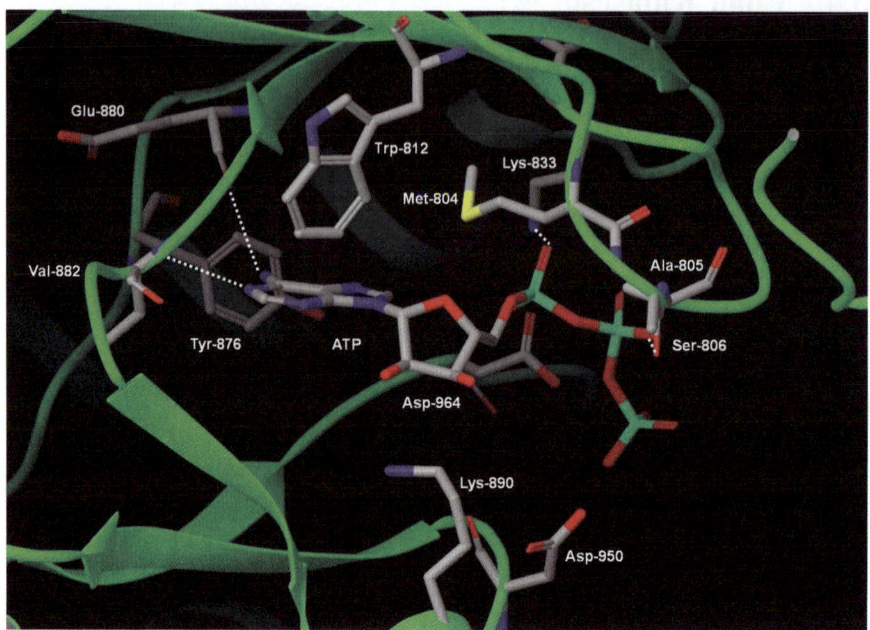

Figure 6.9   Co-crystal structure of ATP bound to PI3K-γ. The binding mode is characterized by the hinge region interaction of the purine ring nitrogen with the Val-882 backbone amide. Additional interactions shown are H-bonds from Ser-806 and Lys-833 to the phosphate; and from the exocyclic amine to Glu-880.

The natural product wortmannin proved unsuitable for clinical use due to poor stability and general toxicity. Oncothyreon has advanced a derivative of wortmannin, PX-866 ((**39**), Figure 6.16), to Phase I clinical trials in patients with advanced solid tumors. The compound displays selectivity for the PI3K-α, -δ, and -γ isoforms relative to PI3K-β and mTOR.[74] A comparison of the behavior of PX-866 and wortmannin in human cancer cell lines was performed; in this study, PX-866 was found to limit cancer cell motility in three-dimensional spheroid cultures at subnanomolar levels.[75]

Co-crystallization of LY294002 (**40**) with PI3K-γ revealed that the morpholine ring in LY294002 partially overlaps the volume occupied by the adenine ring in ATP (Figure 6.11).[73] The backbone NH of Val-882 in PI3K-γ interacts with the morpholine oxygen of LY294002. Like wortmannin, LY294002 engages in additional interactions with the active site. A putative hydrogen bond between the primary amine of Lys-833 and the chromenone ketone in LY294002 was observed. As will be illustrated in the remainder of this section, the combination of the hinge region interaction with Val-882 and additional interactions in the active site is common to many of the ATP competitive inhibitors of PI3K and the PIKKs and forms the basis of the structure guided design of several different classes of inhibitors.

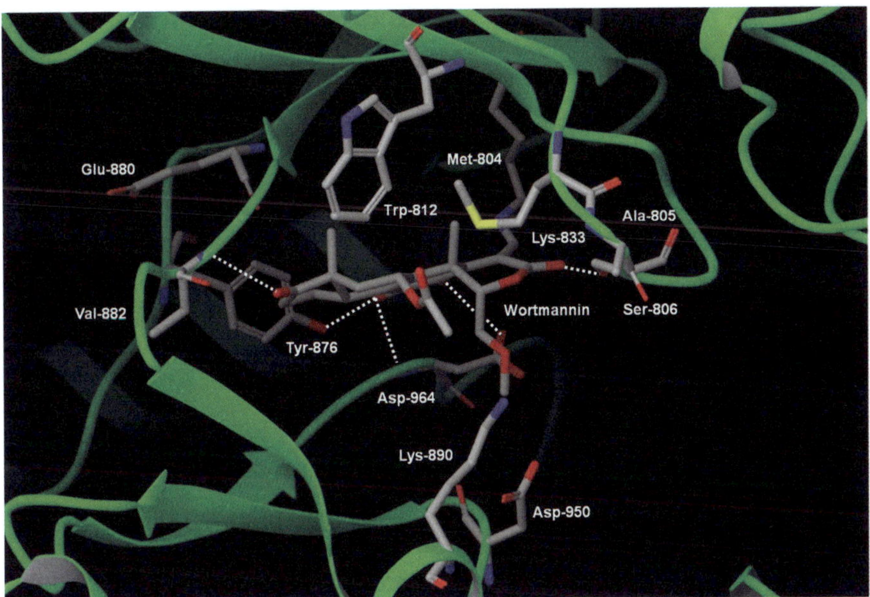

Figure 6.10 Co-crystal structure of wortmannin bound to PI3K-γ. The ketone oxygen of the D ring interacts with the backbone of Val-882, forming the hinge region interaction. The ketone oxygen of the A ring interacts with the hydroxyl of Ser-806, mimicking the interaction of the phosphate in ATP. Lys-833 (which forms the other phosphate interaction in the case of ATP) forms a covalent bond to the furan of wortmannin, which is clearly visible in the crystal structure. As a result of the formation of this covalent bond, the furan ring opens up and the hydroxyl of the B ring interacts with the Asp-964 side chain (this hydroxyl is the ether oxygen of the furan ring in the unreacted wortmannin). Finally, the ketone oxygen of the B ring interacts with the backbone of Asp-964 and with the hydroxyl of Tyr-867.

LY294002 was investigated for its anticancer properties in the early 1990s. However, the compound proved to be an unsuitable therapeutic due to its lack of specificity.[76] SF1126 ((**41**), Figure 6.16, Semafore Pharmaceuticals) consists of a LY294002 derivative conjugated to a tetrapeptide moiety and was designed to specifically bind to integrins within the tumor compartment.[77] SF1126 targets all PI3K class I isoforms as well as other related kinases including DNA-PK, PLK-1, CK2, ATM, and PIM-1. This agent demonstrated *in vitro* efficacy against trastuzumab-resistant, HER-2 over-expressing breast cancer cell lines,[78] and has been in Phase I trials as a single agent since 2007.

In 2008, Genentech disclosed the structure of GDC-0941 ((**42**), Figure 6.16), a pan-PI3K class I inhibitor.[79] This compound was obtained by modification of a phenolic thienopyrimidine (derived from PI-103, (**82**), Figure 6.23), which showed good PI3K activity but poor pharmacokinetics.[80] Modification of the phenol to an indazole and the introduction of an alkylpiperidinyl moiety

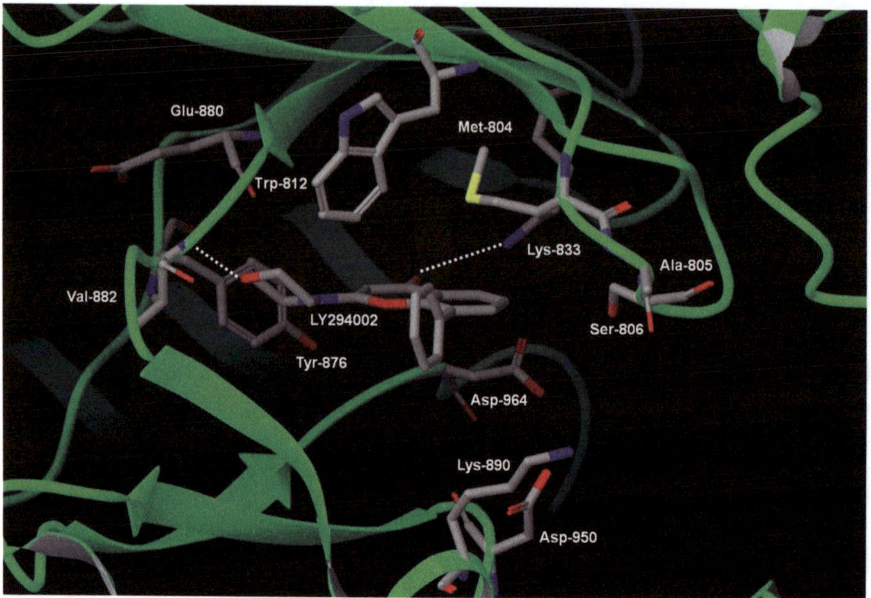

Figure 6.11 Co-crystal structure of LY294002 bound to PI3K-γ. The binding mode mimics that of ATP. Thus, the morpholine oxygen forms the hinge region interaction with the backbone amine of Val-882 (compare the pyrimidine nitrogen in ATP) and the carbonyl interacts with Lys-833 (compare the phosphate in ATP).

resulted in GDC-0941, a compound that displayed dramatically improved clearance, volume of distribution, and oral bioavailability.

A co-crystal structure of GDC-0941 with PI3K-γ showed the same hinge region interaction between the morpholine oxygen and the backbone NH of Val-882 as observed for LY294002 (Figure 6.12).[79] Additional interactions of the indazole nitrogens with the phenol oxygen of Tyr-867 and the carboxyl group of Asp-841 were observed as well. Finally, the oxygens of the sulfonyl group were positioned within H-bonding distance of the side chain of Lys-802 and the amide nitrogen of Ala-805.

Additional studies revealed the efficacy of GDC-0941 in multiple xenograft studies.[81] A PK–PD relationship has been determined for this compound in a breast cancer xenograft model, and a predictive model which correlated reduction in Akt and PRAS40 phosphorylation in these xenografts to growth inhibition was developed.[82] GDC-0941 is currently undergoing multiple Phase I clinical trials. In addition to examination of the safety, tolerability, and pharmacokinetics of GDC-0941 as a single agent in solid tumors or non-Hodgkin's lymphoma (NHL), the compound is under clinical investigation in combination with taxol, carboplatin, and–or bevacizumab in patients with non-small cell lung cancer and metastatic breast cancer. A recent report revealed the antitumor effects of GDC-0941 in models possessing mutations in PI3KCA and hEGFR2 genes were particularly pronounced.[83] A second

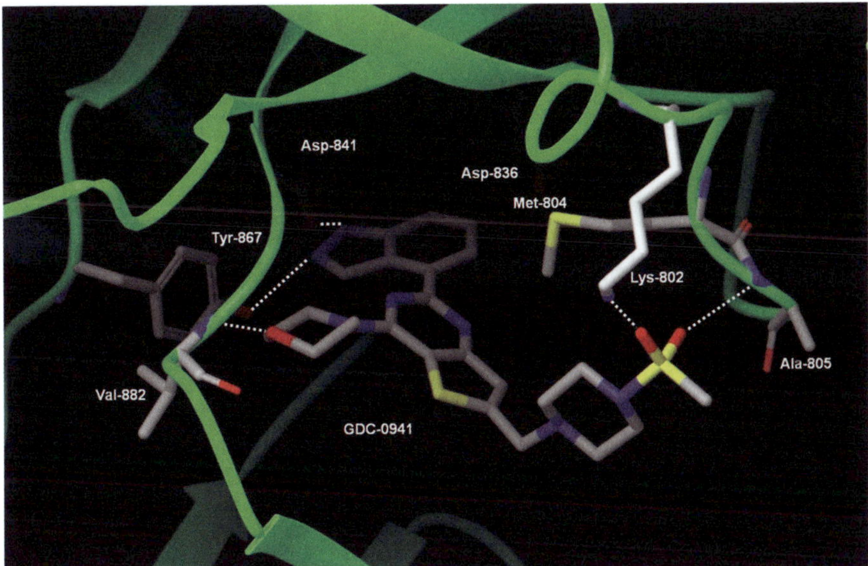

Figure 6.12 Co-crystal structure of GDC-0941 bound to PI3K-γ. The morpholine oxygen forms the hinge region interaction with the backbone NH of Val-882 (compare LY294002); additional interactions shown include H-bonds from the indazole to Tyr-867 and Asp-841; and from the sulfonyl to the Lys-802 side chain and to the Ala-805 backbone amide.

compound from Genentech, GDC-0980 ((**43**), Figure 6.16), has been characterized as a dual PI3K and mTOR inhibitor.[84,85] GDC-0980 is currently in Phase I trials in solid tumors, NHL, and as a combination therapy in metastatic breast cancer. An additional dual PI3K-mTOR inhibitor termed GNE-477 (**44**) has also been disclosed by Genentech.[86] Based on the X-ray structure of a complex of PI3K-γ with a related inhibitor (**45**), the potency of GDC-0980 and GNE-477 likely stems from H-bonds between the amino-pyrimidine NH2 and Asp-836 and Asp-841 in addition to the hinge region interaction between the morpholine oxygen and the backbone NH of Val-882 (Figure 6.13).[87] A water mediated interaction of the pyrimidine nitrogen with Tyr-867 has also been proposed.

The quinoxaline XL147 ((**46**), Figure 6.16, Sanofi-Aventis–Exelixis) first entered the clinic in 2007. This compound displayed a selectivity profile much like PX-866; the PI3K-α, -δ, and -γ isoforms are inhibited to a greater extent than PI3K-β and mTOR.[88] XL147 progressed to Phase II clinical trials in late 2009 in patients with advanced or recurrent endometrial cancer; a second Phase II study involved combination of this agent with trastuzumab or paclitaxel in metastatic breast cancer patients. Sanofi-Aventis and Exelixis are currently conducting a second set of clinical trials with XL765 (**47**). This analog of XL147 has been modified such that it now functions as a dual PI3K–mTOR inhibitor.[89] This compound entered Phase II trials in 2010 in a combination study with letrozole in breast cancer patients.

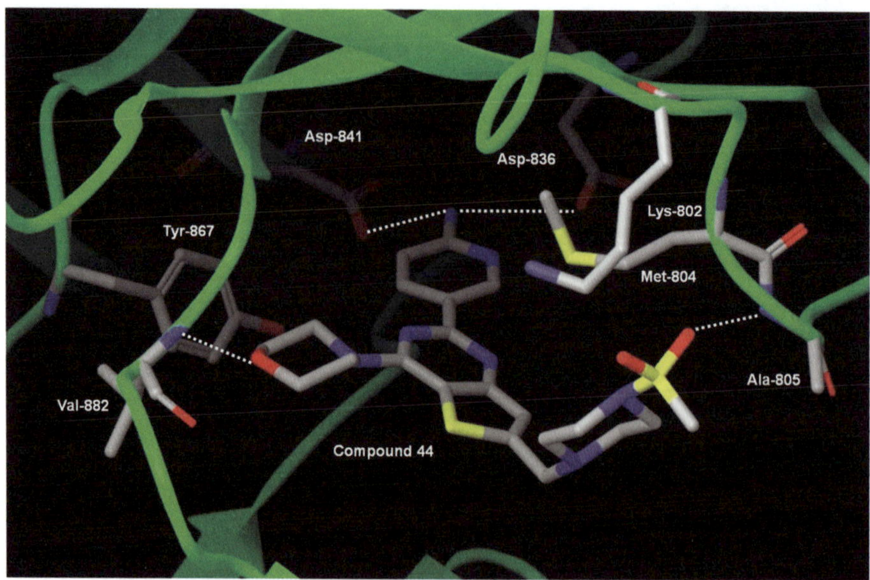

Figure 6.13 Co-crystal structure of an analog of GNE-477 (**45**) bound to PI3K-γ. The morpholine and sulfonyl groups are engaged in similar interactions as observed for GDC-0941. The aminopyridine interacts with Asp-836 and Asp-841.

Researchers at Novartis have been extensively involved in the PI3K–mTOR pathway and have developed both PI3K selective and dual mTOR–PI3K inhibitors. Novartis disclosed the dual mTOR–PI3K inhibitor BEZ235 (NVP-BEZ235, (**48**)).[90] This imidazoquinoline serves as an effective mimic of the adenine moiety of ATP. BEZ235 demonstrated inhibition of VEGF-induced cell proliferation and angiogenesis.[91] Since the allosteric, rapamycin-derived mTOR inhibitor RAD001 did not show this effect, modulation of PI3K was deemed essential in this model. A study directly comparing the effects of rapamycin to those of BEZ235 in renal cell carcinoma models led to the conclusion that dual mTOR–PI3K inhibition resulted in more effective growth inhibition than allosteric mTOR inhibition.[92] BEZ235 is currently in Phase I–II trials for solid tumors. A second compound of undisclosed structure, BGT226, is also at this stage of clinical advancement. In 2008 the pyrimidine derivative BKM120 (**49**) entered the clinic. The structure of BKM120 shares some features with that of GNE-477 and similar interactions with the enzyme can be envisioned. This compound displayed selectivity for class I PI3Ks in cancer cell lines, displaying IC_{50} values in a MEF cell line in the range of 93–242 nM for PI3K isoforms relative to 1800 nM for mTOR.[93,94] A recent presentation disclosed the activity of BKM120 in multiple mouse xenograft models. The compound displayed efficacy in a U87MG glioblastoma model following oral dosing of 30 mg kg^{-1} once daily, a BT474 breast cancer model at 10 mg kg^{-1} once daily, and demonstrated a synergistic effect with the oral

MEK inhibitor selumetinib (AZD6244, AstraZeneca).[94,95] In 2010, BKM120 advanced to a Phase I–II trial as a combination therapy with trastuzumab in breast cancer patients. A second compound, BYL719, began a Phase II trial in late 2010 in patients with solid tumors with a mutation of the PIK3CA gene; however, the properties and biological activities of this compound have yet to be disclosed.

GlaxoSmithKline recently began a dose-escalation study in patients with refractory malignancies with GSK1059615 (**50**). The co-crystal structure of GSK1059615 bound to PI3K-γ revealed that there was a sizable pocket which could be accessed from the 6-position of the quinoline core. Efforts to fill this space, with the aim of increasing potency, resulted in GSK2126458 (**51**), which displayed subnanomolar IC_{50} values against both mTOR and PI3K, and also inhibited the related kinase DNA-PK (IC_{50} = 0.3 nM).[96] The binding mode of GSK2126458 (Figure 6.14) was characterized by the common hinge region interaction with Val-882 and showed an additional charged interaction between the sulfonamide and Lys-833 (*cf.* wortmannin and LY294002) as well as a hydrogen bond between the pyridine nitrogen and a conserved water molecule (*cf.* GNE-477). The compound has also demonstrated efficacy at low doses (3 mg kg^{-1}) in a breast cancer mouse xenograft model and is currently undergoing clinical trials for solid tumors and lymphoma.

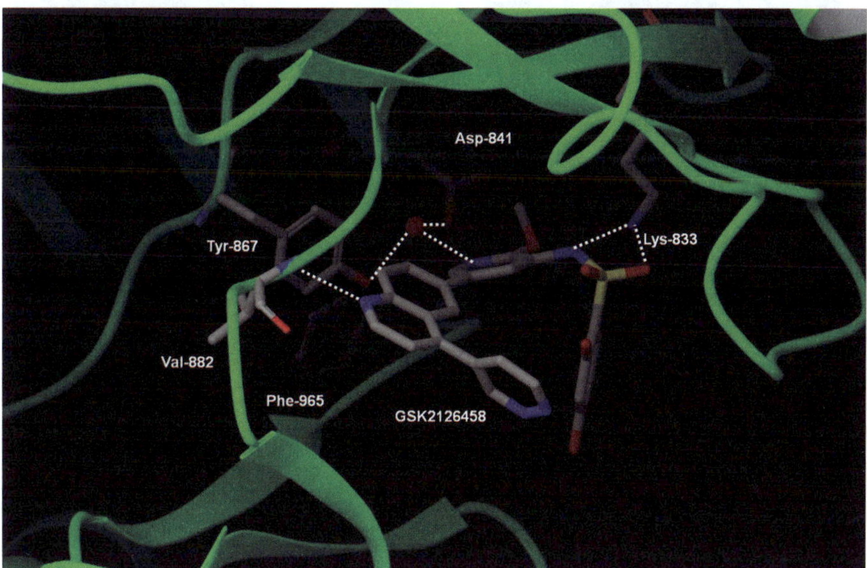

Figure 6.14 Co-crystal structure of GSK2126458 and PI3K-γ. The quinoline nitrogen forms the hinge region H-bond with the backbone NH of Val-882. The pyridine nitrogen forms a H-bond through a conserved water molecule to Tyr-867 and Asp-841. Lys-833 forms a charged interaction with the sulfonamide.

The pyridonepyrimidine PF-04691502 ((**52**), Figure 6.16), developed by Pfizer, displayed dual PI3K and mTOR activity. In an enzymatic assay, the compound displayed K_i values of 0.6 nM *versus* PI3K and 16 nM *versus* mTOR.[97] The 4-methylpyridopyrimidinone scaffold was identified by an HTS campaign. The interactions of this scaffold with PI3K-γ are reminiscent of the binding mode of GSK2126458. Thus, the aminopyrimidine forms a two-point hydrogen bond with the hinge region valine and the methoxypyridine binds the conserved water molecule (Figure 6.15). The methyl group on the pyrimidine ring extends into a small hydrophobic pocket present in PI3K; this has been hypothesized to result in increased selectivity over protein kinases. The compound displayed good PK, with a clearance of 5.2 ml min^{-1} kg^{-1} and a bioavailability of 56%, and entered the clinic at the end of 2009. A second dual inhibitor, PF-04979064, has been recently reported and is currently at the preclinical stage.[98] A triazine-based dual PI3K-mTOR inhibitor, PKI-587–PF-05212384 ((**53**), Figure 6.16, Pfizer), entered Phase I clinical trials for oncology in late 2009. The urea moiety had been previously utilized in the preparation of selective mTOR inhibitors;[99] the presence of two morpholine moieties in PKI-587 resulted in compounds with dual PI3K–mTOR activity. The compound was highly potent in enzymatic assays, inhibiting PI3K-α and mTOR with IC$_{50}$

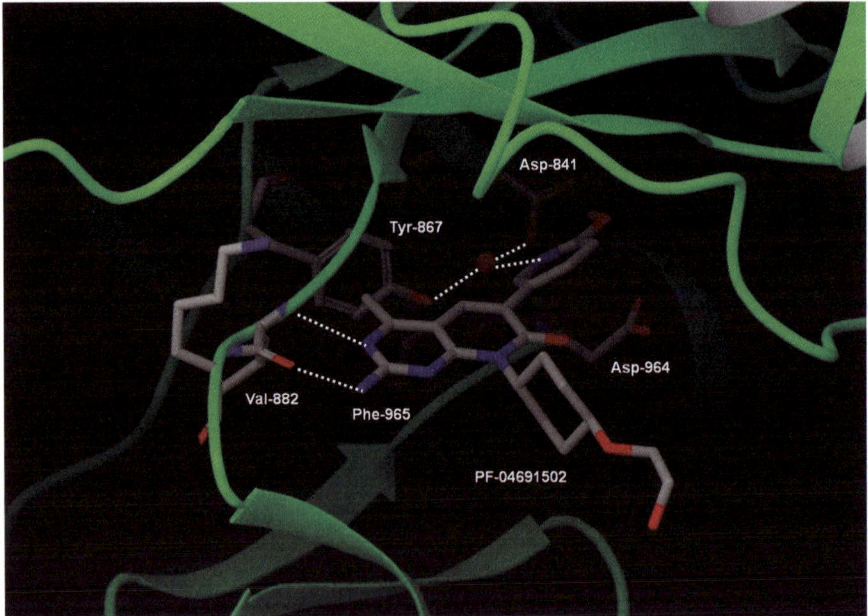

Figure 6.15 Co-crystal structure of PF-04691502 bound to PI3K-γ. The ring nitrogen of the pyridopyrimidinone forms the common H-bond hinge region interaction with the backbone NH of Val-882, while the exocyclic amine forms an additional interaction with the Val-882 backbone carbonyl. The pyridine nitrogen binds to Tyr-867 and Asp-841 through a conserved water molecule.

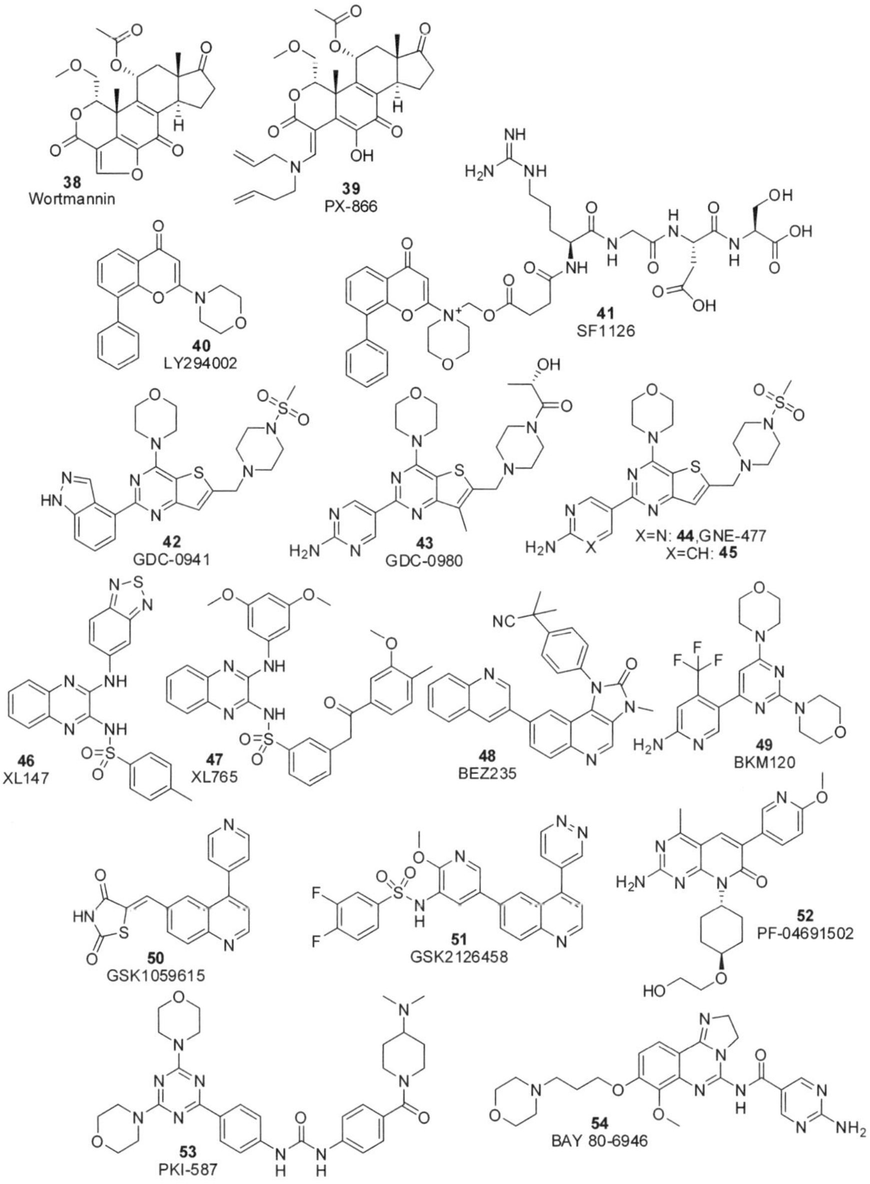

Figure 6.16 PI3K-α Inhibitors, pan-PI3K inhibitors and dual mTOR/PI3K-α Inhibitors.

values of 0.4 nM and 1.6 nM, respectively.[100] A second dual mTOR–PI3K inhibitor utilizing the triazine scaffold, PKI-179, has also been disclosed.[101]

Bayer initiated clinical trials in patients with advanced solid tumors with BAY 80-6946 (**54**) in 2009. This compound displayed an IC_{50} of 0.5 nM against PI3K-α and inhibited mTOR with an IC_{50} of 45 nM, while being selective against a panel of 205 other kinases.[102] Further evaluation of BAY 80-6946 revealed that this compound was particularly effective against tumor cell lines with a PIK3CA mutation, deletion of PTEN or overexpression of HER2. For example, significant tumor growth suppression was observed in a KPL4 breast tumor model in athymic rats at doses as low as 0.5 mg kg^{-1}, given i.v. every other day, whereas doses of 6 mg kg^{-1} i.v. every other day resulted in complete tumor suppression in this model.[103,104]

6.4.1.2 Inhibitors of Other PI3K Isoforms

Differentiation of the role of PI3K-β in disease from other PI3K isoforms, especially PI3K-α, has been accomplished through genetic modifications and the use of isoform selective inhibitors. TGX-221 (PIK-108, (**55**), Figure 6.19, Thrombogenix) is a potent inhibitor of PI3K-β (IC$_{50}$ = 5 nM) with ~1000-fold selectivity over PI3K-α and PI3K-γ (Table 6.1). Its selectivity over PI3K-δ is more modest (~10-fold).[105] Alternative IC$_{50}$ values for TGX-221 have been generated that show a similar trend, but with a 10-fold higher IC$_{50}$ value for PI3K-β and less selectivity *versus* the other isoforms.[106] TGX-221 has been used to show that inhibition of PI3K-β prevents stable integrin αIIbβ3 adhesion contacts leading to defective platelet thrombus formation, establishing PI3K-β as a target for antithrombotic therapy. The *in vivo* effect of TGX-221 in inhibiting thrombus formation without adversely affecting haemodynamic variables and bleeding time has also been demonstrated.[107]

Thrombogenix was later known as Kinacia, which in turn was acquired by Cerylid. Cerylid has also reported the antithrombotic PI3K-β inhibitor CBL1309 (KN309, (**56**)). Recent studies with knockout mice demonstrate different roles for PI3K-α and PI3K-β in cell signaling and tumorigenesis.[108] In the majority of cases, PI3K-α is responsible for carrying the PI3K signal in receptor tyrosine kinase signal transduction. In contrast, PI3K-β responds to GPCRs. Tumors driven by PIK3CA mutations and oncogenic RTK–Ras rely

Table 6.1 IC50 values of inhibitors of PI3K-β, -γ and -δ.

Compound	PI3K-α[a]	PI3K-β[a]	PI3K-δ[a]	PI3K-γ[a]
TGX-221[b] (PIK108, **55**)	5000	5	100	>10 000
TGX-221[c] (PIK108, **55**)	2600	57	260	4100
TGX-286 (**57**)	4500	120	1000	10 000
IC87114 (**58**)	100 000	75 000	500	29 000
PIK-39 (**59**)	>200 000	11 000	180	17 000
PIK-293 (**60**)	>90 000	>90 000	237	10 000
SW13 (**61**)	1240	221	0.7	33
SW30 (**62**)	85 000	740	7	1300
CAL-101 (**63**)	820	565	2.5	89
TG100-115 (**64**)	1300	1200	235	83
AS-252424 (**65**)	940	20 000	20 000	30

[a]IC$_{50}$ (nM), [b]Ref. 104, [c]Ref. 105

on PI3K-α for signaling and growth. PTEN deficient tumorogenisis is mediated by PI3K-β. Using small molecule PI3K isoform-selective inhibitors in a diverse set of breast cancer cell lines, Torbett *et al* demonstrated that PI3K-α inhibitors inhibited phosphorylation of PKB (protein kinase B)–Akt and S6.[109] In contrast, PI3K-β selective inhibitors (TGX-286 (**57**) and TGX-221–PIK-108) only reduced PKB–Akt phosphorylation in PTEN mutant cell lines, and showed less inhibition of S6 phosphorylation. Therefore, isoform selective PI3K inhibitors could have equivalent efficacy in different tumor types depending on their mechanism. This suggests that a higher therapeutic window may potentially be achieved by eliminating off target effects caused by inhibiting non-therapeutically relevant kinases.

PI3K-δ and -γ are expressed mainly in the hematopoietic system and regulate immune cell signaling. PI3K-γ is found in granulocytes, monocytes, and macrophages and PI3K-δ is also expressed in B cells and T cells. In contrast, PI3K-α and -β are ubiquitously expressed and regulate cell survival and metabolism. Because of this difference in expression and function, selective inhibitors of PI3K-γ and–or PI3K-δ have the potential to treat inflammatory and immune conditions. In contrast to pan-PI3K inhibitors and dual PI3K–mTOR inhibitors, several of which are in clinical trials, little is known about the clinical behavior of selective PI3K-γ–δ inhibitors.

The earliest selective PI3K inhibitor, IC87114 (**58**), was discovered in 2003 by ICOS researchers.[110] The PI3K IC_{50} values for IC87114 are: PI3K-δ = 0.5 μM, PI3K-α = 100 μM, PI3K-β = 75 μM, PI3K-γ = 29 μM. Shokat and co-workers synthesized an analog of IC87114, PIK-39 (**59**), with a similar PI3K selectivity profile (IC_{50}s: PI3K-δ = 0.18 μM, PI3K-α > 200 μM, PI3K-β = 11 μM, PI3K-γ = 17 μM).[106] Co-crystallization of PIK-39 with PI3K-γ revealed that residue Met-804 of PI3K-γ, which forms the roof of the ATP binding pocket when bound to non-selective inhibitors or ATP, moves to open up a new pocket, termed the selectivity pocket, that can accommodate the quinazolinone ring of PIK-39 (Figure 6.17). Thus, the structural cause of the selectivity of IC87114 and related analogs, is the movement of Met-804 in PI3K-γ and -δ; such a molecular reorganization is not available in PI3K-α and -β. Further proof of the requirement of a flexible methionine for activity of PIK-39 and IC87114 was the design of resistance mutants in which Met-804 had been mutated to isoleucine or valine. Modeling predicted that these β-branched residues would be static and in the event, neither PIK-39 nor IC87114 was able to inhibit the mutants while the binding affinities of non-selective inhibitors were unaffected.

The crystal structure of the catalytic subunit of PI3K-δ was recently determined along with co-crystal structures of PI3K-δ and selective and non-selective inhibitors.[111] As was also seen in the co-crystal structures of PI3K-γ with PIK-39, the conformationally mobile Met-804 residue moves to expose the selectivity pocket. The co-crystal structure of IC87114 with PI3K-δ shows that the purine moiety hydrogen bonds with the hinge region while the quinazolinone ring projects into the selectivity pocket created by the Met-804

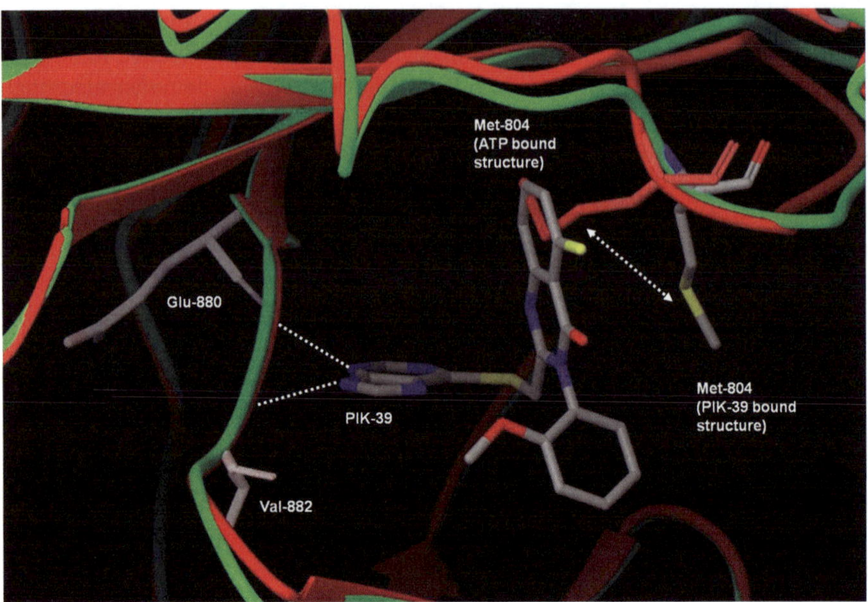

Figure 6.17   Comparison of the PI3K-γ protein co-crystal structures bound to ATP (red) and PIK-39 (green). PIK-39 forms two H-bonds identical to those seen with ATP binding: the hinge region interaction with the backbone NH of Val-882 and a H-bond to the backbone of Glu-880. Binding of PIK-39 results in a significant movement of the Met-804 side chain compared to the ATP-bound conformation.

movement. This selectivity, based on the conformational flexibility differences between closely related kinases, is reminiscent of that seen in imatinib's ability to distinguish between kinases Abl and Src through binding to structurally dynamic regions of the enzyme.[112]

PIK-39 and another analog of IC87114, PIK-293 ((**60**), Figure 6.19) do not occupy the deep, hydrophobic affinity pocket common to all the PI3Ks.[106] Shokat *et al.* designed analogs of PIK-293 that could access the affinity pocket, leading to more potent analogs, without compromising selectivity *versus* PI3K-α and -β.[106,113] Despite the high degree of similarity between the affinity pockets of PI3K-γ and -δ, variable degrees of selectivity for PI3K-δ *versus* PI3K-γ were seen. SW13 (**61**) is the most potent PI3K-δ inhibitor seen to date, and also quite selective *versus* PI3K-α and -β (IC$_{50}$s: PI3K-δ = 0.7 nM, PI3K-α = 1240 nM, PI3K-β = 221 nM, PI3K-γ = 33 nM). An alkynyl affinity pocket group on PIK-293 led to analog SW30 (**62**) with good selectivity *versus* PI3K-γ while maintaining potent PI3K-δ activity (IC$_{50}$ values: PI3K-δ = 7 nM, PI3K-α = 85 000 nM, PI3K-β = 740 nM, PI3K-γ = 1300 nM). Evaluation of these selective PI3K-γ–δ inhibitors in primary tumor cocultures stimulated with proinflammatory agents revealed an unanticipated similarity in the anti-inflammatory properties of these compounds and the glucocorticoids. In addition, pan-PI3K inhibitors demonstrated a decrease in anti-inflammatory effects relative to the PI3K-γ–δ selective inhibitors.

Calistoga Pharmaceuticals, founded by former Icos scientists when Icos was acquired by Eli Lilly, has continued Icos' work in this field. CAL-101 (**63**), also a derivative of IC87114, is an orally active, selective PI3K-δ inhibitor currently under clinical evaluation in patients with B-cell malignancies.[114] CAL-101 is 40- to 300-fold more selective for PI3K-δ relative to other PI3Ks (IC$_{50}$ values: PI3K-δ = 2.5 nM; PI3K-α = 820 nM, PI3K-β = 565 nM, PI3K-γ = 89 nM). Selectivity (400- to 4000-fold) was also seen against related kinases C2b, hVPS34, DNA-PK and mTOR. No significant activity was observed against a panel of 402 diverse kinases at 10 μM.

TargeGen claims compound TG100-115 (**64**) to be a PI3K-γ–δ dual inhibitor (IC$_{50}$ values: PI3K-γ = 83 nM, PI3K-δ = 235 nM, PI3K-β = 1200 nM, PI3K-α = 1300 nM) that was selective *versus* 136 protein kinases also tested.[115] TG100-115 contains two metabolically labile phenol groups, making it suitable for treatment of illnesses such as myocardial infarction, which require treatments with quick onset of action and rapid clearance. In a rat model of myocardial infarction, an i.v. bolus of TG100-115 (0.5 mg kg^{-1}) 60 minutes post-reperfusion gave a 50% reduction in the infarct area after 22 h.[116] This compound entered the clinic in 2005 for the treatment of patients who suffer a heart attack and then undergo angioplasty.

Researchers at Serono in 2006 published an article on the discovery of AS-252424 (**65**), a potent and selective PI3K-γ inhibitor (IC$_{50}$: PI3K-γ = 30 nM, PI3K-α = 940 nM, PI3K-β = 20 000 nM, PI3K-δ = 20 000 nM).[117] This compound was also selective *versus* 79 of 80 protein kinases tested (CK2 IC$_{50}$ = 20 nM). X-Ray co-crystallization with PI3K-γ revealed that the binding relied on a key salt bridge between the enzyme (Lys-833, protonated under physiological conditions) and the NH of the thiazolidinedione core (which is expected to be deprotonated) (Figure 6.18). In addition, the hydroxy group of the phenol forms the hinge region interaction with the backbone amide of Val-882. It should be noted that these two interactions are again similar to those observed for the prototypical inhibitors wortmannin and LY294002. Despite its short half-life and high clearance, oral dosing of AS-252424 at 10 mg kg^{-1} produced a decrease in leukocyte infiltration (35%) in a murine peritonitis model similar to that seen in PI3K-γ deficient mice.

6.4.1.3 mTOR Inhibitors

As mentioned, due to the homology between mTOR and PI3K, many inhibitors will affect both enzymes. However, since mTOR is located downstream of PI3K it has been postulated that selective inhibition of mTOR may also lead to effective cancer therapeutics while minimizing the side effects associated with pan-PI3K inhibition. Selective mTOR inhibitors may be divided into two classes. The first class consists of the allosteric inhibitors derived from rapamycin, which selectively inhibit mTORC1 by binding to FKBP12 and

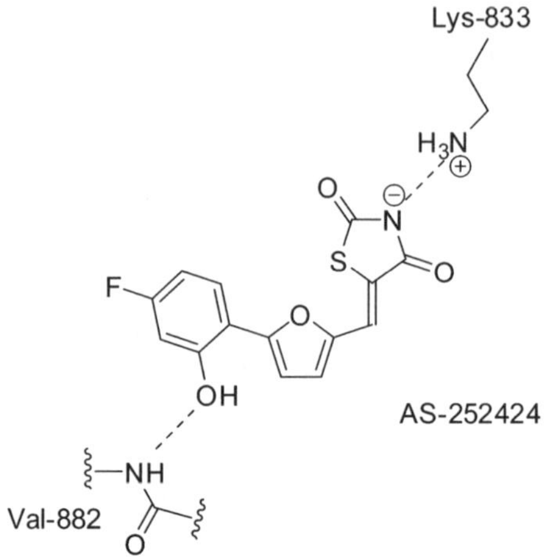

Figure 6.18   Schematic representation of the key interactions involved in binding of AS-252424 to PI3K-γ. The phenol forms the hinge region interaction with the Val-882 backbone NH. A charged interaction between Lys-833 and the thiazolidinedione contributes to potency.

preventing the association of FKBP12 with the FRB domain of mTOR. Two members of this class are currently approved drugs: everolimus (Novartis) for renal-cell carcinoma (RCC), pancreatic neuroendocrine tumors and subependymal giant cell astrocytomas (SEGAs) associated with tuberous sclerosis patients, and temsirolimus (Pfizer) for RCC and mantle-cell lymphoma. These two compounds, as well as rapamycin (sirolimus, Pfizer), ridaforolimus (Merck–Ariad), and ABI-009 (Abraxis Bioscience), are currently undergoing clinical trials for multiple cancers. The use of rapamycin and its analogs as anticancer drugs has been extensively reviewed.[118,119]

The second class of mTOR inhibitors is comprised of small molecules that directly interact at the ATP-binding site of mTOR. In contrast to rapamycin and its analogs, these compounds inhibit both functional complexes of mTOR, mTORC1 and mTORC2. The dual inhibition of both complexes may have potential clinical benefits, as inhibition solely of mTORC1 has been correlated with a negative feedback mechanism,[120] and there is evidence that certain mTORC1 functions are not fully inhibited by sirolimus.[121] Relative to the rapalogs, the members of this second class of mTOR compounds are a newer entry to the field and have only recently begun clinical trials.

Researchers at KuDOS Pharmaceuticals Ltd (now AstraZeneca) have developed pyridopyrimidines that function as selective mTOR inhibitors. A compound from this class, AZD8055 ((**66**), Figure 6.21, AstraZeneca), entered the clinic in 2008 and has been studied in advanced solid tumors as well as hepatocellular carcinoma. AZD8055 exhibited more than 1000-fold selectivity for mTOR over all PI3K isoforms, and demonstrated no appreciable activity against a panel of 260 kinases.[122] A Phase I trial of a second mTOR inhibitor, AZD2014 (AstraZeneca, structure not disclosed), was initiated at the end of 2009.

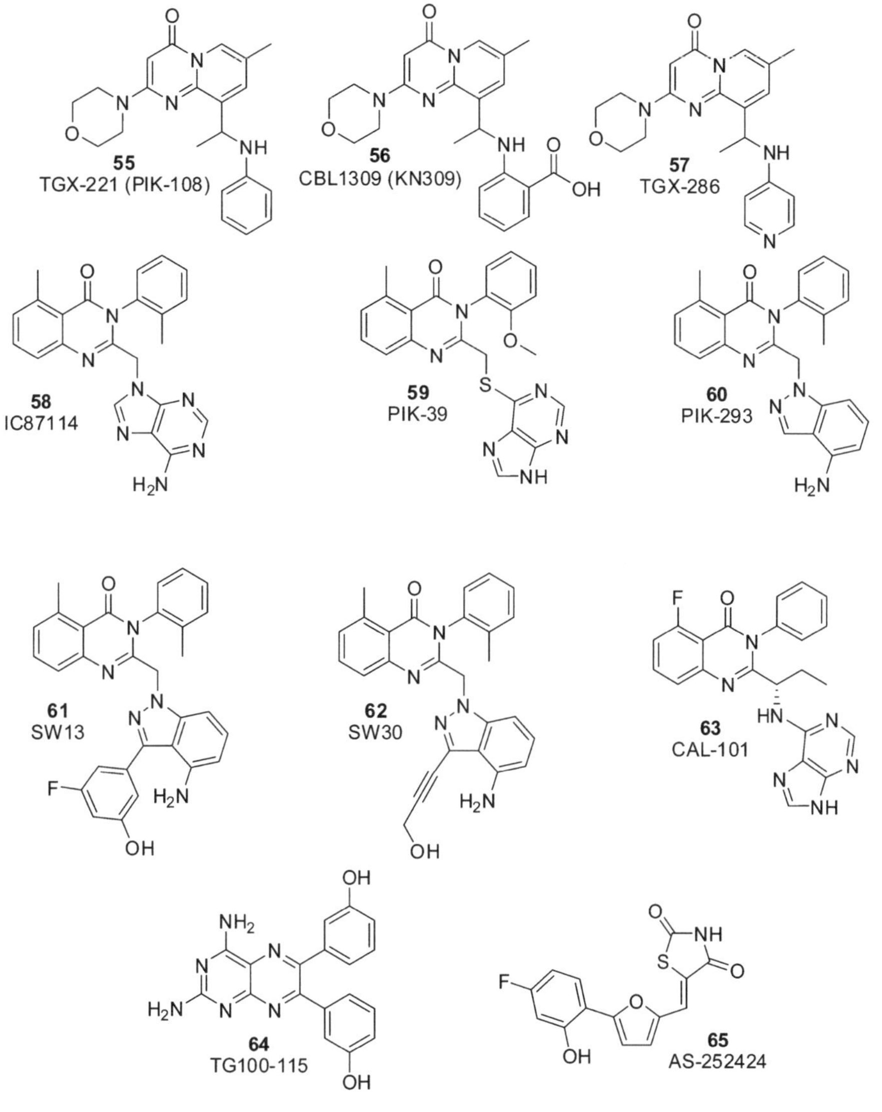

Figure 6.19 Inhibitors of other PI3K isoforms.

The selective mTOR inhibitor OSI-027 (**67**) from OSI Pharmaceuticals (now Astellas) is undergoing a Phase I clinical trial in patients with advanced solid tumors or lymphoma.[123] A comparison study of this compound with rapamycin showed that while the mTORC1 inhibitor rapamycin only partially inhibited phosphorylation of 4E-BP1, OSI-027 fully inhibited this process. OSI-027 also proved more effective than rapamycin in induction of apoptosis in a variety of tumor cell lines.[124] Other studies demonstrated that OSI-027 is effective in chronic myeloid leukemia (CML) patients with a particular

BCR-ABL mutation, providing evidence for the involvement of TORC1 and TORC2 in survival and growth of such cells.[125]

A third selective mTOR inhibitor that is currently being investigated in the clinic is INK128 (Intellikine, structure not disclosed), which entered Phase I trials in late 2009; a second study in multiple myeloma began in 2010. This compound inhibits mTOR with an IC_{50} value of 1.6 nM and displays moderate to excellent selectivity for mTOR over the various PI3K isoforms (IC_{50} values of 172 to 296 nM for PI3Kα, PI3Kβ and PI3Kγ, and 6000 nM for PI3Kδ). INK128 demonstrated *in vivo* efficacy in mice following once-daily oral doses of 1 to 3 mg kg^{-1}, and displayed *in vivo* inhibition of mTOR signaling at 8 h following a 1 mg kg^{-1} dose.[126] A recent presentation led to the conclusion that INK128 displayed more potent *in vitro* anti-leukemic efficacy than rapamycin.[127]

A final mTOR inhibitor in clinical trials for cancer is CC-223 (Celgene). No data on this compound has been disclosed to date. Although all mTOR inhibitors discussed above are under investigation for oncology, inhibition of this pathway is also being investigated in at least one other therapeutic area. Palomid 529 ((**68**), Paloma Pharmaceuticals) appears to disrupt the complex formation of both mTORC1 and mTORC2.[128] A clinical trial of Palomid 529 for age-related macular degeneration began in 2010. This compound also provided a synergistic effect *in vitro* in combination with radiotherapy in a prostate cancer model.[129]

In addition to the selective mTOR inhibitors described above which have progressed to the clinical stage, a number of additional compounds have been investigated preclinically.[130,131] The structure of one such compound, Torin1 (**69**), has been recently disclosed.[132] A screening campaign identified a quinoline derivative which was subsequently modified to a tricyclic scaffold, resulting in compounds with dramatically improved mTOR activity. This inhibitor displayed potency against mTOR with an IC_{50} value of 0.3 nM, showed 1000-fold selectivity over class I PI3Ks, and had a minimum efficacious dose of 20 mg kg^{-1} in a U87MG mouse xenograft model while displaying inhibition of relevant mTOR biomarkers.

Researchers at Wyeth (now Pfizer) have described highly selective inhibitors of mTOR. The incorporation of bridged morpholine derivatives on a pyrazolopyrimidine core resulted in compounds such as WYE-125132 (**70**) which inhibited mTOR with an IC_{50} value of 0.2 nM and was more than 5000-fold selective for mTOR over all PI3K isoforms.[133] The selectivity of this class of compounds was explained by the presence of a leucine moiety in mTOR adjacent to the hinge region valine; this residue is a phenylalanine in PI3K-α. WYE-125132 demonstrated efficacy in tumor xenograft mouse models for multiple cancers, and displayed a synergistic effect when co-administered with the anti-VEGF drug bevacizumab.[134] The co-crystal structure of WYE-125132 analog (**71**) with PI3K-γ revealed that the potency of this class of inhibitors stems from the formation of three hydrogen bonds between the ureido group and the side chains of Lys-833 (*cf.* wortmannin, LY294005, GSK2126458,

AS-252424) and Asp-841 (*cf.* GDC-0941, GNE-477), in addition to the hinge region interaction between the morpholine oxygen and the backbone NH of Val-882 (Figure 6.20, panel A).[99] The co-crystal structure also revealed that binding of an arylureido group induced a conformational shift in the protein, exposing the 4-position of the arylureido moiety to solvent (Figure 6.20, panel B). Introduction of water-solubilizing groups at this position resulted in an enhancement in cellular potency, giving compounds that showed subnanomolar inhibition of proliferation of MDA361 breast cancer and LNCAP prostate cancer cell lines.[135]

The pyrazolopyrimidine inhibitor PP242 ((**72**), Figure 6.21) is a member of a third class of selective mTOR inhibitors that has recently been described. PP242 inhibited mTOR with an IC_{50} value of 8 nM, and also displayed selectivity for mTOR over PI3K-α (IC_{50} = 2 μM).[136] This compound exhibited activity against certain mouse and human leukemia cell lines that were resistant to sirolimus.[137] Such findings provide evidence that dual mTORC1–mTORC2 inhibitors behave significantly differently than rapamycin and its analogs, highlighting the importance of continued studies with such compounds.

6.4.1.4 Other PIKKs

The phosphoinositide 3-kinase (PI3K) related kinases (PIKKs) are atypical serine–threonine protein kinases whose ATP binding sites are structurally similar to the PI3Ks. The members of this family of kinases are DNA-dependent protein kinase (DNA-PK), ataxia telangiectasia mutated (ATM), ataxia telangiectasia and rad3 related (ATR), hSMG-1 and the mammalian target of rapamycin (mTOR). As a result of the homology between the ATP binding pockets of the two classes of enzymes, design of inhibitors of the PI3Ks must take into account selectivity *versus* PIKKs. As mentioned above, a number of PI3K inhibitors recently entering the clinic have potent mTOR activity. After mTOR, the most active area for inhibitor discovery has been DNA-PK. Ionizing radiation and DNA damaging drugs (*e.g.* topoisomerase inhibitors) create double strand breaks that lead to cell death. DNA damage results in upregulation of DNA-PK, ATM and ATR. DNA-PK controls the repair process of double stranded DNA breaks, a process called non-homologous end joining, leading to cell survival. ATM and ATR activate the checkpoint kinases Chk1 and Chk2 leading to cell cycle delay and increasing the opportunity for cells to carry out DNA repair processes. Thus, use of inhibitors of DNA-PK, ATM and ATR would act as radio- and chemo-sensitizers and potentially require lower dosing of these therapies resulting in increased tolerability.[138]

Several research groups have designed DNA-PK inhibitors that are analogs of the broad-spectrum kinase inhibitor LY294002, a low μM IC_{50} inhibitor of the class I PI3Ks, mTOR and DNA-PK, but not ATM or ATR. In a study of arylmorpholines as PI3K inhibitors, Shokat's group at UCSF identified

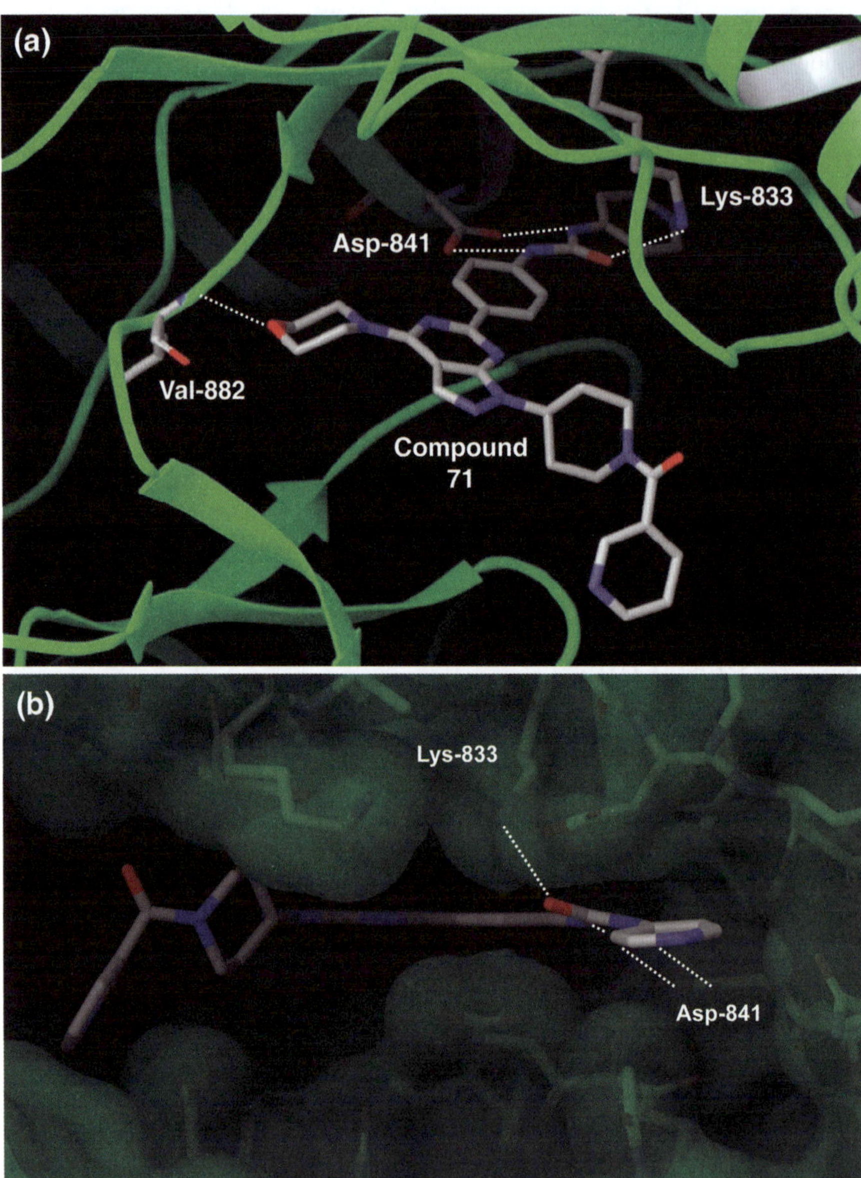

Figure 6.20   Co-crystal structure of WYE-125132 analog **71** bound to PI3K-γ. Panel A: The morpholine oxygen forms the hinge region interaction with the backbone NH of Val-882, while the urea group engages in a triple H-bond with the Lys-833 and Asp-841 side chains. Panel B: alternate view (rotated approximately 90 degrees from panel A) with surface representation, illustrating how binding of the pyridinyl urea induced a conformational shift in the enzyme, exposing the 4-position of the pyridinyl ring to solvent.

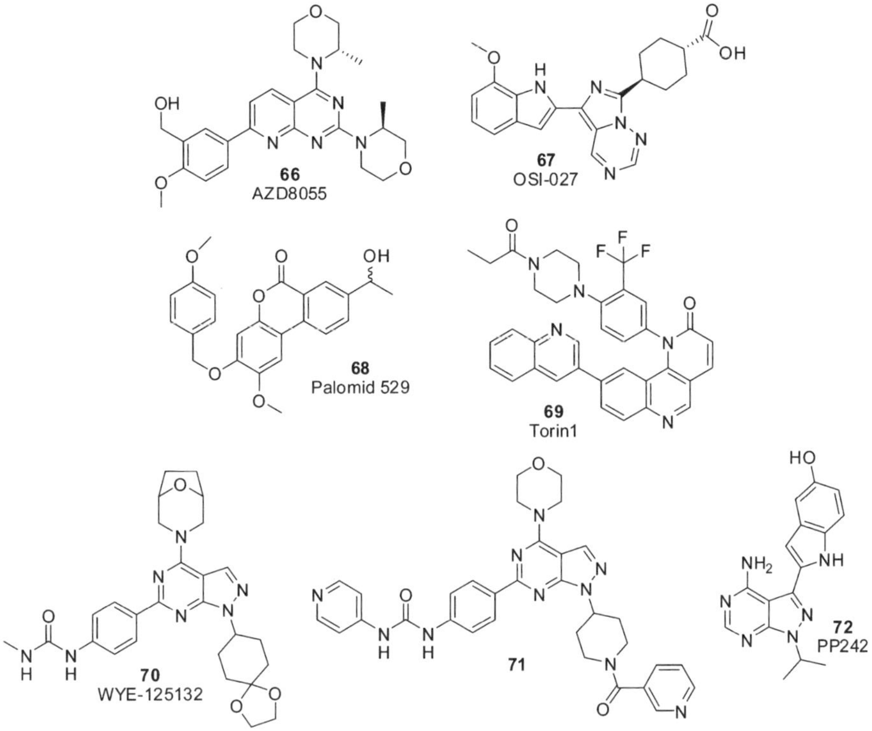

Figure 6.21 Structures of selective mTOR inhibitors.

AMA37 ((**73**), Figure 6.22) as a selective DNA-PK inhibitor with 10-fold specificity relative to p110β and 100-fold specificity relative to the other class I PI3Ks, ATM, ATR and mTOR.[139] Researchers at University of Newcastle and KuDOS Pharmaceuticals elaborated the chromenone ring system found in LY294002, and in a series of publications describe the development of SAR and the ultimate discovery of NU7441 (**74**), a potent and selective DNA-PK inhibitor.[140] NU7441 has an $IC_{50} = 14$ nM for DNA-PK and IC_{50} values of > 1 μM for mTOR, 5 μM for PI3K, and > 100 μM for ATM and ATR. The high DNA-PK selectivity of NU7441 was also confirmed against a commercial panel of 60 diverse kinases, with no inhibitory activity being observed at an inhibitor concentration of 10 μM. Preclinical evaluation in *in vitro* and *in vivo* models of chemosensitization and radiosensitization demonstrated sufficient proof of principle for the use of DNA-PK inhibitors in the clinic. However, the limited solubility and bioavailability of NU7441 precluded its further development.[141] Further work to improve the physicochemical properties of NU7441 through incorporation of water solubilizing groups resulted in (**75**). SAR studies showed that the chromenone ring in NU7441 could be replaced with a 4H-pyrido[1,2-a]pyrimidin-4-one ring and that a water solubilizing group could be attached at the 1-position of the dibenzothiophene ring. Compound (**75**) had an $IC_{50} = 8$ nM for DNA-PK and at 100 nM exhibited

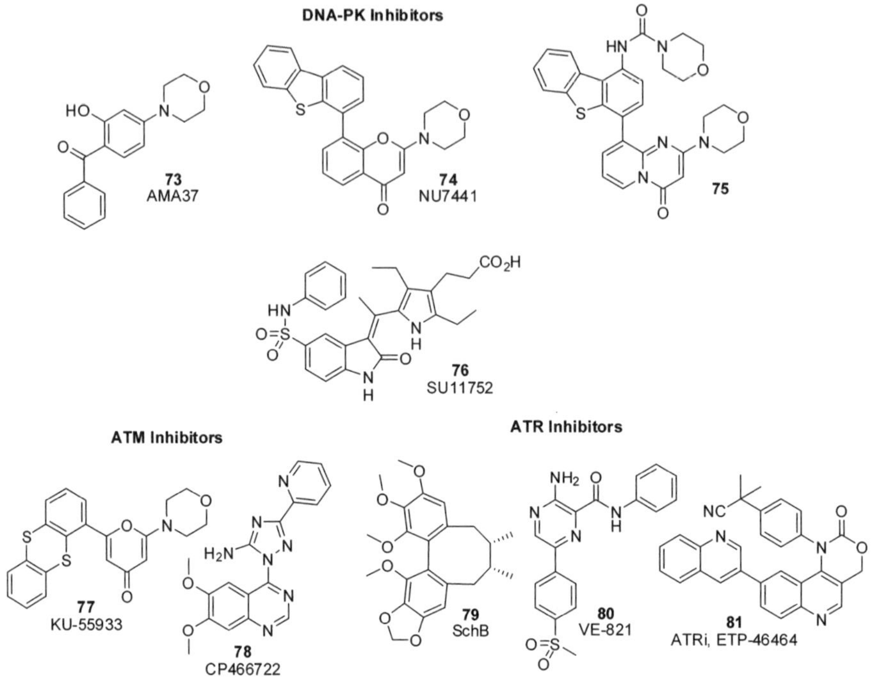

Figure 6.22   Other PIKK inhibitors.

excellent potentiation of ionizing radiation (IR) induced cytotoxicity in the HeLa tumor cell line (dose modification ratio (DMR) = 3.5 at IR = 2 Gy).[142] Researchers at Sugen identified SU11752 (**76**), a selective DNA-PK inhibitor structurally different from AMA37 or NU7441. SU11752 inhibited DNA-PK with an IC_{50} = 140 nM and showed selectivity *versus* PI3K and ATM.[143]

KuDOS Pharmaceuticals researchers screened a combinatorial library based around the nonspecific PI3K and DNA-PK inhibitor LY294002 to find KU-55933 (**77**), a potent and selective inhibitor of ATM. KU-55933 inhibits ATM with an IC_{50} of 13 nmol L^{-1} and a K_i of 2.2 nmol L^{-1}. KU-55933 shows significant specificity with respect to inhibition of other PIKKs (DNA-PK, ATM, ATR, mTOR) and PI3K. Cellular inhibition of ATM by KU-55933 was demonstrated by the reduction of ionizing radiation-dependent phosphorylation of a range of ATM targets. Exposure of cells to KU-55933 resulted in a significant sensitization to the cytotoxic effects of ionizing radiation and to the DNA double-strand break-inducing chemotherapeutic agents etoposide, doxorubicin, and camptothecin.[144] Recent studies with KU-55933 show that in addition to inhibiting nuclear ATM, it also inhibits cytoplasmic ATM resulting in inhibition of the phosphorylation of Akt and leading to inhibition of cellular proliferation through induction of G1 arrest. KU-55933 also inhibits the feedback phosphorylation of Akt induced by treatment with the allosteric mTORC1 inhibitor rapamycin. Combination of KU-55933 and rapamycin led to apoptosis and increased inhibition of cellular proliferation.[145] Researchers at St. Jude Children's Hospital and Pfizer screened a targeted library of 1500 compounds and identified CP466722 (**78**), an inhibitor of ATM that acts rapidly and reversibly. CP466722 did not inhibit PI3K or other PIKKs. As with the other ATM inhibitors, CP466722 inhibited ATM signal transduction, disrupted cell cycle checkpoint function, and sensitized tumor cells to ionizing radiation.[146]

More recently, inhibitors of ATR have been reported. Schisandrin B (SchB, (**79**)), a natural product isolated from Fructus schisandrae, was shown to inhibit ATR activity with an IC_{50} = 7.25 μM and not to inhibit ATM, Chk1, PI3K, DNA-PK and mTOR. SchB in ATM-deficient cells reduced UV-induced phosphorylation of p53 and Chk1.[147] Vertex has disclosed ATR inhibitors,[148] including VE-821 (**80**), which inhibited ATR with an IC_{50} of 26 nM while being selective over ATM (> 8 μM) and DNA-PK (4.3 μM). VE-821 showed strong synergy with genotoxic agents as well as ionizing radiation.[149] Toledo *et al.* described a cell-based screen for the identification of ATR inhibitors and showed that the dual PI3K–mTOR inhibitor BEZ235 ((**48**), Figure 6.16) and a related compound (ETP-46464, ATRi, (**81**), Figure 6.22) also potently inhibited ATR.[150]

6.4.2 Other Phosphatidylinositol Kinases

6.4.2.1 *Inositol 1,4,5-trisphosphate 3-kinase (Itpk)*

Itpk converts the second messenger inositol 1,4,5-trisphosphate (IP_3) into the soluble inositol-polyphosphate inositol 1,3,4,5-tetrakisphosphate (IP_4). Relatively little is known about the physiological role of IP_4. Selective and specific ItpkB inhibitors have been proposed for the treatment of B-cell-dependent immune disorders such as systemic lupus erythematosus (SLE) or allergies, and would be expected to give fewer side effects than currently available immunosuppressants.[151] However, the design of selective inhibitors may be hard to achieve due to the conserved ATP binding region.[152] The limited efforts to design inhibitors have been reviewed.[153]

Most efforts have centered on inositol analogs as pharmacological tools from the groups of Barry Potter,[153–155] and Sung-Kee Chung.[156–158] Micro-molar inhibitors were identified when a purine library was screened for compounds that targeted the ATP binding site rather than the inositol binding site.[159] One of these inhibitors was later also identified as an inositol hexakisphosphate inhibitor.[160] Polyphenolic compounds were also disclosed as pharmacological tools, constituting the most potent inhibitors to date.[161] These compounds also inhibit DNA polymerase and Topo II.

The patent literature contains several additional applications. Exelixis has claimed methods for modulating the IGFR pathway by controlling Itpk activity and described screens for the identification of modulators of Itpk.[162] IRM LLC has published three patent applications for ItpkB selective inhibitors. Limited activity data was provided but selected compounds were claimed to have IC_{50} values between 0.5 and 5 nM.[163–165]

6.4.2.2 *Phosphatidylinositol 4-kinase (PI4K)*

The first committed step in the synthesis of polyphosphoinositides is the formation of phosphatidylinositol 4-phosphate catalysed by phosphatidylinositol 4-kinase. Young and co-workers have described inhibitors of human erythrocyte PI4K based on adenine derivatives ((**82**), Figure 6.23, IC_{50}: 3.7 μM),[166] as well as inositol derivatives.[167] Shi *et al.* from Eli Lilly prepared echiguanines (**83**) as inhibitors of PI4K.[168] Additional echiguanine derivatives were reported.[169,170] Ribofuranosyl derivatives of the echiguanines have also been described, although the presence of the ribose group led to decreased activity.[171] The physiological relevance of PI4K inhibition is not yet fully established.

6.4.2.3 *Phosphatidylinositol 4-phosphate 5-kinase (PIP5K)*

PIP5K produces the versatile phospholipid phosphatidylinositol 4,5-bisphosphate (PI4,5P(2)). The physiological functions of the various PIP5K isoforms are not yet fully characterized. In an abstract presented at the AACR in 2009, Treinies reported on a novel PIP5K assay developed by CRT Discovery

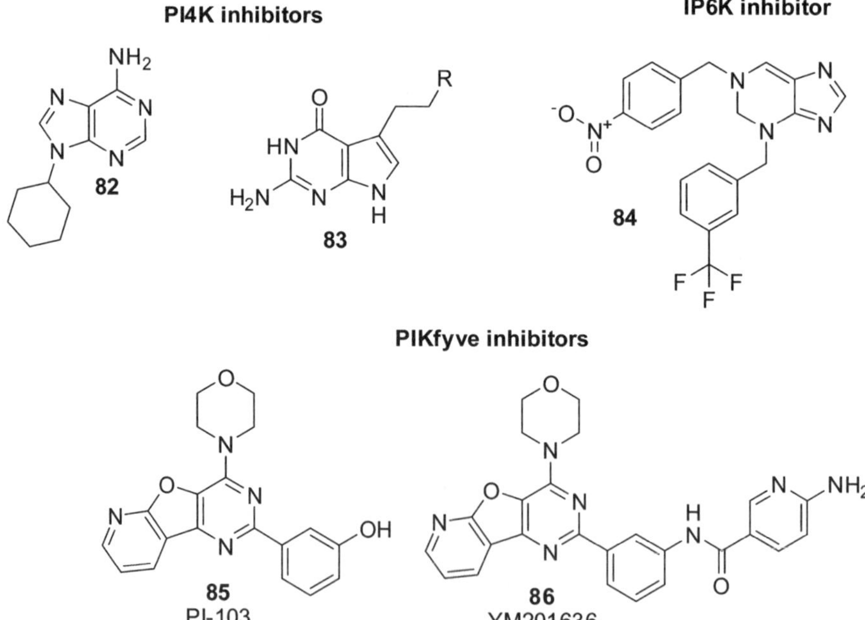

Figure 6.23   Inhibitors of other phosphatidyl inositol kinases.

Laboratories.[172] A high throughput screen led to the identification of two distinct chemical series that have since entered a hit to lead program. The compounds inhibited PIP5K with nanomolar potencies and were selective against a panel of protein kinases including PI3K.

6.4.2.4   Inositol Hexakisphosphate Kinase (IP6K)

IP6K phosphorylates inositol hexakisphosphate (InsP$_6$) to diphosphoinositol pentakisphosphate, a "high energy" candidate regulator of cellular trafficking. Compound (**84**), originally identified as an Itpk inhibitor, was shown to be a relatively selective and reversible inhibitor of IP6K.[160]

6.4.2.5   PIKfyve

PIKfyve is a mammalian class III phosphatidylinositol phosphate kinase which synthesizes phosphatidylinositol 3,5-bisphosphate. The physiological role of this pathway is not yet clear. The PI3K inhibitor PI-103 (**85**) also inhibits PIKfyve. A structural analog, YM201636 (**86**), was shown to be selective for PIKfyve.[173,174]

6.4.3   Other Lipid Kinases

6.4.3.1   Diacylglycerol Kinase (DGK)

Diacylglycerol kinases (DGKs) attenuate diacylglycerol-induced protein kinase C activation during stimulated phosphatidylinositol turnover and initiate phosphatidylinositol resynthesis. Janssen Pharmaceutica (now Johnson & Johnson) developed two inhibitors that have found widespread use as pharmacological tools: R59022,[175] and R59949.[176] The selectivity of R59949 for various diacylglycerol kinase subtypes has been investigated.[177] The effects of another inhibitor, stemphone, on mouse portal vein contraction were described and the use of DGK inhibitors as novel medications for hypervascular contraction was proposed.[178] The therapeutic potential of DGK inhibitors for various diseases has been reviewed.[179]

6.4.3.2   Choline Kinase (CHK)

Choline kinase phosphorylates choline to give a phosphocholine and participates in glycine, serine and threonine metabolism and glycerophospholipid metabolism. Hemicholinium-7 is the prototypical tool compound used to inhibit CHK. Based on inhibitor studies, it has been proposed that CHK is important for the regulation of cell proliferation. Inhibition of choline kinase is also used to target plasmodium and develop novel antimalarials. A series of papers on pyridinium based inhibitors have been published,[180–197] but no disclosures of more drug-like molecules have been made.

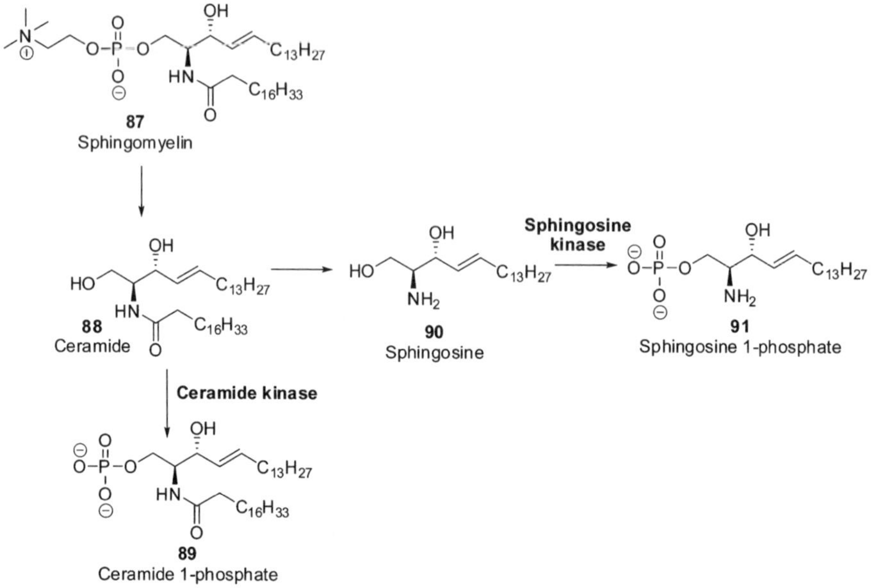

Figure 6.24 Biosynthesis of sphingosine- and ceramide-phosphate.

6.4.3.3 *Ceramide Kinase (CERK)*

Ceramide ((**88**), Figure 6.24) is a bioactive metabolite of sphingomyelin (**87**) and plays important roles in the regulation of cell proliferation, survival and apoptosis. Phosphorylation of ceramide by CERK gives ceramide 1-phosphate (**89**), a mediator of mast cell degranulation. The biological role of CERK and its potential as a drug target have been reviewed.[198,199]

Some sphingosine kinase inhibitors show moderate affinity for CERK. Kim *et al.* explored analogs of the sphingosine kinase inhibitor F-12509A ((**95**), Figure 6.26) and discovered K1 ((**92**), Figure 6.25), a novel CERK inhibitor which has no effects on sphingosine kinase and diacylglycerol kinase activity.[200–203] They also showed that CERK inhibition suppresses IgE–antigen-induced mast cell degranulation, suggesting therapeutic potential of CERK inhibitors for the treatment of allergic diseases.

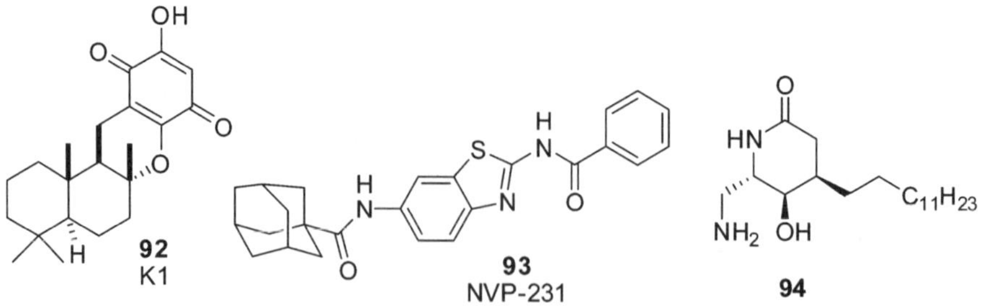

Figure 6.25 CERK inhibitors.

A team at Exelixis developed an HTS assay to identify CERK inhibitors and reported the first inhibitors with nanomolar activity.[204] They identified four scaffolds, with three of them having inhibitors with IC_{50} values of ~50 nM. This included a non-ATP-competitive inhibitor as well. No structures were disclosed.

Researchers at Novartis discovered the potent CERK inhibitor NVP-231 ((**93**), IC_{50} = 12 nM),[205,206] and also described a secondary CERK assay.[207] Mathew *et al.* explored conformationally restricted ceramide analogs as inhibitors and reported a sub-micromolar inhibitor (**94**).[208] Method of use patent applications have also appeared, but no inhibitors were disclosed.[209,210]

6.4.3.4 Sphingosine Kinase Inhibitors

Sphingosine kinase resides in the same signaling pathway as ceramide kinase (Figure 6.24), and converts sphingosine (**90**) into the important signaling molecule sphingosine 1-phosphate (**91**). Two sphingosine kinase isoforms are known, SK1 and SK2. Although both have divergent biological roles, each isoform is believed to have an anti-apoptotic role and inhibition of either isoform has received significant attention for the development of anticancer drugs. The biological role and therapeutic potential of the sphingosine kinases have recently been reviewed.[211,212]

Kono *et al.* identified several natural products as micromolar sphingosine kinase inhibitors. F-12509a ((**95**), Figure 6.26) was isolated from the fungus *Trichopezizella barbata.* This inhibitor was shown to be competitive with respect to sphingosine.[213] As mentioned, F-12509a also had weak affinity for CERK and was used as a starting point for the development of CERK inhibitors. B5354c (**96**) was isolated from the marine bacterium SANK 71896, and inhibited sphingosine kinase in a non-competitive manner with respect to sphingosine.[214] S-15183a (**97**) and S-15183b (**98**) were isolated from the fungus *Zopfiella inermis* and were shown to be cell permeable SK inhibitors with selectivity over PKC, PI3K and neutral sphingomyelinase.[215]

Synthetic inhibitors of sphingosine kinase have also been prepared. Several sphingosine analogs were found to be inhibitors. For instance, *N,N*-di-methyl sphingosine (DMS, (**99**)) selectively inhibited sphingosine kinase, but not protein kinase C.[216] This compound has found widespread use as a pharmacological tool to elucidate the role of the sphingosine kinases. De Jonghe *et al.* explored sphingosine analogs in which the alkyl chain was shortened. They found that replacement of the alkyl chain by an aromatic residue or replacement of the 3-hydroxyl group with a fluorine atom led to sphingosine kinase inhibitors that were more potent than DMS (*cf.* compounds (**100**), (**101**), (**102**)).[217] Based on these findings, Paugh *et al.* designed SK1-I (sphingosine kinase 1-inhibitor, also known as BML-258, (**103**)); not to be confused with SKI-I discussed below.[218] The stereochemistry of the compound as depicted in Figure 6.26 reflects the stereochemistry reported in the original publication (2*R*,3*S*). However, it should be noted that the original publication

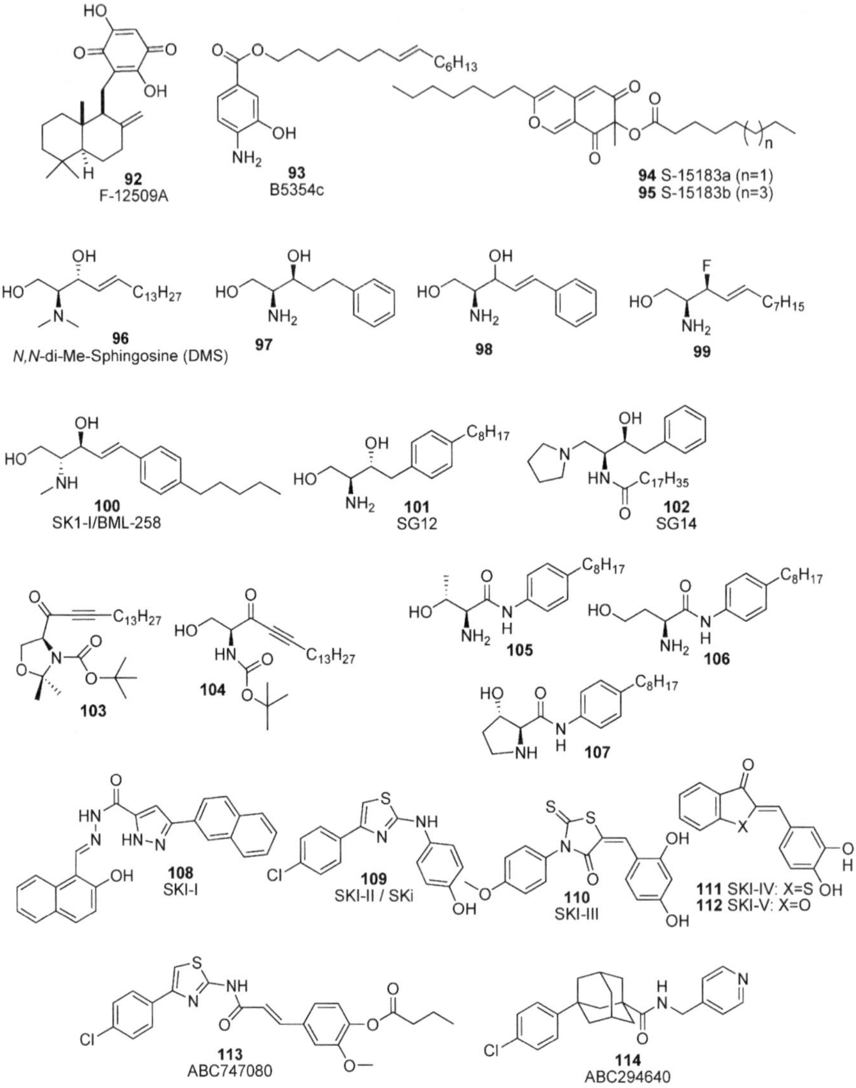

Figure 6.26 Inhibitors of sphingosine kinase.

depicts the same stereochemistry for sphingosine (2R,3S), whereas the natural product's configuration is (2S,3R). Therefore it is unclear at this time whether SK1-I is the reported (2R,3S) compound or has the same stereochemistry as sphingosine (2S,3R). SK1-I selectively inhibited SK1 (K_i 10 μM) while not inhibiting SK2 or ceramide kinase. In addition, SK1-I did not significantly inhibit PKCα and PKCδ, PKA, Akt1, ERK1, EGFR, CDK2, IKKβ, or CamKIIβ. SK1-I inhibited growth of human leukemia cells and induced apoptosis through reduction of sphingosine 1-phosphate levels. Administration of

SK1-I (ip, 20 mg kg^{-1} day^{-1}) to nude mice carrying U937 human leukemia xenografts resulted in reduced tumor size.

Kim *et al.* prepared 16 sphingoid analogs and screened them for inhibition of SK1 and SK2.[219] Two of the prepared compounds selectively inhibited SK2 to a greater extent than SK1. SG12 (**104**) possessed an IC$_{50}$ of 22 μM, while SG14 (**105**) had an IC$_{50}$ of 4 μM against SK2. SG14 was selective over PKC as well, and inhibited SK2 in cellular assays. Alkyne containing sphingosine analogs have also been explored.[220] The Boc-protected oxazolidine (**106**) had an IC$_{50}$ of 3.3 μM against sphingosine kinase 1, whereas the Boc-protected aminoethanol (**107**) had an IC$_{50}$ of 1.2 μM against this isozyme. Compound (**106**) was selective for the sphingosine kinase 1 isozyme, whereas (**107**) inhibited both isoforms.

Replacement of the aminodiol headpiece of SK1-I with a serine amide led to a novel scaffold.[221] Introduction of an extra methyl group (*i.e.* replacing serine with threonine) resulted in the first sub-micromolar sphingosine kinase inhibitor. Compound (**108**) had an IC$_{50}$ of 0.65 μM. Extending the distance between the terminal alcohol and the amide functionality (*i.e.* by replacing serine with homoserine) led to a further increase in potency. Thus, *S* enantiomer (**109**) had an IC$_{50}$ of 40 nM. The stereochemistry was crucial, as demonstrated by the significantly decreased inhibitory potency of the *R* isomer (2.2 μM). A conformationally restricted analog ((**110**), resulting from replacement of serine with 3-hydroxyproline) also demonstrated good potency (IC$_{50}$ 62 nM).

As outlined, most sphingosine kinase inhibitors reported to date have been derived from the natural lipid substrate. French *et al.* screened a library of ~16 000 compounds for non-lipid inhibitors of SK1.[222] Four distinct chemotypes were identified, as exemplified by SKI-I (**111**), -II (**112**), -III (**113**) and -IV (**114**) (Sphingosine Kinase Inhibitor -I, -II, -III and -IV). An analog of SKI-IV was synthesized as well (SKI-V, (**115**)). These novel inhibitors were not competitive inhibitors of the ATP-binding site of sphingosine kinase. Of these five inhibitors, SKI-II showed moderately potent inhibition of sphingosine kinase (IC$_{50}$ 0.5 μM) while exhibiting selectivity over ERK2, PI3K and PKCα. All five compounds inhibited sphingosine kinase activity in tumor cells, inhibited tumor cell proliferation and induced apoptosis. SKI-V was evaluated in an *in vivo* cancer model and was shown to inhibit tumor growth. SKI-II is now commercially available (often referred to as SKi, Sphingosine Kinase inhibitor) and is widely used as a research tool.

A team at Apogee identified two novel orally bioavailable sphingosine kinase inhibitors, ABC747080 (**116**) and ABC294640 (**117**), with *in vivo* SK inhibitory activity.[223] Analysis of ~140 aryladamantanes, analogs of a library screen hit, revealed that ABC294640 had excellent solubility and oral absorption and this compound was selected for further characterization.[224] ABC294640 selectively inhibited SK2 with a K_i of 9.8 μM while being selective over SK1 and a panel of 20 kinases. This compound was a competitive inhibitor with respect to sphingosine. ABC294640 also showed *in vivo* antitumor activity in a syngeneic tumor model that uses the mouse JC mammary adenocarcimona cell line in immunocompetent mice. An IND for the treatment of solid tumors with this compound was approved in 2010.

6.4.4 Conclusion

The majority of drug discovery work on lipid kinases has focused on the development of inhibitors of PI3K and the structurally related serine threonine protein kinase mTOR. Although details of the PI3K–mTOR signaling pathway have only recently been elucidated, we already have seen a large number of inhibitors enter clinical trials for the treatment of cancer. This approach has been validated by the approved cancer therapeutics temsirolimus and everolimus, allosteric inhibitors of mTOR. The physiological roles of other phosphatidylinositol kinases are not yet fully known. As more details about the role of these kinases come to light, it is conceivable that some of these enzymes will become attractive drug targets. Likewise, the physiological roles of other lipid kinases have not been fully elucidated. Nonetheless, several lipid kinases have been proposed as therapeutic targets. The interest in these kinases is likely to increase further as we gain more detailed knowledge about their role in cell signaling.

6.5 Other Non-Protein Kinases

6.5.1 Mevalonate Kinase

Mevalonate ((**118**), Figure 6.27) is a key intermediate, and mevalonate kinase is a key early enzyme, in isoprenoid and sterol synthesis. Inhibitors of this enzyme have potential applications for treatment of cardiovascular disease and cancer. Mevalonate kinase activity is controlled post-transcriptionally *via* competitive inhibition at the ATP site by prenyl phosphates, such as geranyl diphosphate (**119**). A bifunctional inhibitor with micromolar IC_{50} values against mevalonate kinase and mevalonate 5-diphosphate decarboxylase (a

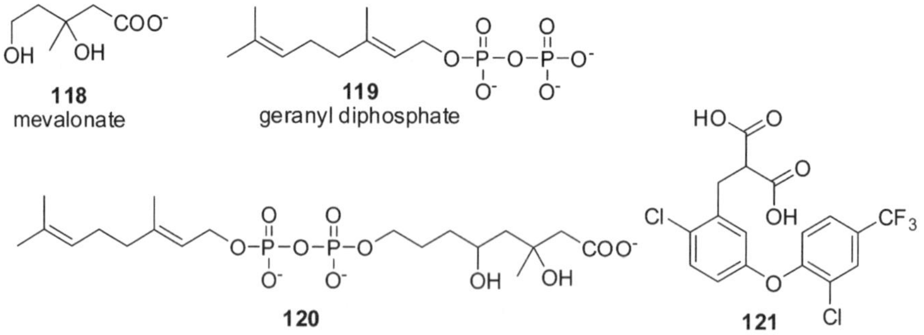

Figure 6.27 Mevalonate kinase: substrates and inhibitors.

downstream enzyme in the same cholesterol biosynthesis pathway) was reported (**120**).[225] This compound combines the structure of the post-translational regulator geranyl diphosphate with that of the substrate mevalonate. Mevalonate kinase inhibitors have also been claimed for fungicide applications; the structure of one of these analogs has been disclosed (**121**).[226,227]

6.5.2 Pyruvate Kinase (PK)

Pyruvate kinase plays an essential role in glycolysis, by catalyzing the transfer of a phosphate group from phosphoenolpyruvate (((**122**), Figure 6.28) to ADP, resulting in the formation of ATP and pyruvate (**123**). Tumor cells express the pyruvate kinase subtype M2, which can exist in the dimeric form (low affinity for its substrate PEP) or tetrameric form (high affinity for PEP). Glycolysis rates are controlled by the transition between dimeric and tetrameric forms of M2-PK, affecting tumor cell proliferation and survival. The dimeric form of the M2 isoenzyme is known as Tumor M2-PK and is a clinically used biomarker to screen for cancer.

Synthetic peptide aptamers were used to bind to M2-PK and shift the isoenzyme into the low affinity dimeric conformation.[228] A high throughput

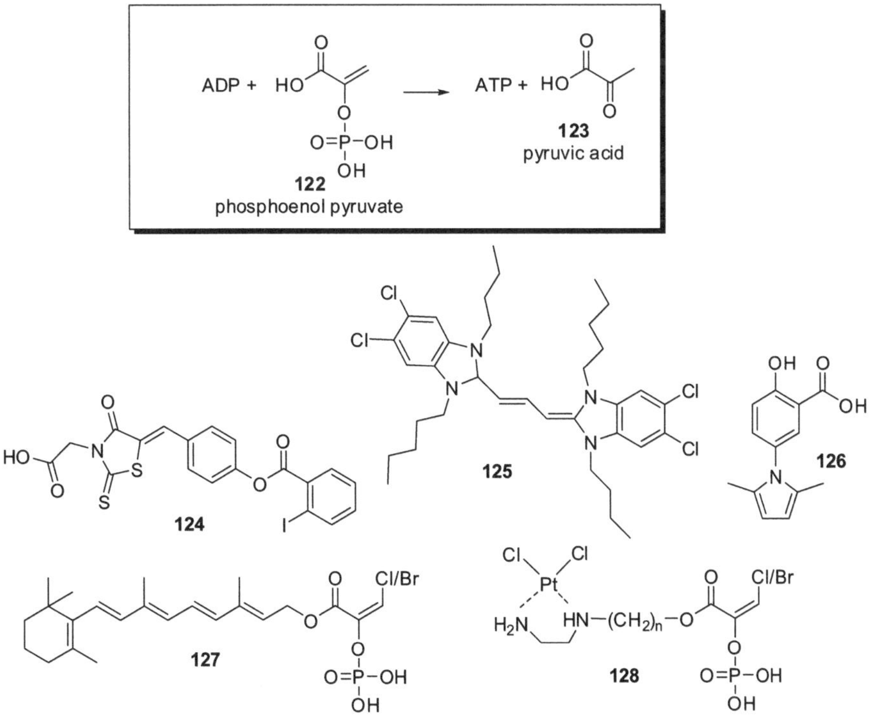

Figure 6.28 Pyruvate kinase: reaction and inhibitors.

screen of one million compounds led to the identification of three classes of inhibitors with some selectivity over the M1 isoform (**124 – 126**). Treatment of tumor cells with these inhibitors led to decreased glycolysis and increased cell death.[229,230] Virginia Commonwealth University and Allos Therapeutics have claimed allosteric inhibitors with micromolar to millimolar IC_{50} values.[231,232] Darwin Pharma has claimed micromolar inhibitors based on substrate analogs, some of which released other biologically active species (*e.g.* retinoic acid in the case of (**127**) or *cis*-platinum in the case of (**128**)).[233]

An opposite approach was followed by Jiang *et al.*, who were looking for activators (instead of inhibitors) of M2-PK with the aim of returning cancer cells to a metabolic state characteristic of normal cells.[234]

6.5.3 Pantothenate Kinase

Pantothenate kinase phosphorylates Vitamin B6, the first and rate controlling step in Coenzyme A biosynthesis. Due to the lack of pantothenate kinase inhibitors and activators, the exact physiological role of this enzyme in metabolism and disease is not yet known. In a recent study aimed at the discovery of antimicrobial agents, a set of novel inhibitors of *S. aureus* pantothenate kinase were disclosed that also inhibited mammalian (murine) pantothenate kinase with high micromolar IC_{50} values.[235] A team at St. Jude Children's Hospital studied a library of known bioactive compounds to find inhibitors of pantothenate kinase. Several inhibitors with IC_{50} values below 10 µM were identified.[236]

6.6 Conclusion

In the previous sections, we have discussed examples of inhibitors of non-protein kinases for the development of novel therapeutics. Due to their important role in sugar metabolism, sugar kinase inhibitors have potential utility for the treatment of cancer, metabolic disorders–diabetes and cardiovascular disease. Nucleoside kinase inhibitors are established oncology targets that operate by inhibition of the salvage pathway. Adenosine kinase inhibitors and lipid kinase inhibitors are highly significant in the development of cancer therapeutics.

A potential advantage of targeting non-protein kinases for drug discovery lies in the structural diversity and small size of their substrates, which sets these kinases apart from protein kinases and may facilitate the design of selective inhibitors. Many of the earlier inhibitors, mostly substrate analogs, were indeed selective over protein kinases. However, the majority of these compounds were hampered by low potency and–or poor drug-like properties. More recently, highly potent and selective ATP competitive inhibitors of non-protein kinases have been developed, particularly against the lipid kinase PI3K. Many of these more recent inhibitors are currently moving through clinical development.

Details about the physiological role of a large number of non-protein kinases and their products continue to emerge. With this new information, the number of non-protein kinases pursued as targets for inhibitor design is only expected to increase. In addition, several older target classes (*e.g.* sugar kinases and nucleoside kinases) may see a renewed interest when modern drug discovery technologies (HTS, structure based drug design, ADME principles) are applied to identify ATP-competitive inhibitors rather than the previously explored substrate analogs. If the successes seen in the development of PI3K inhibitors are reproduced for other non-protein kinases, the inhibition of non-protein kinases may prove to be a very attractive approach for drug discovery.

Acknowledgement

The authors are grateful to Helena Kocis for conducting the literature searches.

References

1. O. Warburg, *München. Med. Wchnschr.,* 1961, **103**, 2504–2506.
2. J. G. Buchanan, D. M. Clode and N. Vethaviyasar, *J. Chem. Soc., Perkin Trans. 1,* 1976, 1449–1453.
3. I. Riviere-Alric; M. Willson and J. Perie, *Phosphorus, Sulfur and Silicon and the Related Elements,* 1991, **56**, 71–80.
4. K. S. Akerfeldt and P. A. Bartlett, *J. Org. Chem.,* 1991, **56**, 7133–7144.
5. M. Kurtoglu, J. C. Maher and T. J. Lampidis, *Antioxid. Redox Signaling,* 2007, **9**, 1383–1390.
6. E. A. Coats, K. A. Skau, C. A. Capcrelli and D. Solomacha, *J. Enzyme Inhib. ,*1993, **6**, 271–282.
7. A. Tiwari, N. S. Krishna, K. Nanda and A. Chugh, *Expert Opin. Invest. Drugs,* 2005, **14**, 1359–1372.
8. G. Giorgioni, S. Ruggieri, A. Di Stefano, P. Sozio, B. Cinque, L. Di Marzio, G. Santoni and F. Claudi, *Bioorg. Med. Chem. Lett.,* 2008, **18**, 2445–2450.
9. M. Matteucci, H. E. Selick and P. Rao, Threshold Pharmaceuticals, Inc.,WO 2006/010077 .
10. J. J. R. Mertens, Mallinckrodt Medical, Inc., WO 1994/14477.
11. Y. H. Ko, P. L. Pedersen and J. F. Geschwind, *Cancer Lett.,* 2001, **173**, 83–91.
12. J.-F. H. Geschwind, Y. H. Ko, M. S. Torbenson, C. Magee and P. L. Pedersen, *Cancer Res.,* 2002, **62**, 3909–3913.
13. H. J. Jae, J. W. Chung, H. S. Park, M. J. Lee, K. C. Lee, H. C. Kim, J. H. Yoon, H. Chung and J. H. Park, *Korean J. Radiol.,* 2009, **10**, 596–603.
14. T. Wang and J. Kang, *Electrophoresis,* 2009, **30**, 1349–1354.

15. R. J. Johnson, T. Nakagawa, M .S. Segal and Y. Y. Sautin, University of Florida Research Foundation Inc., WO 2008/024902.
16. T. Hirata, M. Watanabe, S. Miura, K. Ijichi, M. Fukasawa and R. Sakakibara, *Biosci., Biotechnol., Biochem.*, 2000, **64**, 2047–2052.
17. B. Clem, S. Telang, A. Clem, A. Yalcin, J. Meier, A. Simmons, M. A. Rasku, S. Arumugam, W. L. Dean, J. Eaton, A. Lane, J. O. Trent and J. Chesney, *Mol. Cancer Ther.*, 2008, **7**, 110–120.
18. M. S. Shaikh, A. Mittal and P. V. Bharatam, *J. Mol. Graphics Modell.*, 2008, **26**, 900–906.
19. K. J. Wierenga, K. Lai, P. Buchwald and M. Tang, *J. Biomol. Screening*, 2008, **13**, 415–423.
20. K. Lai, K. J. Wierenga and M. Tang, University of Miami, WO 2009/023773.
21. D. A. Carson, J. Kaye and J. E. Seegmiller, *Proc. Natl. Acad. Sci. U. S. A.*, 1977, **74**, 5677–5681.
22. D. A. Carson, J. Kaye and D. B. Wasson, *J. Immunol.*, 1981, **126**, 348–352.
23. N. K. Ahmed, *Int. J. Biochem.*, 1982, **14**, 259–262.
24. M.-I. Lim, J. D. Moyer, R. L. Cysyk and V. E. Marquez, *J. Med. Chem.*, 1984, **27**, 1536–1538.
25. R. Schibli, M. Netter, L. Scapozza, M. Birringer, P. Schelling, C. Dumas, J. Schoch and P. A. Schubiger, *J. Organomet. Chem.*, 2003, **668**, 67–74.
26. M.-J. Pérez-Pérez, A.-I. Hernández, E.-M. Priego, F. Rodríguez-Barrios, F. Gago, M.-J. Camarasa and J. Balzarini, *Curr. Top. Med. Chem.*, 2005, **5**, 1205–1219.
27. M.-J. Pérez-Pérez, E.-M. Priego, A.-I. Hernández, O. Familiar, M.-J. Camarasa, A. Negri, F. Gago and J. Balzarini, *Med. Res. Rev.*, 2008, **28**, 797–820.
28. J. Balzarini, I. Van Daele, A. Negri, N. Solaroli, A. Karlsson, S. Liekens, F. Gago and S. Van Calenbergh, *Mol. Pharmacol.*, 2009, **75**, 1127–1136.
29. S. Van Poecke, A. Negri, F. Gago, I. Van Daele, N. Solaroli, A. Karlsson, J. Balzarini and S. Van Calenbergh, *J. Med. Chem.*, 2010, **53**, 2902–2912.
30. X.-C. Yu, M. Miranda, Z. Liu, S. Patel, N. Nguyen, K. Carson, Q. Liu and J. C. Swaffield, *J. Biomol. Screening*, 2010, **15**, 72–79.
31. T. C. Jessop, J. E. Tarver, M. Carlsen, A. Xu, J. P. Healy, A. Heim-Riether, Q. Fu, J. A. Taylor, D. J. Augeri, M. Shen, T. R. Stouch, R. V. Swanson, L. W. Tari, M. Hunter, I. Hoffman, P. E. Keyes, X.-C. Yu, M. Miranda, Q. Liu, J. C. Swaffield, S. D. Kimball, A. Nouraldeen, A. G. E. Wilson, A. M. Digeorge-Foushee, K. Jhaver, R. Finch, S. Anderson, T. Oravecz and K. G. Carson, *Bioorg. Med. Chem. Lett.*, 2009, **19**, 6784–6787.
32. D. J. Augeri, M. Carlsen, K. G. Carson, Q. Fu, A. Heim-Riether, T. C. Jessop, J. Tarver and J. A. Taylor, Lexicon Pharmaceuticals Inc., WO 2008/076778.

33. T. C. Jessop, J. E. Tarver, A. M. Digeorge-Foushee, K. C. Jhaver, J. A. Taylor, X. Xu, Q. Fu, M. Carlsen, M. Shen, X.-C. Yu, Z. Liu, M. L. Miranda, A. Nouraldeen, Z.-Q. Yu, L. W. Tari, M. J. Hunter, I. D. Hoffman, A. G. E. Wilson, S. J. Anderson, T. Oravecz, T. R. Stouch, D. J. Augeri and K. G. Carson, 236th ACS National Meeting, Philadelphia, PA, 2008. Abstract Medi 361. http://www.acsmedchem. org/mediabstract2008.pdf.

34. J. E. Tarver, T. C. Jessop, M. Carlsen, D. J. Augeri, Q. Fu, J. P. Healy, A. Heim-Riether, A. Xu, J. A. Taylor, M. Shen, P. E. Keyes, S. D. Kimball, X.-C. Yu, M. Miranda, Q. Liu, J. C. Swaffield, A. Nouraldeen, A. G. E. Wilson, R. Finch, K. Jhaver, A. M. DiGeorge Foushee, S. Anderson, T. Oravecz and K. G. Carson, *Bioorg. Med. Chem. Lett.*, 2009, **19**, 6780–6783.

35. All X-Ray structures were taken from the PDB. Graphics were created using Maestro (Version 1.9.207), from Schrodinger.

36. A. Gomtsyan and C.-H. Lee, *Curr. Pharm. Des.*, 2004, **10**, 1093–1103.

37. S. McGaraughty, M. Cowart, M. F. Jarvis and R. F. Berman, *Curr. Top. Med. Chem.*, 2005, **5**, 43–58.

38. E. A. Kowaluk and M. F. Jarvis, *Expert Opin. Invest. Drugs,* 2000, **9**, 551–564.

39. C.-H. Lee, J. F. Daanen, M. Jiang, H. Yu, K. L. Kohlhaas, K. Alexander, M. F. Jarvis, E. L. Kowaluk and S. S. Bhagwat, *Bioorg. Med. Chem. Lett.*, 2001, **11**, 2419–2422.

40. P. J. Hajduk, A. Gomtsyan, S. Didomenico, M. Cowart, E. K. Bayburt, L. Solomon, J. Severin, R. Smith, K. Walter, T. F. Holzman, A. Stewart, S. McGaraughty, M. F. Jarvis, E. A. Kowaluk and S. W. Fesik, *J. Med. Chem*, 2000, **43**, 4781–4786.

41. A. Gomtsyan, S. Didomenico, C.-H. Lee, M. A. Matulenko, K. Kim, E. A. Kowaluk, C. T. Wismer, J. Mikusa, H. Yu, K. Kohlhaas, M. F. Jarvis and S. S. Bhagwat, *J. Med. Chem.,* 2002, **45**, 3639–3648.

42. S. W. Muchmore, R. A. Smith, A. O. Stewart, M. D. Cowart, A. Gomtsyan, M. A. Matulenko, H. Yu, J. M. Severin, S. S. Bhagwat, C.-H. Lee, E. A. Kowaluk, M. F. Jarvis and C. L. Jakob, *J. Med. Chem.,* 2006, **49**, 6726–6731.

43. S. Bhutoria and N. Ghoshal, *J. Mol. Graphics Modell.,* 2010, **28**, 577–591.

44. C.-H. Lee, M. Jiang, M. Cowart, G. Gfesser, R. Perner, K. H. Kim, Y. G. Gu, M. Williams, M. F. Jarvis, E. A. Kowaluk, A. O. Stewart and S. S. Bhagwat, *J. Med. Chem.,* 2001, **44**, 2133–2138

45. M. F. Jarvis, H. Yu, K. Kohlhaas, K. Alexander, C.-H. Lee, M. Jiang, S. S. Bhagwat, M. Williams and E. A. Kowaluk, *J. Pharmacol. Exp. Ther.,* 2000, **295**, 1156–1164.

46. E. A. Kowaluk, J. Mikusa, C. T. Wismer, C. Z. Zhu, E. Schweitzer, J. J. Lynch, C.-H. Lee, M. Jiang, S. S. Bhagwat, A. Gomtsyan, J. McKie, B. F. Cox, J. Polakowski, G. Reinhart, M. Williams and M. F. Jarvis, *J. Pharmacol. Exp. Ther.,* 2000, **295**, 1165–1174.

47. M. Cowart, C.-H. Lee, G. A. Gfesser, E. K. Bayburt, S. S. Bhagwat, A. O. Stewart, H. Yu, K. L. Kohlhaas, S. McGaraughty, C. T. Wismer, J. Mikusa, C. Zhu, K. M. Alexander, M. F. Jarvis and E. A. Kowaluk, *Bioorg. Med. Chem. Lett.*, 2001, **11**, 83–86.

48. G. Z. Zheng, C.-H. Lee, J. K. Pratt, R. J. Perner, M. Q. Jiang, A. Gomtsyan, M. A. Matulenko, Y. Mao, J. R. Koenig, K. H. Kim, S. Muchmore, H. Yu, K. Kohlhaas, K. M. Alexander, S. McGaraughty, K. L. Chu, C. T. Wismer, J. Mikusa, M. F. Jarvis, K. Marsh, E. A. Kowaluk, S. S. Bhagwata and A. O. Stewart, *Bioorg. Med. Chem. Lett.*, 2001, **11**, 2071–2074.

49. G. Z. Zheng, Y. Mao, C.-H. Lee, J. K. Pratt, J. R. Koenig, R. J. Perner, M. D. Cowart, G. A. Gfesser, S. McGaraughty, K. L. Chu, C. Zhu, H. Yu, K. Kohlhaas, K. M. Alexander, C. T. Wismer, J. Mikusa, M. F. Jarvis, E. A. Kowaluk and A. O. Stewart, *Bioorg. Med. Chem. Lett.*, 2003, **13**, 3041–3044.

50. A. Gomtsyan, S. Didomenico, C.-H. Lee, A. O. Stewart, S. S. Bhagwat, E. A. Kowaluk and M. F. Jarvis, *Bioorg. Med. Chem. Lett.*, 2004, **14**, 4165–4168.

51. R. J. Perner, C.-H. Lee, M. Jiang, Y.-G. Gu, S. DiDomenico, E. K. Bayburt, K. M. Alexander, K. L. Kohlhaas, M. F. Jarvis, E. L. Kowaluk and S. S. Bhagwat, *Bioorg. Med. Chem. Lett.*, 2005, **15**, 2803–2807.

52. M. A. Matulenko, C.-H. Lee, M. Jiang, R. R. Frey, M. D. Cowart, E. K. Bayburt, S. DiDomenico, Jr., G. A. Gfesser, A. Gomtsyan, G. Z. Zheng, J. A. McKie, A. O. Stewart, H. Yu, K. L. Kohlhaas, K. M. Alexander, S. McGaraughty, C. T. Wismer, J. Mikusa, K. C. Marsh, R. D. Snyder, M. S. Diehl, E. A. Kowaluk, M. F. Jarvis and S. S. Bhagwat, *Bioorg. Med. Chem.*, 2005, **13**, 3705–3720.

53. M. A. Matulenko, E. S. Paight, R. R. Frey, A. Gomtsyan, S. DiDomenico, Jr., M. Jiang, C.-H. Lee, A. O. Stewart, H. Yu, K. L. Kohlhaas, K. M. Alexander, S. McGaraughty, J. Mikusa, K. C. Marsh, S. W. Muchmore, C. L. Jakob, E. A. Kowaluk, M. F. Jarvis and S. S. Bhagwat, *Bioorg. Med. Chem.*, 2007, **15**, 1586–1605.

54. S. H. Boyer, M. D. Erion and B. G. Ugarkar, Gensia Inc., WO 1996/040705.

55. B. G. Ugarkar, M. D. Erion, J. E. Gomez-Galeno, A. J. Castellino and C. E. Browne, Gensia Inc., WO 1996/040706.

56. G. S. Firestein, *Drug Dev. Res.*, 1996, **39**, 371–376.

57. M. D. Erion, B. G. Ugarkar, J. DaRe, A. J. Castellino, J. M. Fujitaki, R. Dixon, J. R. Appleman and J. B. Wiesner, *Nucleosides, Nucleotides Nucleic Acids,* 1997, **16**, 1013–1021.

58. B. G. Ugarkar, J. M. DaRe, J. J. Kopcho, C. E. Browne, III, J. M. Schanzer, J. B. Wiesner and M. D. Erion, *J. Med. Chem.*, 2000, **43**, 2883–2893.

59. B. G. Ugarkar, A. J. Castellino, J. M. DaRe, J. J. Kopcho, J. B. Wiesner, J. M. Schanzer and M. D. Erion, *J. Med. Chem.*, 2000, **43**, 2894–2905.

60. B. G. Ugarkar, A. J. Castellino, J. S. DaRe, M. Ramirez-Weinhouse, J. J. Kopcho, S. Rosengren and M. D. Erion, *J. Med. Chem.*, 2003, **46**, 4750–4760.

61. B. C. Bookser, M. C. Matelich, K. Ollis and B.G. Ugarkar, *J. Med. Chem.*, 2005, **48**, 3389–3399.

62. S. H. Boyer, B. G. Ugarkar, J. Solbach, J. Kopcho, M. C. Matelich, K. Ollis, J. E. Gomez-Galeno, R. Mendonca, M. Tsuchiya, A. Nagahisa, M. Nakane, J. B. Wiesner and M. D. Erion, *J. Med. Chem.*, 2005, **48**, 6430–6441.

63. B. C. Bookser, B. G. Ugarkar, M. C. Matelich, R. H. Lemus, M. Allan, M. Tsuchiya, M. Nakane, A. Nagahisa, J. B. Wiesner and M. D. Erion, *J. Med. Chem.*, 2005, **48**, 7808–7820.

64. P. Singh, R. Kumar and B. K. Sharma, *J. Enzyme Inhib.*, 2003, **18**, 395–402.

65. M. Bauser, G. Delapierre, M. Hauswald, T. Flessner, D. D'Urso, A. Hermann, B. Beyreuther, J. De Vry, P. Spreyer, E. Reissmüller and H. Meier, *Bioorg. Med. Chem. Lett.*, 2004, **14**, 1997–2000.

66. K. Gomes Franchini, M. J. Abdalla Saad, R. Rittner Neto, R. M. Marin and S. Rocco Aparecido, Universidade Estadual de Campinas – Unicamp, WO 2005/085213.

67. J. Park, G. Vaidyanathan, B. Singh and R.S. Gupta, *Protein J.*, 2007, **26**, 203–212.

68. S. Butini, S. Gemma, M. Brindisi, G. Borrelli, A. Lossani, A. M. Ponte, A. Torti, G. Maga, L. Marinelli, V. La Pietra, I. Fiorini, S. Lamponi, G. Campiani, D. M. Zisterer, S.-M. Nathwani, S. Sartini, C. La Motta, F. Da Settimo, E. Novellino and F. Focher *J. Med. Chem.*, 2011, **54**, 1401–1420.

69. Y. Lee, N. Bharatham, K. Bharatham and K. W. Lee, *Bull. Korean Chem. Soc.*, 2007, **28**, 561–566.

70. M. C. Long and W. B. Parker, *Biochem. Pharmacol.*, 2006, **71**, 1671–1682.

71. M. C. Long, P. W. Allan, M.-Z. Luo, M.-C. Liu, A. C. Sartorelli and W. B. Parker, *J. Antimicrob. Chemother.*, 2007, **59**, 118–121.

72. R. Gupta, J. Park and B. Singh, McMaster University, CA 2495134.

73. E. H. Walker, M. E. Pacold, O. Perisic, L. Stephens, P. T. Hawkins, M. P. Wymann and R. L. Williams, *Mol. Cell*, 2000, **6**, 909–919.

74. N. T. Ihle, R. Williams, S. Chow, W. Chew, M. I. Berggren, G. Paine-Murrieta, D. J. Minion, R. J. Halter, P. Wipf, R. Abraham, L. Kirkpatrick and G. Powis, *Mol. Cancer Ther.*, 2004, **3**, 763–772.

75. A. L. Howes, G. G. Chiang, E. S. Lang, C. B. Ho, G. Powis, K. Vuori and R. T. Abraham, *Mol. Cancer Ther.*, 2007, **6**, 2505–2514.

76. S. I. Gharbi, M. J. Zvelebil, S. J. Shuttleworth, T. Hancox, N. Saghir, J. F. Timms and M. D. Waterfield, *Biochem. J.*, 2007, **404**, 15–21.

77. J. R. Garlich, P. De, N. Dey, J. D. Su, X. Peng, A. Miller, R. Murali, Y. Lu, G. B. Mills, V. Kundra, H.-K. Shu, Q. Peng and D. L. Durden, *Cancer Res.*, 2008, **68**, 206–215.

78. T. Ozbay, D. L. Durden, T. Liu, R. M. O'Regan and R. Nahta, *Cancer Chemother. Pharmacol.*, 2010, **65**, 697–706.

79. A. J. Folkes, K. Ahmadi, W. K. Alderton, S. Alix, S. J. Baker, G. Box, I. S. Chuckowree, P. A. Clarke, P. Depledge, S. A. Eccles, L. S. Friedman, A. Hayes, T. C. Hancox, A. Kugendradas, L. Lensum, P. Moore, A. G. Olivero, J. Pang, S. Patel, G. H. Pergl-Wilson, F. I. Raynaud, A. Robson, N. Saghir, L. Salphati, S. Sohal, M. H. Ultsch, M. Valenti, H. J. A. Wallweber, N. C. Wan, C. Wiesmann, P. Workman, A. Zhyvoloup, M. J. Zvelebil and S. J. Shuttleworth, *J. Med. Chem.*, 2008, **51**, 5522–5532.

80. M. Hayakawa, H. Kaizawa, H. Moritomo, T. Koizumi, T. Ohishi, M. Okada, M. Ohta, S. Tsukamoto, P. Parker, P. Workman and M. Waterfield, *Bioorg. Med. Chem.*, 2006, **14**, 6847–6858.

81. F. I. Raynaud, S. A. Eccles, S. Patel, S. Alix, G. Box, I. Chuckowree, A. Folkes, S. Gowan, A. De Haven Brandon, F. Di Stefano, A. Hayes, A. T. Henley, L. Lensun, G. Pergl-Wilson, A. Robson, N. Saghir, A. Zhyvoloup, E. McDonald, P. Sheldrake, S. Shuttleworth, M. Valenti, N. C. Wan, P. A. Clarke and P. Workman, *Mol. Cancer Ther.*, 2009, **8**, 1725–1738.

82. L. Salphati, H. Wong, M. Belvin, D. Bradford, K. A. Edgar, W. W. Prior, D. Sampath and J. J. Wallin, *Drug Metab. Dispos.*, 2010, **38**, 1436–1442.

83. C. O'Brien, J. J. Wallin, D. Sampath, D. GuhaThakurta, H. Savage, E. A. Punnoose, J. Guan, L. Berry, W. W. Prior, L. C. Amler, M. Belvin, L. S. Friedman and M. R. Lackner, *Clin. Cancer Res.*, 2010, **16**, 3670–3683.

84. K. Walker and M. Padhiar, *IDrugs*, 2010, **13**, 10–12.

85. D. P. Sutherlin, M. Belvin, L. Bao, L. Berry, M. Berry, G. Castanedo, K. Edgar, A. Folkes, L. Friedman, T. Heffron, S. Patel, A. Olivero, J. Lesnick, C. Lewis, J. Marsters, Jr., J. Nonomiya, J. Pang, W. W. Prior, L. Salphati, D. Sampath, V. Tsui, J. Wallin, B. Q. Wei, C. Weismann and B.-Y. Zhu, 102nd Annual Meeting of the American Association for Cancer Research, April 2-6, 2011, Orlando, Florida, Abstract 2787.

86. T. P. Heffron, M. Berry, G. Castanedo, C. Chang, I. Chuckowree, J. Dotson, A. Folkes, J. Gunzner, J. D. Lesnick, C. Lewis, S. Mathieu, J. Nonomiya, A. Olivero, J. Pang, D. Peterson, L. Salphati, D. Sampath, S. Sideris, D. P. Sutherlin, V. Tsui, N. C. Wan, S. Wang, S. Wong and B. Zhu, *Bioorg. Med. Chem. Lett.*, 2010, **20**, 2408–2411.

87. D. P. Sutherlin, D. Sampath, M. Berry, G. Castanedo, Z. Chang, I. Chuckowree, J. Dotson, A. Folkes, L. Friedman, R. Goldsmith, T. Heffron, L. Lee, J. Lesnick, C. Lewis, S. Mathieu, J. Nonomiya, A. Olivero, J. Pang, W. W. Prior, L. Salphati, S. Sideris, Q. Tian, V. Tsui, N. C. Wan, S. Wang, C. Wiesmann, S. Wong and B.-Y. Zhu, *J. Med. Chem.*, 2010, **53**, 1086–1097.

88. P. G. Foster, AACR-NCI-EORTC International Conference on Molecular Targets and Cancer Therapeutics, 2007, Abs. C199.

89. P. LoRusso, B. Markman, J. Tabarnero, R. Shazer, L. Nguyen, E. Heath, A. Patnaik and K. Papadopoulos, *ASCO Annual Meeting,* 2009, Abs. 3502.
90. S-M. Maira, F. Stauffer, J. Brueggen, P. Furet, C. Schnell, C. Fritsch, S. Brachmann, P. Chene, A. De Pover, K. Schoemaker, D. Fabbro, D. Gabriel, M. Simonen, L. Murphy, P. Finan, W. Sellers and C. Garcia-Echeverria, *Mol. Cancer Ther.,* 2008, **7**, 1851–1863.
91. C. R. Schnell, F. Stauffer, P. R. Allegrini, T. O'Reilly, P. M. J. McSheehy, C. Dartois, M. Stumm, R. Cozens, A. Littlewood-Evans, C. Garcia-Echeverria and S-M. Maira, *Cancer Res.*, 2008, **68**, 6598–6607.
92. D. C. Cho, M. B. Cohen, D. J. Panka, M. Collins, M. Ghebremichael, M. B. Atkins, S. Signoretti and J. W. Mier, *Clin. Cancer Res.,* 2010, **16**, 3628–3638.
93. C. F. Voliva, S. Pecchi, M. Burger, T. Nagel, C. Schnell, C. Fritsch, S. Brachmann, D. Menezes, M. Knapp, K. Shoemaker, M. Wiesmann, K. Huh, I. Zaror, M. Dorsch, W. R. Sellers, C. Garcia-Echeverria and M. Maira, *AACR National Meeting,* 2010, Abs. 4498.
94. V. L. Mason, *IDrugs,* 2010, **13**, 360–362.
95. M. Maira, D. Menezes, S. Pecchi, K. Shoemaker, M. Burger, C. Schnell, C. Fritsch, S. Brachmann, T. Nagel, W. R. Sellers, C. Garcia-Echeverria, M. Wiesmann and C. F. Voliva, *AACR National Meeting,* 2010, Abs. 4497.
96. S. D. Knight, N. D. Adams, J. L. Burgess, A. M. Chaudhari, M. G. Darcy, C. A. Donatelli, J. I. Luengo, K. A. Newlander, C. A. Parrish, L. H. Ridgers and M. A. Sarpong, *ACS Med. Chem. Lett.*, 2010, **1**, 39–43.
97. H.-M. Cheng, S. Bagrodia, S. Bailey, M. Edwards, J. Hoffman, Q.-Y. Hu, R. Kania, D. R. Knighton, M. A. Marx, S. Ninkovic, S.-X. Sun and E. Zhang, *Med. Chem. Commun.*, 2010, **1**, 139–144.
98. H.-M. Cheng, S. Bagrodia, S. Bailey, S. M. Baxi, L. Goulet, L. Guo, J. Hoffman, T. O. Johnson, T. W. Johnson, C. Li, J. Li, K. Liu, Z. Liu, M. A. Marx, G. Paderes, S. Sun, M. Walls, P. A. Wells, A. Yang, M.-J. Yin and J. Zhu, *240th ACS National Meeting,* 2010, Abs. MEDI-493.
99. A. Zask, J. C. Verheijen, K. Curran, J. Kaplan, D. J. Richard, P. Nowak, D. J. Malwitz, N. Brooijmans, J. Bard, K. Swenson, J. Lucas, L. Toral-Barza, W. G. Zhang, I. Hollander, J. J. Gibbons, R. T. Abraham, S. Ayral-Kaloustian, T. S. Mansour and K. Yu, *J. Med. Chem.*, 2009, **52**, 5013–5016.
100. A. M. Venkatesan, C. M. Dehnhardt, E. Delos Santos, Z. Chen, O. Dos Santos, S. Ayral-Kaloustian, G. Khafizova, N. Brooijmans, R. Mallon, I. Hollander, L. Feldberg, J. Lucas, K. Yu, J. Gibbons, R. T. Abraham, I. Chaudhary and T. S. Mansour, *J. Med. Chem.*, 2010, **53**, 2636–2645.
101. A. M. Venkatesan, Z. Chen, O. Dos Santos, C. Dehnhardt, E. Delos Santos, S. Ayral-Kaloustian, R. Mallon, I. Hollander, L. Feldberg, J. Lucas, K. Yu, I. Chaudhary and T. S. Mansour, *Bioorg. Med. Chem. Lett.*, 2010, **20**, 5869–5873.

102. W. J. Scott, M. Hentemann, B, Rowley, C. Bull, S. Jenkins, A. M. Bullion, J. Johnson, A. Redman, M. Brands, W. Esler, P. Fracasso, T. Garrison, M. Hamilton, N. Liu, M. Michels, J. E. Wood, D. Wilkie and H. Xiao, *22nd EORTC-NCI-AACR symposium*, 2010, Abstract 444/poster185.

103. N. Liu, B. Rowley, C. Schneider, C. Bull, J. Hoffmann, S. Kaekoenen, M. Hentemann, W. Scott, D. Mumberg and K. Ziegelbauer, *101st Annual Meeting of the American Association for Cancer Research*, April 17–21, 2010, Washington, DC, Abstract 4476.

104. N. Liu, A. Haegebarth, C. Bull, C. Schatz, S. Wiehr, B.J. Pichler, P. Hauff, D. Mumberg, S. Jenkins, T. Schwarz, K. Ziegelbauer, *101st Annual Meeting of the American Association for Cancer Research*, April 17–21, 2010, Washington, D.C., Abstract 4478.

105. S. P Jackson, S. M. Schoenwaelder, I. Goncalves, W. S. Nesbitt, C. L. Yap, C. E. Wright, V. Kenche, K. E. Anderson, S. M. Dopheide, Y. Yuan, S. A. Sturgeon, H. Prabaharan, P. E. Thompson, G. D. Smith, P. R. Shepherd, N. Daniele, S. Kulkarni, B. Abbott, D. Saylik, C. Jones, L. Lu, S. Giuliano, S. C. Hughan, J. A. Angus, A. D. Robertson and H. H. Salem, *Nat. Med.*, 2005, **11**, 507–514.

106. Z. A. Knight, B. Gonzalez, M. E. Feldman, E. R. Zunder, D. D. Goldenberg, O. Williams, R. Loewith, D. Stokoe, A. Balla, B. Toth, T. Balla, W. A. Weiss, R. L. Williams and K. M. Shokat, *Cell*, 2006, **125**, 733–747.

107. S. A. Sturgeon, C. Jones, J. A. Angus and C. E. Wright, *Eur. J. Pharmacol.*, 2008, **587**, 209–215.

108. S. Jia, T. M. Roberts and J. J. Zhao, *Curr. Opin. Cell Biol.*, 2009, **21**, 199–208

109. N. E. Torbett, A. Luna-Moran, Z. A. Knight, A. Houk, M. Moasser, W. Weiss, K. M. Shokat and D. Stokoe, *Biochem. J.*, 2008, **415**, 97–110.

110. C. Sadhu, B. Masinovsky, K. Dick, C. G. Sowell and D. E. Staunton, *J. Immunol.*, 2003, **170**, 2647–2654.

111. A. Berndt, S. Miller, O. Williams, D. D. Le, B. T. Houseman, J. I. Pacold, F. Gorrec, W.-C. Hon, P. Ren, Y. Liu, C. Rommel, P. Gaillard, T. Ruckle, M. K. Schwarz, K. M. Shokat, J. P. Shaw and R. L. Williams, *Nat. Chem. Biol.*, 2010, **6**, 117–124.

112. T. Schindler, W. Bornmann, P. Pellicena, W. T. Miller, B. Clarkson and J. Kuriyan, *Science,* 2000, **289**, 1938–1942.

113. O. Williams, B. T. Houseman, E. J. Kunkel, B. Aizenstein, R. Hoffman, Z. A. Knight and K. M. Shokat, *Chem. Biol.*, 2010, **17**, 123–134.

114. B. J. Lannutti, S. A. Meadows, S. E. M. Herman, A. Kashishian, B. Steiner, A. J. Johnson, J. C. Byrd, J. W. Tyner, M. M. Loriaux, M. Deininger, B. J. Druker, K. D. Puri, R. G. Ulrich and N. A. Giese, *Blood*, 2011, **117**, 591–594.

115. M. S. S. Palanki, E. Dneprovskaia, J. Doukas, R. M. Fine, J. Hood, X. Kang, D. Lohse, M. Martin, G. Noronha, R. M. Soll, W. Wrasidlo, S. Yee and H. Zhu, *J. Med. Chem.*, 2007, **50**, 4279–4294.

116. J. Doukas, W. Wrasidlo, G. Noronha, E. Dneprovskaia, R. Fine, S. Weis, J. Hood, A. DeMaria, R. Soll and D. Cheresh, *Proc. Natl. Acad. Sci. U. S. A.* , 2006, **103**, 19866–19871.

117. V. Pomel, J. Klicic, D. Covini, D. D. Church, J. P. Shaw, K. Roulin, F. Burgat-Charvillon, D. Valognes, M. Camps, C. Chabert, C. Gillieron, B. Francon, D. Perrin, D. Leroy, D. Gretener, A. Nichols, P.-A. Vitte, S. Carboni, C. Rommel, M. K. Schwarz and T. Ruckle, *J. Med. Chem.*, 2006, **49**, 3857–3871.

118. R. T. Abraham, J. J. Gibbons and E. I. Graziana in *The Enzymes,* ed. M. N. Hall and F. Tamanoi, Elsevier, San Diego, 2010, pp 329–366.

119. J. Dancey, *Nat. Rev. Clin. Oncol.*, 2010, **7**, 209–219.

120. J. Huang and B. D. Manning, *Biochem. Soc. Trans.*, 2009, **37**, 217–222.

121. C. C. Thoreen, S. A. Kang, J. W. Chang, Q. Liu, J. Zhang, Y. Gao, L. J. Reichling, T. Sim, D. M. Sabatini and N. S. Gray, *J. Biol. Chem.*, 2009, **284**, 8023–8032.

122. C. M. Chresta, B. R. Davies, I. Hickson, T. Harding, S. Cosulich, S. E. Critchlow, J. P. Vincent, R. Ellston, D. Jones, P. Sini, D. James, H. Z. Howard, P. Dudley, G. Hughes, L. Smith, S. Maguire, M. Hummersone, K. Malagu, K. Menear, R. Jenkins, M. Jacobsen, G. C. M. Smith, S. Guichard and M. Pass, *Cancer Res.*, 2010, **70**, 288–298.

123. P. C. Gokhale, S. V. Bhagwat, A. P. Crew, A. Cooke, C. Mantis, J. C. Workman, D. Landfair, M. Bittner, D. M. Epstein, J. A. Pachter and R. Wild, *AACR National Meeting,* 2010, Abs. 4486.

124. S. V. Bhagwat, A. P. Crew, P. C. Gokhale, Y. Yao, J. Kahler, D. M. Epstein, R. Wild and J. A. Pachter, *AACR National Meeting,* 2010, Abs. 4487.

125. N. Carayol, E. Vakana, A. Sassano, S. Kaur, D. J. Goussetis, H. Glaser, B. J. Druker, N. J. Donato, E. K. Altman, S. Barr and L. C. Platanias, *Proc. Natl. Acad. Sci. U. S. A.*, 2010, **107**, 12469.

126. K. Jessen, S. Wang, L. Kessler, X. Guo, J. Kucharski, J. Staunton, L. Lan, M. Elia, J. Stewart, J. Brown, L. Li, K. Chan, M. Martin, P. Ren, C. Rommel and Y. Liu, *AACR-NCI-EORTC International Conference: Molecular Targets and Cancer Therapeutics,* 2009, Abs. B148.

127. D. A. Fruman, M. R. Janes, J. J. Limon, L. So, J. Chen, M. B. Martin, P. Ren, Y. Liu and C. Rommel, *AACR National Meeting,* 2010, Abs. 1798.

128. Q. Xue, B. Hopkins, C. Perruzzi, D. Udayakumar, D. Sherris and L. E. Benjamin, *Cancer Res.*, 2008, **68**, 9551–9557.

129. R. Diaz, P. A. Nguewa, J. A. Diaz-Gonzalez, E. Hamel, O. Gonzalez-Marino, R. Catena, D. Serrano, M. Redrado, D. Sherris and A. Calvo, *Br. J. Cancer,* 2009, **100**, 932–940.

130. D. J. Richard, J. C. Verheijen and A. Zask, *Curr. Opin. Drug Discovery Dev.*, 2010, **13**, 428–440.

131. C. Garcia-Echeverria, *Bioorg. Med. Chem. Lett.*, 2010, **20**, 4308–4321.

132. Q. Liu, J. W. Chang, J. Wang, S. A. Kang, C. C. Thoreen, A. Markhard, W. Hur, J. Zhang, T. Sim, D. M. Sabatini and N. S. Gray, *J. Med. Chem.*, 2010, **53**, 7146–7155.

133. A. Zask, J. Kaplan, J. C. Verheijen, D. J. Richard, K. Curran, N. Brooijmans, E. M. Bennett, L. Toral-Barza, I. Hollander, S. Ayral-Kaloustian and K. Yu, *J. Med. Chem.*, 2009, **52**, 7942–7945.

134. K. Yu, C. Shi, L. Toral-Barza, J. Lucas, B. Shor, J. E. Kim, W.-G. Zhang, R. Mahoney, C. Gaydos, L. Tardio, S. K. Kim, R. Conant, K. Curran, J. Kaplan, J. Verheijen, S. Ayral-Kaloustian, T. S. Mansour, R. T. Abraham, A. Zask and J. J. Gibbons, *Cancer Res.*, 2010, **70**, 621–631.

135. D. J. Richard, J. C. Verheijen, K. Curran, J. Kaplan, L. Toral-Barza, I. Hollander, J. Lucas, K. Yu and A. Zask, *Bioorg. Med. Chem. Lett.*, 2009, **19**, 6830–6835.

136. M. E. Feldman, B. Apsel, A. Uotila, R. Loewith, Z. A. Knight, D. Ruggero and K. M. Shokat, *PLos Biol.*, 2009, **7**, 371–383.

137. M. R. Janes, J. J. Limon, L. So, J. Chen, R. J. Lim, M. A. Chavez, C. Vu, M. B. Lilly, S. Mallya, S. T. Ong, M. Konopleva, M. B. Martin, P. Ren, Y. Liu, C. Rommel and D. A. Fruman, *Nat. Med.*, 2010, **16**, 205–213.

138. R. Plummer, *Clin. Cancer Res.*, 2010, **16**, 4527–4531.

139. Z. A. Knight, G. G. Chiang, P. J. Alaimo, D. M. Kenski, C. B. Ho, K. Coan, R. T. Abraham and K. M. Shokat, *Bioorg. Med. Chem.*, 2004, **12**, 4749–4759.

140. J. J. J. Leahy, B. T. Golding, R. J. Griffin, I. R. Hardcastle, C. Richardson, L. Rigoreau and G. C. M. Smith, *Bioorg. Med. Chem. Lett.*, 2004, **14**, 6083–6087.

141. Y. Zhao, H. D. Thomas, M. A. Batey, I. G. Cowell, C. J. Richardson, R. J. Griffin, A. H. Calvert, D. R. Newell, G. C. M. Smith and N. J. Curtin, *Cancer Res.*, 2006, **66**, 5354–5362.

142. C. Cano, O. R. Barbeau, C. Bailey, X.-L. Cockcroft, N. J. Curtin, H. Duggan, M. Frigerio, B. T. Golding, I. R. Hardcastle, M. G. Hummersone, C. Knights, K. A. Menear, D. R. Newell, C. J. Richardson, G. C. M. Smith, B. Spittle and R. J. Griffin, *J. Med. Chem.*, 2010, **53**, 8498–8507.

143. I. H. Ismail, S. Martensson, D. Moshinsky, A. Rice, C. Tang, A. Howlett, G. McMahon and O. Hammarsten, *Oncogene*, 2004, **23**, 873–882.

144. I. Hickson, Y. Zhao, C. J. Richardson, S. J. Green, N. M. B. Martin, A. I. Orr, P. M. Reaper, S. P. Jackson, N. J. Curtin and G. C. M. Smith, *Cancer Res.*, 2004, **64**, 9152–9159.

145. Y. Li and D.-Q. Yang, *Mol. Cancer Ther.*, 2010, **9**, 113–125.

146. M. D. Rainey, M. E. Charlton, R. V. Stanton and M. B. Kastan, *Cancer Res.*, 2008, **68**, 7466–7474.

147. H. Nishida, N. Tatewaki, Y. Nakajima, T. Magara, K. M. Ko, Y. Hamamori and T. Konishi, *Nucleic Acids Res.*, 2009, **37**, 5678–5689.

148. J.-D. Charrier, S. J. Durrant, J. M. C. Golec, D. P. Kay, R. M. A. Knegtel, S. MacCormick, M. Mortimore, M. E. O'Donnell, J. L.Pinder,

P. M. Reaper, A. P. Rutherford, P. S. H. Wang, S. C. Young and J. R. Pollard, *J. Med. Chem.*, 2011, **54**, 2320–2330.

149. P. M. Reaper, M. R. Griffiths, J. M. Long, J.-D. Charrier, S. MacCormick, P. A. Charlton, J. M. C. Golec and J. R. Pollard, *Nat. Chem. Biol.*, 2011, **7**, 428–430.

150. L. I. Toledo, M. Murga, R. Zur, R. Soria, A. Rodriguez, S. Martinez, J. Oyarzabal, J. Pastor, J. R. Bischoff and O. Fernandez-Capetillo, *Nat. Struct. Mol. Biol.*, 2011, **18**, 721–727.

151. Y. H. Huang, K. Hoebe and K. Sauer, *Expert Opin. Ther. Targets,* 2008, **12**, 391–413.

152. M. J. Schell, *Cell. Mol. Life Sci.,* 2010, **67**, 1755–1778.

153. S. T. Safrany, D. A. Sawyer, S. R. Nahorski and B. V. L. Potter, *Chirality*, 1992, **4**, 415–422.

154. C. Liu, A. M. Riley, X. Yang, S. B. Shears and B. V. L. Potter, *J. Med. Chem.,* 2001, **44**, 2984–2989

155. A. Poinas, K. Backers, A. M. Riley, S. J. Mills, C. Moreau, B. V. L. Potter and C. Erneux, *ChemBioChem*, 2005, **6**, 1449–1457.

156. G. Choi, Y.-T. Chang, S.-K. Chung and K. Y. Choi, *Bioorg. Med. Chem. Lett.,* 1997, **7**, 2709–2714.

157. S.-K. Chung and Y.-T. Chang, *Bioorg. Med. Chem. Lett.*, 1997, **7**, 2715–2718.

158. Y.-U. Kwon, J. Im, G. Choi, Y.-S. Kim, K. Y. Choi and S.-K. Chung, *Bioorg. Med. Chem. Lett.*, 2003, **13**, 2981–2984

159. Y.-T. Chang, G. Choi, Y.-S. Bae, M. Burdett, H.-S. Moon, J. W. Lee, N. S. Gray, P. G. Schultz, L. Meijer, S.-K. Chung, K. Y. Choi, P.-G. Suh and S. H. Ryu, *ChemBioChem,* 2002, **3**, 897–901.

160. U. Padmanabhan, D. E. Dollins, P. C. Fridy, J. D. York and C. P. Downes, *J. Biol. Chem.,* 2009, **284**, 10571 10582.

161. G. W. Mayr, S. Windhorst and K. Hillemeier, *J. Biol. Chem.*, 2005, **280**, 13229–13240; corrected in *J. Biol. Chem.*, 2007, **282**, 35424–35431.

162. L. Friedman, H. Francis-Lang, A. L. Parks, K. J. Shaw, L. M. Bjerke and T. S. Heuer, Exelixis, Inc., WO 2005/072475.

163. S. Pan, Y. Liu, Y. F. Xie, D. Cheng, Y. Wan, D. Han, Y. Yang, W. Gao, J. Jiang, B. Bursulaya, P. Chamberlain, D. S. Karanewsky and X. Wang, IRM, Llc., WO 2008/011611.

164. B. Bursulaya, D. Cheng, J. Jiang, D.S. Karanewsky, Y. Liu, S. Pan, Y. Wan, X. Wang, Y. F. Xie and Y. Yang, IRM Llc., WO 2008/157210.

165. Y. Wan, S. Pan, G. Zhang, X. Wang, Y. F. Xie, J. Jiang, D. Phillips and Y. Yang, IRM Llc., WO 2009/123948.

166. R. C. Young, M. Jones, K. J. Milliner, K. K. Rana and J. G. Ward, *J. Med. Chem.,* 1990, **33**, 2073–2080.

167. R. C. Young, C. P. Downes, M. Jones, K. J. Milliner, K. K. Rana and J. G. Ward, *Eur. J. Med. Chem.,* 1994, **29**, 537–549.

168. C. Shih and Y. Hu, *Tetrahedron Lett.,* 1994, **35**, 4677–4680.

169. Y. Saito and K. Umezawa, *Bioorg. Med. Chem. Lett.,* 1997, **7**, 861–864.

170. Y. Saito, K. Kato and K. Umezawa, *Tetrahedron*, 1998, **54**, 4251–4266.
171. K. Sanpei, Y. Saito, M. Imoto and K. Umezawa, *Bioorg. Med. Chem. Lett.*, 1996, **6**, 2487–2490.
172. I. Treinies, S. Lachmann, O. Kern, J. Smith, T. Hunter, M. Jimenez Quesada, L. Bill, E. Trivier, S. Klee, A. C. Wong, M. Swarbrick, L. Tonkin, M. Pittman, L. Pang, R. Wood, T. Hammonds, D. Jones, N. Divecha, C. Foxton and J. Roffey, *100th AACR Annual Meeting, 2009*, Abs. 1782.
173. H. B. J. Jefferies, F. T. Cooke, P. Jat, C. Boucheron, T. Koizumi, M. Hayakawa, H. Kaizawa, T. Ohishi, P. Workman, M. D. Waterfield and P. J. Parker, *EMBO Reports*, 2008, **9**, 164–170.
174. O. C. Ikonomov, D. Sbrissa and A. Shisheva, *Biochem. Biophys. Res. Commun.*, 2009, **382**, 566–570.
175. D. de Chaffoy de Courcelles, P. Roevens and H. Van Belle, *J. Biol. Chem.*, 1985, **260**, 15762–15770.
176. D. de Chaffoy de Courcelles, P. Roevens, H. Van Belle, L. Kennis, Y. Somers and F. De Clerck, *J. Biol. Chem.*, 1989, **264**, 3274–3285.
177. Y. Jiang, F. Sakane, H. Kanoh and J. P. Walsh, *Biochem. Pharmacol.*, 2000, **59**, 763–772.
178. K. Nobe, M. Miyatake, H. Nobe, Y. Sakai, J. Takashima and K. Momose, *Br. J. Pharmacol.*, 2004, **143**, 166–178.
179. F. Sakane, S.-I Imai, M. Kai, S. Yasuda and H. Kanoh, *Curr. Drug Targets*, 2008, **9**, 626–640.
180. R. Hernández-Alcoceba, L. Saniger, J. Campos, M. C. Núñez, F. Khaless, M. Angel Gallo, A. Espinosa and J. C. Lacal, *Oncogene*, 1997, **15**, 2289–2301.
181. R. Hernández-Alcoceba, F. Fernandez and J. C. Lacal, *Cancer Res.*, 1999, **59**, 3112–3118.
182. J. Campos, M. del Carmen Núñez, V. Rodríguez, M. A. Gallo and A. Espinosa, *Bioorg. Med. Chem. Lett.*, 2000, **10**, 767–770.
183. J. Campos, M. C. Núñez, A. Conejo-García, R. M. Sánchez-Martín, R. Hernández-Alcoceba, A. Rodríguez-González, J. C. Lacal, M. A. Gallo and A. Espinosa, *Curr. Med. Chem.*, 2003, **10**, 1095–1112.
184. B. Jimeñez, L. del Peso, S. Montaner, P. Esteve and J. C. Lacal, *J. Cell. Biochem.*, 1995, **57**, 141–149.
185. J. C. Lacal, *IDrugs*, 2001, **4**, 419–426.
186. A. Rodríguez-González, A. Ramírez de Molina, F. Fernández, M. Angeles Ramos, M. del Carmen Núñez, J. Campos and J. C. Lacal, *Oncogene*, 2003, **22**, 8803–8812.
187. A. Conejo-García, J. M. Campos, R. M. Sánchez-Martín, M. A. Gallo and A. Espinosa, *J. Med. Chem.*, 2003, **46**, 3754–3757.
188. A. Conejo-García, J. Campos, R. M. Sánchez, A. Rodríguez-González, J. C. Lacal, M. A. Gallo and A. Espinosa, *Eur. J. Med. Chem.*, 2003, **38**, 109–116.

189. A. Conejo-García, M. Báñez-Coronel, R. M. Sánchez-Martín, A. Rodríguez-González, A. Ramos, A. Ramírez de Molina, A. Espinosa, M. A. Gallo, J. M. Campos and J. C. Lacal, *J. Med. Chem.*, 2004, **47**, 5433–5440.

190. A. Conejo-García, A. Entrena, J. M. Campos, R. M. Sánchez-Martín, M. A. Gallo and A. Espinosa, *Eur. J. Med. Chem.*, 2005, **40**, 315–319.

191. S. Janardhan, P. Srivani and G. N. Sastry, *QSAR Comb. Sci.*, 2006, **25**, 860–872.

192. R. Sánchez-Martín, J. M. Campos, A. Conejo-García, O. Cruz-López, M. Báñez-Coronel, A. Rodríguez-González, M. A. Gallo, J. C. Lacal and A. Espinosa, *J. Med. Chem.*, 2005, **48**, 3354–3363.

193. L. Milanese, A. Espinosa, J. M. Campos, M. A. Gallo and A. Entrena, *ChemMedChem*, 2006, **1**, 1216–1228.

194. J. M. Campos, R. M. Sánchez-Martín, A. Conejo-García, A. Entrena, M. A. Gallo and A. Espinosa, *Curr. Med. Chem.*, 2006, **13**, 1231–1248.

195. A. Conejo-García, J. M. Campos, A. Entrena, R. M. Sánchez-Martín, M. A. Gallo and A. Espinosa, *J. Org. Chem.*, 2005, **70**, 5748–5751.

196. M. C. Núñez, A. Conejo-García, R. M. Sánchez-Martín, M. A. Gallo, A. Espinosa and J. M. Campos, *Curr. Comput.-Aided Drug Des.*, 2007, **3**, 297–313.

197. P. Srivani and G. N. Sastry, *J. Mol. Graphics Modell.*, 2009, **27**, 676–688.

198. N. F. Lamour and C.E. Chalfant, *Curr. Drug Targets*, 2008, **9**, 674–682.

199. S. Saxena, M. Banerjee, R. K. Shirumalla and A. Ray, *Curr. Opin. Invest. Drugs*, 2008, **9**, 455–462.

200. J.-W. Kim, Y. Inagaki, S. Mitsutake, N. Maezawa, S. Katsumura, Y.-W. Ryu, C.-S. Park, M. Taniguchi and Y. Igarashi, *Biochim. Biophys. Acta*, 2005, **1738**, 82–90.

201. JP 2006/298849

202. J.-W. Kim, Y. Inagaki, S. Mitsutake, N. Maezawa, S. Katsumura, Y.-W. Ryu, C.-S. Park, M. Taniguchi and Y. Igarashi, *Biochim. Biophys. Acta*, 2007, **1771**, 1262.

203. JP 2009/051743

204. N. Munagala, S. Nguyen, W. Lam, J. Lee, A. Joly, K. McMillan and W. Zhang, *Assay Drug Dev. Technol.*, 2007, **5**, 65–73.

205. C. Graf, M. Klumpp, M. Habig, P. Rovina, A. Billich, T. Baumruker, B. Oberhauser and F. Bornancin, *Mol. Pharmacol.*, 2008, **74**, 925–932.

206. F. Bornancin and B. Oberhauser, Novartis Pharma Gmbh, WO 2007/112914.

207. C. Graf, P. Rovina and F. Bornancin, *Anal. Biochem.*, 2009, **384**, 166–169.

208. T. Mathew, M. Cavallari, A. Billich, F. Bornancin, P. Nussbaumer, G. De Libero and A. Vasella, *Chem. Biodiversity*, 2009, **6**, 1688–1715.

209. S. Kossida, J. Encinas and E. Takao, Bayer Aktiengesellschaft, US 2004/0192580.

210. C. E. Chalfant, Y. A. Hannun, B. J. Pettus and A. Bielawska, US 2006/0030537.

211. M. R. Pitman and S. M. Pitson, *Curr. Cancer Drug Targets*, 2010, **10**, 354–367

212. O. Cuvillier, *Anti-Cancer Drugs*, 2007, **18**, 105–110.

213. K. Kono, M. Tanaka, T. Ogita, T. Hosoya and T. Kohama, *J. Antibiot.*, 2000, **53**, 459–466.

214. K. Kono, M. Tanaka, T. Mizuno, K. Kodama, T. Ogita and T. Kohama, *J. Antibiot.*, 2000, **53**, 753–758.

215. K. Kono, M. Tanaka, Y. Ono, T. Hosoya, T. Ogita and T. Kohama, *J. Antibiot.*, 2001, **54**, 415–420.

216. L. C. Edsall, J. R. Van Brocklyn, O. Cuvillier, B. Kleuser and S. Spiegel, *Biochemistry*, 1998, **37**, 12892–12898.

217. S. De Jonghe, I. Van Overmeire, S. Poulton, C. Hendrix, R. Busson, S. Van Calenbergh, D. De Keukeleire, S. Spiegel and P. Herdewijn, *Bioorg. Med. Chem. Lett.*, 1999, **9**, 3175–3180.

218. S. W. Paugh, B. S. Paugh, M. Rahmani, D. Kapitonov, J. A. Almenara, T. Kordula, S. Milstien, J. K. Adams, R. E. Zipkin, S. Grant and S. Spiegel, *Blood*, 2008, **112**, 1382–1391.

219. J.-W. Kim, Y.-W. Kim, Y. Inagaki, Y.-A. Hwang, S. Mitsutake, Y.-W. Ryu, W. K. Lee, H.-J. Ha, C.-S. Parka and Y. Igarashi, *Bioorg. Med. Chem.*, 2005, **13**, 3475–3485.

220. L. Wong, S. S. L. Tan, Y. Lam and A. J. Melendez, *J. Med. Chem.*, 2009, **52**, 3618–3626.

221. Y. Xiang, G. Asmussen, M. Booker, B. Hirth, J. L. Kane, Jr., J. Liao, K. D. Noson and C. Yee, *Bioorg. Med. Chem. Lett.*, 2009, **19**, 6119–6121.

222. K. J. French, R. S. Schrecengost, B. D. Lee, Y. Zhuang, S. N. Smith, J. L. Eberly, J. K. Yun and C. D. Smith, *Cancer Res.*, 2003, **63**, 5962–5969.

223. L. W. Maines, L. R. Fitzpatrick, K. J. French, Y. Zhuang, Z. Xia, S. N. Keller, J. J. Upson and C. D. Smith, *Dig. Dis. Sci.*, 2008, **53**, 997–1012.

224. K. J. French, Y. Zhuang, L. W. Maines, P. Gao, W. Wang, V. Beljanski, J. J. Upson, C. L. Green, S. N. Keller and C. D. Smith, *J. Pharmacol. Exp. Ther.*, 2010, **333**, 129–139.

225. Y. Qiu and D. Li, *Org. Lett.*, 2006, **8**, 1013–1016.

226. A. Freund, T. Lacour and J. Rether, BASF Aktiengesellschaft, WO 2004/070038.

227. R. Suelmann, E. Koopmann, A. Trautwein, K.-H. Kuck, M. Adamczewski, B. Grimiig and R. Kirsten, Bayer Cropscience Aktiengesellschaft, WO 2005/003355.

228. G. A. Spoden, S. Mazurek, D. Morandell, N. Bacher, M. J. Ausserlechner, P. Jansen-Dürr, E. Eigenbrodt and W. Zwerschke, *Int. J. Cancer*, 2008, **123**, 312–321.

229. M. G. Vander Heiden, H. R. Christofk, E. Schuman, A. O. Subtelny, H. Sharfi, E. E. Harlow, J. Xian and L. C. Cantley, *Biochem. Pharmacol.*, 2010, **79**, 1118–1124.

230. L. C. Cantley, M. G. Vander Heiden and H. R. Christoffk, Beth Israel Deaconess Medical Center, WO 2008/019139.
231. D. Abraham, R. Danso-Danquah, T. Boyiri, I. Nnamani, C. Wang, M. Gerber, S. Hoffman and G. Joshi, Allos Therapeutics, Inc.; Virginia Commonwealth University, WO 1999/048490.
232. D. J. Abraham, C. Wang, R. Danso-Danquah, J. C. Burnett, G. S. Joshi and S. J. Hoffman, US 2001/0046997
233. B. R. Smith, T.-H. Chan and B. Leyland-Jones, Darwin Pharma Inc., WO 2006/125323.
234. J.-K. Jiang, M. B. Boxer, M. G. Vander Heiden, M. Shen, A. P. Skoumbourdis, N. Southall, H. Veith, W. Leister, C. P. Austin, H. W. Park, J. Inglese, L. C. Cantley, D. S. Auld and C. J. Thomas, *Bioorg. Med. Chem. Lett.*, 2010, **20**, 3387–3393.
235. K. G. Virga, Y.-M. Zhang, R. Leonardi, R. A. Ivey, K. Hevener, H.-W. Park, S. Jackowski, C. O. Rock and R. E. Lee, *Bioorg. Med. Chem.*, 2006, **14**, 1007–1020.
236. R. Leonardi, Y.-M. Zhang, M.-K. Yun, R. Zhou, F.-Y. Zeng, W. Lin, J. Cui, T. Chen, C. O. Rock, S. W. White and S. Jackowsk, *Chem. Biol.*, 2010, **17**, 892–902.

CHAPTER 7

The Drug Discovery and Development of Kinase Inhibitors Outside of Oncology

A. J. RATCLIFFE

Cellzome Ltd, Chesterford Research Park, Little Chesterford, Cambridge, CB10 1XL, UK

7.1 Introduction

Over 20 years ago the kinome tree consisted of < 20 kinases. Today the tree details an extensive network of 518 distinct human kinases, which form the basis of a Smorgasbord of drug discovery targets. Testament to the interest in kinase drug discovery is the projected US $75 000 000 000 plus spent to date on research and development. The return from such investment has witnessed 10 small molecule inhibitors reach the market (excluding the rapalogs). Of these, nine fall into the oncology arena,[1] with fasudil the exception, being approved in Japan as an i.v. therapy to treat cerebral vasospasm and ischemia following a subarachnoid hemorrhage. The drive towards an oncology endpoint is also apparent from a 2011 protein kinase inhibitor oncology pipeline update, which shows 1500 developmental projects towards this indication.[2] Aligned with this, analysis of kinase patents between 2006–2009 supports the view that kinase research has a high oncology inertia, with >40% claiming inhibitors against only 10 targets, the majority of which are linked to a role in the disease.[3]

RSC Drug Discovery Series No. 19
Kinase Drug Discovery
Edited by Richard A. Ward and Frederick Goldberg
© Royal Society of Chemistry 2012
Published by the Royal Society of Chemistry, www.rsc.org

The reasons for this *status quo* may stem from several factors, one being the hope of fulfilling the expectations generated by the remarkable ascendancy of imatinib (Gleevec®), the first FDA approved kinase inhibitor in 2001, which found utility in the treatment of chronic myelogenous leukemia (CML) and reached blockbuster status, breaking global sales of over US $1 billion. A second may revolve around the significant challenges associated with the drug design of kinase inhibitors. A hotly debated subject is the degree of kinase inhibitor selectivity required in order to deliver clinical efficacy against the intended disease in concert with imparting a favorable therapeutic window. Most kinase inhibitors compete with the common ATP binding site. As a consequence targeting high specificity and potency against a single kinase within the kinome tree presents a daunting and difficult task which is rarely achieved, in particular when in addition the need for optimization of physicochemical, ADME and general compound specific drug safety (hERG, Ames, phospholipidosis *etc.*) is required in order to generate robust drug candidates. Using toxicity readouts at both the *in vitro* and *in vivo* level in tandem with kinase selectivity profiles, research informaticians, in collaboration with nonclinical drug safety and medicinal chemistry, are beginning to identify kinase inhibitor profiles that represent a high risk factor to exhibiting unacceptable safety.[4] As a general trend the severity of organ toxicity appears to be associated with increased kinase promiscuity. However, more detailed analysis is leading to the suggestion that inhibition of unique kinase signatures, involving set combination of specific kinases rather than individual kinases *per se*, are the trigger for toxicity. The situation is further complicated with evidence suggesting that in some cases clinical efficacy, or superior clinical efficacy, may be dependent on polypharmacology and inhibition of multiple kinase targets.[5] Under this scenario the kinase drug discovery process becomes a more complex and challenging interplay between identification of disease relevant target combinations, and whether a single molecule can be efficiently crafted to inhibit multiple selected targets, which may be unrelated by protein sequence, whilst avoiding triggering a toxic kinase signature.

Overlaying the question of kinase selectivity is the complexity of the signaling network environments that kinases control, and whether inhibition of a set signaling pathway can result in activation of compensatory signaling processes that manifest in unwanted toxicity. As a consequence of the above kinase drug discovery landscape it is not surprising, in particular in the early days of the field where assessment of kinome wide selectivity was lacking, kinase specific structural information was in its infancy, and toxicity and side effect profiles dominated thinking, that oncology presented itself as the initial disease area of most interest (and success to market), where medicines of the day relied chiefly on cytotoxic chemotherapeutics.

Despite this background there is a small band of kinase inhibitors currently undergoing human clinical evaluation outside the oncology franchise, with several in late Phase status (II–III). Therapeutic areas being investigated include inflammation, autoimmune, respiratory, cardiovascular, pain, trans-

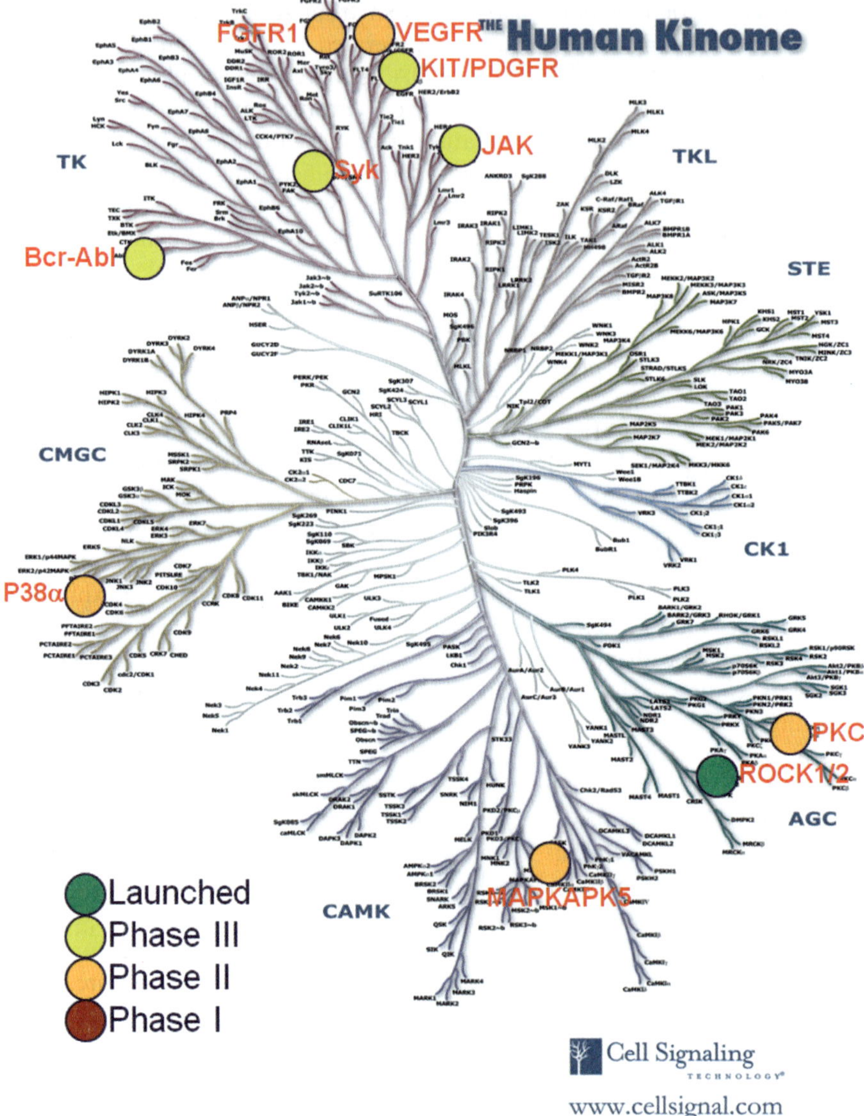

Figure 7.1   Kinase Map of Targets and Clinical Status of most Advanced Inhibitors.

plantation and ophthalmology. The purpose of this chapter is to review such kinase inhibitors. A map of the kinase targets in question (Figure 7.1) reveals collectively the targets are spread over several kinase classes, although the tyrosine kinase (TK) group of kinases appears to represent a hot spot. The clinical status of the most advanced inhibitors is also highlighted.

7.2 Inhibitors of Rho Kinase – fasudil

Rho associated kinase (ROCK), a member of the AGC (PKA–PKG–PKC) family of serine-threonine protein kinases, acts downstream of small GTPase Rho proteins.[6,7] At the human level two ROCK isoforms, ROCK1 and ROCK2, have been detected ubiquitously. Higher expression levels of ROCK1 are mapped to liver, lung, spleen, kidney and testis compared to ROCK2, which is enriched in the brain and heart. At the catalyst kinase domain level high protein sequence homology exists between the two isoforms (92%).[7]

ROCK activity promotes phosphorylation of a number of targets that include the LIM kinases, regulatory myosin light chain (MLC) and the myosin-binding subunit (MYPT1) of the MLC phosphatase. Taken together these events manifest in controlling a wide spectrum of biological processes that include vascular smooth muscle regulation and endothelial barrier function, actin cytoskeleton organization and neurite retraction.[6]

The role of ROCK in the vascular system has been a driver to discover and develop ROCK inhibitors to treat cardiovascular diseases. Fasudil (**7.1**), the only clinically approved kinase inhibitor outside the oncology arena, is effective in reversing blood vessel spasm and constriction resulting from bleeding into the subarachnoid space around the brain.[6] If left untreated the subarachnoid hemorrhage (SAH) can lead to stroke or at the extreme, death. Biochemical and cell based profiling of fasudil has suggested that the clinical effects observed against SAH may reside from inhibition of ROCK, although potency against both isoforms is low μM, with cell based activities often in the 10–50 μM range. Selectivity profiling has also highlighted inhibition of other kinases at a similar level, in particular other members of the AGC subfamily, including PRK2, MSK1, RSK1, RSK2 and PKA, raising the overall question whether the clinical efficacy observed with fasudil is solely driven through inhibition of the ROCK pathway.[8] Nevertheless, post marketing surveillance studies in over 1400 SAH patients have found fasudil to be well tolerated and safe. Clinical trials with fasudil have been conducted in other indications connected to the cardiovascular system, but as of 2010 does not rank highly within pipelines of companies. Of note, Asahi Kasei Pharma Corp. recently reported that Phase III studies for acute cerebral infarction failed to meet its primary endpoint and is considering whether to continue development.

Fasudil has been used, along with Y-27632 (**7.2**), a sub-μM and slightly more selective pan ROCK inhibitor, as prototypes in establishing potential impact of ROCK inhibition in other disease settings.[7] The results have spurred pharmaceutical companies to discover more potent and selective ROCK inhibitors, in particular targeted to the treatment of glaucoma, where pharmacology studies have suggested impact on the disease at several levels; reduction of intraocular pressure and increase of aqueous outflow by relaxation of trabecular meshwork tissue;[9] improving blood flow to the optic nerve;[10] protection of healthy ganglion cells;[11] and reduced subconjunctival scanning after surgery.[12]

Several companies are competing for the glaucoma opportunity and a worldwide market worth annually of the order of US $4.5 billion. Santen

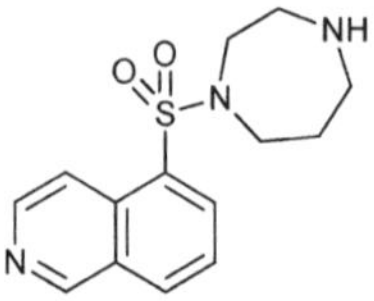

(7.1)

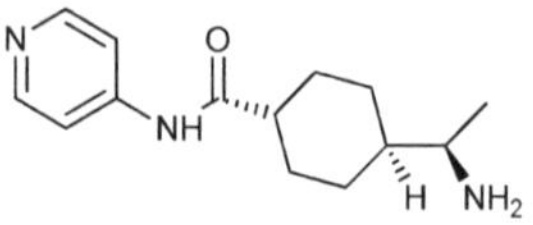

(7.2)

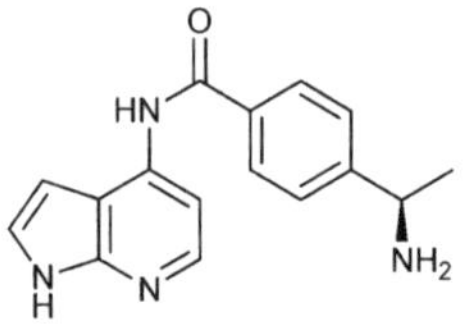

(7.3)

Figure 7.2 Structures of ROCK Inhibitors.

Pharmaceutical Co. Ltd. in partnership with UBE industries are currently in Phase I–II with DE-104 (structure undisclosed). Senju Pharmaceutical Co. Ltd., under licence from Mitsubishi Pharmaceutical Corp., is pursuing Y-39983 (**7.3**) (ROCK1 (K_i) 28 nM; ROCK2 (K_i) 2.4 nM) in Phase II studies, and has presented Phase I data for topical administration of an ophthalmic solution (SNJ-1656) to healthy volunteers.[13,14] Single instillation of SNJ-1656 over a concentration range dose dependently reduced intracocular pressure (IOP) without any significant adverse findings, either systemic or ocular. Inspire Pharmaceuticals have reported mild IOP effects with INS117548 (structure undisclosed. ROCK1 (K_i) 14.4 nM; ROCK2 (K_i) 5.8 nM) in a Phase I trial, claiming INS117548 demonstrated superior selectivity to Y-39983.[13] When tested in a KinaseProfiler[TM] system of 226 human kinases INS117548 (10 µM) only inhibited 28 kinases by $\geqslant 70\%$. In comparison Y-39983 inhibited 61 kinases under the same conditions. Aerie Pharmaceuticals Inc., a biotechnology company dedicated to the discovery and development of novel treatments for glaucoma, have reported Phase IIa results with AR-12286 (structure undisclosed). Top line data from a study of 88 patients demonstrated statistically significant IOP change from baseline, with no serious side effects reported. Based on the positive results Phase IIb trials have been initiated.

7.3   Inhibitors of p38α

In terms of drug discovery and development activities p38α represents one of the most intensely investigated kinase targets within the industry. There are several principal reasons for this, one being the pivotal role the kinase plays in the regulation of pro-inflammatory cytokines, including TNF-α, IL-1β and IL-6, inhibition of which using biologics has proved an effective strategy to treat several chronic inflammatory diseases, including rheumatoid arthritis (RA), psoriasis and inflammatory bowel disease (IBD).[15,16] Although four isoforms exist in the p38 family (p38α, p36β, p38δ, p38γ), analysis of expression and activities, in particular from disease patients, has highlighted p38α as the preferred target to inhibit.

A second reason lies in the drug design perspective, where the use of structural information has aided the generation of potent p38α inhibitors demonstrating exquisite selectivity, that dosed as oral agents deliver proven pre-clinical efficacy in a range of animal models, including chronic inflammatory, at low doses. In addition, the chemical and drug space of p38α inhibitors sits with imparting favorable compound properties required for candidates to transition into development (*e.g.* lack of genotoxicity, drug-drug interactions, hERG *etc.*). However, despite this success clinical benefit in chronic inflammatory diseases, in particular RA, has been disappointing, with many compounds failing to progress beyond Phase II. ARRY-797, BIRB-796, GSK-856553 (losmapimod), KC706, PH-797804, RO4402257 (pamapimod), SCIO-469, VX-745 and VX-702 fall into this category.[15–17] Factors causing discontinuation range from observed liver and CNS toxicity to a general lack of efficacy.

Several RA trials noted a reduction of C-reactive protein (CRP) levels, a biomarker for disease severity, in the early stages of dosing that failed to be sustained throughout the duration of the trial. This has led to the proposal that prolonged p38α inhibition may cause upregulation of compensatory inflammatory pathways.[18] In support of this concept inhibition of p38α has been suggested to dampen feedback loops that switch off the activity associated with upstream TAK1 and MLKs (MSK1 and MSK2). It is speculated that the resulting hyperactivation of these kinases kickstarts other pro-inflammatory pathways, such as those leading to the activation of JNK and IKKβ. The hyperactivation of JNK may also offer an explanation for the observed liver problems reported.[18]

As a consequence of the high failure rate in RA trials a focus for the development of p38α inhibitors has shifted away from a chronic inflammatory scene to acute pathologies or diseases where drugs can be delivered locally. Currently, only BMS-582949 (**7.4**) remains listed in RA clinical trials. The compound exhibits IC_{50} values of 13 and 50 nM for the inhibition of p38α and LPS induced TNF-α production in hPBMCs respectively.[19] Although more potent analogues have been described by BMS at the *in vitro* level, BMS-582949 proved to possess superior efficacy in both acute and chronic animal disease models as a result of its highly favorable pharmacokinetic (PK) properties. In a rat adjuvant arthritis model dosed under a therapeutic regimen BMS-582949 displayed dose dependent reduction in paw swelling, with significant effects observed at 10 mg kg^{-1} q.d. Moving to b.i.d. dosing resulted in improved efficacy at doses as low as 1 mg kg^{-1}. In line with the general p38α field the compound displays high selectivity, showing > 2000 fold selectivity against a diverse panel of 57 kinases, including p38γ and p38δ. In contrast only

5-fold selectivity is witnessed against p38β, with JNK2 and Raf registering 450- and 190-fold, respectively.[19]

A Phase I single ascending dose study comprised of monitoring safety, pharmacodynamics (PD) and PK. Using an ex-vivo LPS induced TNF-α protocol the highest dose, at 600 mg q.d., completely suppressed TNF-α inhibition.[15] In a second Phase I study RA patients received BMS-582949 co-dosed with methotrexate. Although no firm details of the trial have been published, grey information suggests that the highest dose, 300 mg q.d., tends towards improving Disease Activity Scores (DAS), a measure of disease remission.[15] Currently, the compound is being dosed in a 12-week Phase II study at 300 mg q.d. in methotrexate treated RA patients who are subject to an inadequate response from the disease-modifying anti-rheumatic drug (DMARD). The BMS p38α program is in collaboration with Ligand Pharmaceuticals Inc. Based on the Ligand pipeline BMS-582949 is associated with additional Phase II trials in patients with moderate to severe plaque psoriasis and atherosclerosis. There is accumulating evidence that activated p38 may play a key role in facilitating atherosclerosis, through influencing increased monocyte adhesion to endothelial cells lining the vascular endothelium and promoting other aspects of endothelial dysfunction.[20,21] Left unchecked atherosclerosis can lead to cardiovascular complications.

ARRY-797 (structure unknown) from Array Biopharma is listed as having completed successful Phase II trials for acute inflammatory pain. Signaling through the p38 pathway has been shown to play a role in the regulation of PGE2 production, inhibition of which in human whole blood has been demonstrated as a predictor of therapeutic analgesic effects in humans.[22,23] To assess an analgesic effect in the clinic, two placebo controlled postsurgical dental pain studies, in which two or more molars were removed, were conducted.[24] In the first study ARRY-797, given at doses of 200 mg pre and post surgery, and 400 mg post surgery only, produced a significant reduction in postsurgical pain intensity. In the second study ARRY-797 was dosed at 200, 400 and 600 mg post operation head to head with the 400 mg maximum labeled dose of the COX2 inhibitor celecoxib. Again, total pain relief over 6 h reached significance, with the 400 mg and 600 mg dosing regimen of ARRY-797 providing similar onset, magnitude and duration of analgesia as provided by 400 mg celecoxib. Both ARRY-797 clinical studies revealed a low incidence of adverse events. The p38α activity of ARRY-797 has been disclosed as IC_{50} 4.5 nM, with no reported activity against 200 kinases at 1 μM.[23] Whether the panel of kinases tested included the other isoforms of p38 (β, δ and γ) is unknown. Further development options of ARRY-797 are currently under evaluation.

Microglial cells play a pivotal role in the pathology of neuropathic pain. Accumulating evidence has linked p38 kinase activation to signal transduction in microglia and development and maintenance of neuropathic pain.[25] In support of this concept several p38α inhibitors have significantly reduced neuropathic pain symptoms in various animal models. Currently PH-797804

(7.5), a p38α inhibitor designed and under development from Pfizer, is in Phase II clinical trials for pain resulting from post herpetic neuralgia.

The compound is also in Phase II for COPD. Several lines of evidence have placed p38 kinase as a target of interest for COPD. Analysis of p38 activation in COPD patients has shown increases in key cells and tissues within the airways that closely associates with the degree of lung function impairment.[26] In a tobacco smoke animal model designed to present the pulmonary inflammation observed in COPD patients, dosing of a p38α inhibitor in either prophylactic or therapeutic mode led to marked pulmonary anti-inflammatory effects.[27] Data for PH-797804 from the Phase II trial has recently been reported at the 2010 European Respiratory Society Annual Congress.[28] Oral PH-797804 at doses of 3, 6 and 10 mg once daily for 6 weeks met the primary endpoint of significant improvements of trough fixed expiratory volume in the first second (FEV1) over placebo. In addition, analysis of the 3 and 6 mg dose groups using a Transitional Dyspnoea Index conveyed clinical meaningful results. Significant effects were also noted on secondary endpoints and CRP levels, which remained consistently lower across the treatment groups compared to placebo. The observations regarding CRP levels are interesting in light of the experiences encountered during RA trials, and it will be intriguing to see whether PH-797804 is further progressed as an oral agent into longer term COPD studies.

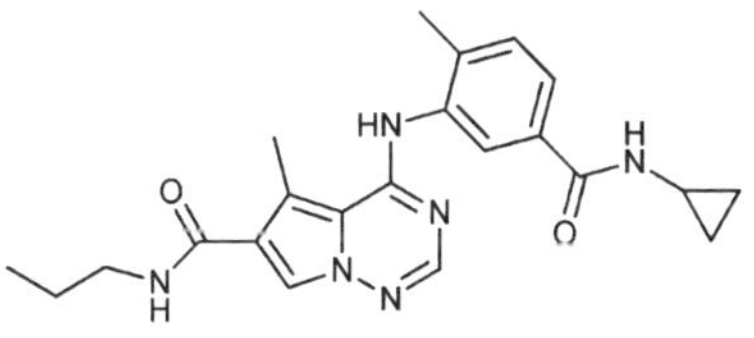

(7.4)

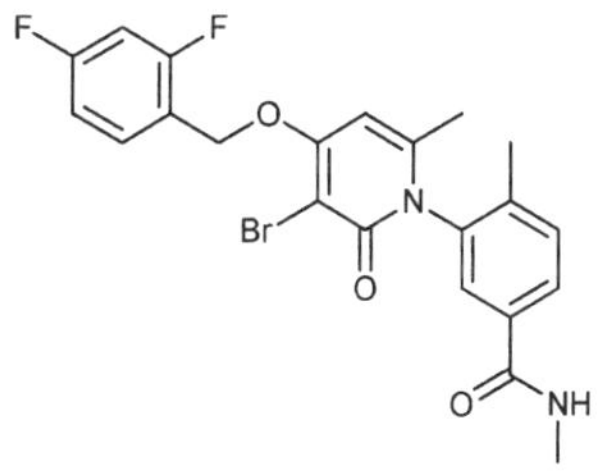

(7.5)

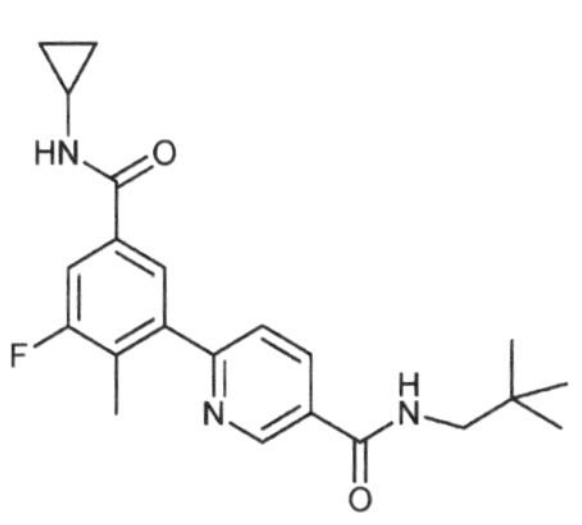

(7.6)

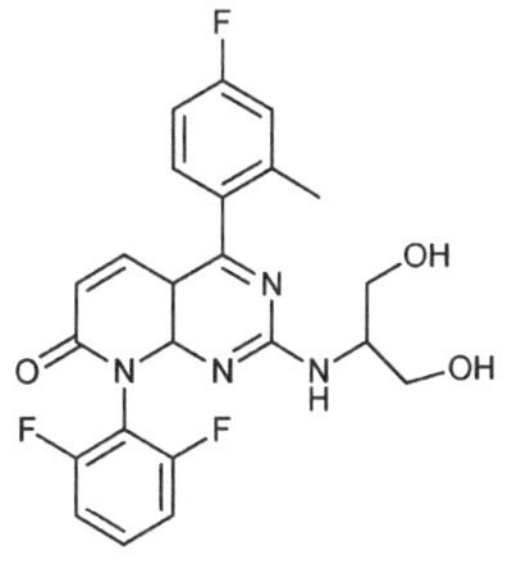

(7.7)

Figure 7.3   Structures of p38 Inhibitors.

PH-797804 is a sterically congested N-phenyl pyridone which exists as atropisomers. X-Ray crystallography has confirmed predictions from modeling, that it is the (a*S*) atropisomer that conveys high p38α potency, with a K_i of 5.8 nM.[29] Of the remaining p38 isoforms inhibition against p38β is approximately 5 fold lower, with K_i for both p38δ and p38γ > 200 μM. From profiling studies against a further 58 other kinases PH-797804 exerts exceptional selectivity, with windows of 500 fold or greater. A structural bioinformatics analysis has rationalized the selectivity at a molecular level.[29]

GSK has a number of clinical p38α inhibitors in various disease indications. However, little information is in the public domain regarding clinical data from such studies. The most advanced compound is oral losmapimod (**7.6**), which based on GSK's 2010 pipeline is in Phase II for COPD, depression, neuropathic pain and cardiovascular indications. With respect to COPD a 12-week study using losmapimod 7.5 mg b.i.d. to assess efficacy, safety and anti-inflammatory effects in subjects with COPD has been completed, with an extended 24 week study initiated in October 2010. In terms of cardiovascular indications Phase II studies focusing on atherosclerosis aspects have been completed. In addition, a 12 week study is currently recruiting to evaluate the effects of losmapimod on inflammatory markers, infarct size and cardiac function in patients with myocardiac infarction. Recent research has suggested that p38 may control some of the intracellular signaling following infarction that leads to diminished contractility, apototosis, fibrosis and heart failure.[30]

The SAR optimization of losmapimod has been described.[31] The compound exhibits a K_i against p38α of 7.9 nM, with approximately 3-fold selectivity over p38β. No significant inhibition was reported against a panel of 67 human kinases. Details of the analgesic and cardioprotective effects of losmapimod in relevant animal models have recently been disclosed.[31,32]

A second GSK drug candidate that has been extensively investigated in the clinic is dilmapimod (**7.7**). Based on GSK's 2010 pipeline the compound is only listed in Phase I for acute lung injury and acute respiratory distress syndrome by the i.v. route, with recruitment ongoing to support a Phase II trial. A role for p38 inhibitors in acute lung injury and acute respiratory distress syndrome is based on the findings linking p38 inhibition to a reduction of endotoxin induced lung inflammation, bronchoconstriction and neutrophil influx.[33,34]

Although COPD and cardiovascular disease endpoints are no longer of interest for dilmapimod, some clinical data conducted in these areas has been released that underpin the current clinical trials being undertaken with losmapimod. In a Phase I study in COPD patients a single oral dose of 7.5 mg and 25 mg of dilmapimod significantly suppressed whole blood biomarkers of inflammation.[35] It is well documented that patients who have undergone a percutaneous coronary intervention (PCI) to treat atherosclerosis driven stenotic coronary arteries of the heart often experience post inflammatory responses, in which elevated CRP levels are a biomarker for risk of clinical restenosis.[36] In a 28 day Phase II study, oral dilmapimod 7.5 mg given daily to patients three days before PCI and 25 days post PCI significantly attenuated post PCI inflammatory responses, as measured by a reduction of CRP levels relative to placebo throughout the duration of the study.[36] The sustained reduction of CRP levels noted above, coupled with similar observations in the PH-797804 COPD Phase II study, compared to RA trial data, throw up some interesting questions as to whether such differences reflect the complex and different signaling networks p38 plays in each disease. GSK also have a third p38 inhibitor in development in GSK610677 (structure undisclosed). The compound, targeting a COPD endpoint, has completed a Phase I study in healthy volunteers by the inhaled route. One of the potential advantages of going inhaled may lie in an increased efficacy and safety window. The results from the trial, and whether investment is made in further progression, are awaited.

7.4   Inhibitors of SYK

Spleen tyrosine kinase (Syk) is a cytoplasmic protein tyrosine kinase predominantly expressed in hematopoietic cells. Unchecked activation of Syk and its role in orchestrating downstream immunoreceptor signaling is thought to contribute to the pathology of several allergic and autoimmune diseases.[37,38] In allergic driven asthma, rhinitis and atopic dermatitis enhanced IgE production and binding to mast cell Fcε-R1 receptors trigger Syk activation, which leads to the synthesis and release of mediators, such as histamine, prostaglandins, leukotrienes and cytokines, that can instigate an acute allergic reaction.[37] Syk activity is upregulated in several autoimmune diseases, such as RA, multiple sclerosis and systemic lupus erythematosus *via* antigen driven T and B cell signaling, leading to the production of pro-inflammatory mediators which can further amplify and promote the pathogenesis of the disease.[37,38] For these reasons Syk has become an interesting and attractive target to inhibit and a target actively being pursued by many pharamaceutical and biotech companies.[39]

The most advanced compound is fostamatinib (**7.8**), with Phase III studies initiated in RA in Sept 2010. The Syk inhibitor was originally discovered by Rigel, but recently licensed to AstraZeneca in early 2010 under a structuring deal worth potentially over US $1 billion if specified development, regulatory and commercial sales are achieved. Fostamatinib is a phosphate prodrug of R406 (**7.9**), the active Syk component, which inhibits the kinase with a K_i = 30 nM and prevents Fcε-R1 dependent mast cell activation with an EC_{50} = 43 nM.[40] Pharmacology data on R406 showing low oral dose beneficial effects in models of airway hyper-responsiveness, airway inflammation and rodent collagen and immune complex mediated RA models have been published.[40–42]

R406 is stated to be relatively selective for Syk over other kinases, although full details have not been published.[42] Initially, off target kinases significantly inhibited by R406 were identified from classical biochemical screening, and then more fully assessed at the cellular level *via* inhibition of known phosphorylation targets of the respective kinases. From this exercise Flt3,

Lck and AK1–AK3 emerged as kinases inhibited by R406 at some 5–100-fold lower potency than Syk inhibition, with Flt3 being most potently inhibited.[42] Given the role these kinases play in inflammation it is likely that their inhibition adds to the beneficial effects observed.

The most recent published clinical data stems from a 6 month Phase IIb trial involving 457 RA patients who failed to respond to methotrexate treatment.[43] Patients receiving fostamatinib 150 mg q.d. or 100 mg b.i.d. achieved statistically higher American College Rheumatology (ACR) response rates than placebo. In addition both drug dosed groups delivered a significantly higher DAS28 remission rate compared to the placebo group. During the trial patients remained on methotrexate therapy. Common drug related adverse events corresponded to diarrhea, upper respiratory infection, neutropenia and hypertension, the latter of which occurred with a high frequency and responded to conventional anti-hypertensive medications. Data from the forthcoming Phase III program called OSKIRA (Oral Syk Inhibition in RA), involving around 2000 patients enrolled across two 12 month studies to further examine efficacy against patients who fail to respond to current DMARDs, and a 6 month study to assess effects against anti-TNFα non-responders, is awaited with great interest. Successful Phase III studies could lead to regulatory filings in 2013.

Little information is in the public domain regarding a second Rigel Syk inhibitor R343, believed to be structure (**7.10**).[37] Pfizer licensed the compound from Rigel at the preclinical stage for development as an inhaled drug for the treatment of allergic disorders. To date the compound has completed Phase I

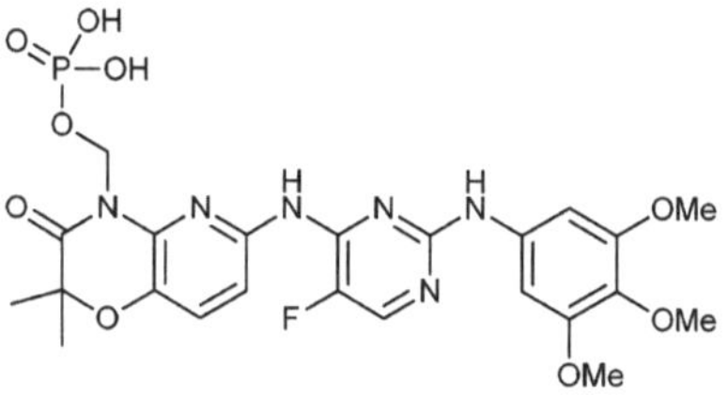

(7.8)

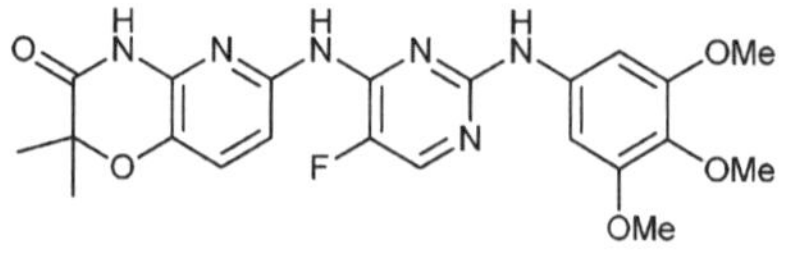

(7.9)

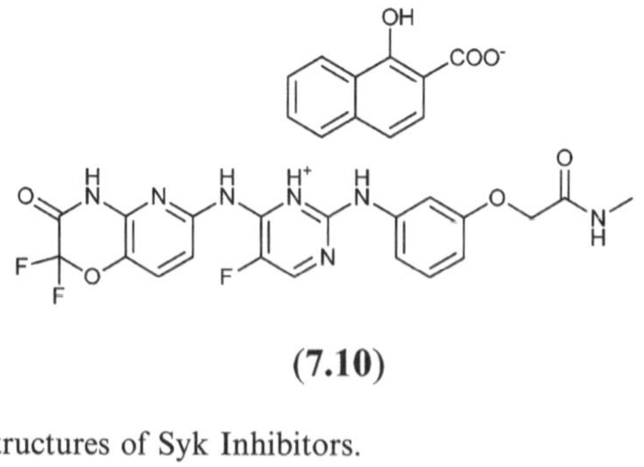

(7.10)

Figure 7.4 Structures of Syk Inhibitors.

studies to assess effects on lung function following an allergic challenge to asthmatic subjects. Results from the study are pending and will dictate whether Pfizer progresses to Phase II.

Portola Pharmaceuticals Inc, a US based biopharmaceutical company, are currently undertaking a Phase I study with an in-house oral Syk inhibitor PRT-062607 (structure unknown) to evaluate safety, PK and PD in healthy volunteers, ahead of investigating the drug in chronic inflammatory disease indications. Very scant information is known about the compound, except it is claimed to be highly selective for Syk from profiling in a broad panel of *in vitro* kinase and cellular assays.

7.5 Inhibitors of PKC

The protein kinase C (PKC) family of serine–threonine kinases plays an important role in the regulation of immune cell functions.[44] Although the family consists of 10 isotypes that share sequence and structural homology, the isotypes can be divided into three sub-groups based on their need for co-factors Ca^{2+} and diacylglycerol (DAG). One group containing "classical" isotypes α, β and γ require both co-factors. In contrast, "novel" isotypes δ, ε, η and θ depend only on DAG, while "atypical" isotypes ι/λ and ξ function without any co-factor involvement.[45] Through principally gene ablation approaches, the role of the different PKC isotypes involved in T-cell signaling and function has been unraveled. Key PKC isotypes revolve around PKCθ and PKCα. PKCθ appears critical to inducing T-cell receptor cytokine secretion and proliferation, and development of a robust Th17–Th2 dependent immune response. PKCα is essential for eliciting an immune response controlled by Th1, and expression of interferon-γ.[44] Thus, small molecule blockade of PKCθ and PKCα may inhibit T-cell activation and provide an approach to T-cell immunomodulation in autoimmune diseases and transplantation. To this end Novartis have discovered and are developing sotrastaurin (**7.11**).[45] The compound is currently in Phase II studies for a number of indications, including kidney and liver transplantation, ulcerative colitis and psoriasis.

Data from a small proof of concept study in psoriasis has been published.[46] A total of 32 patients received sotrastaurin at 25, 100, 200 and 300 mg b.i.d. for 2 weeks. Over this period a dose dependent improvement in lesions was observed. Assessing clinical severity using the psoriasis area severity index (PASI), the 300 mg b.i.d. gave a mean PASI score of 69%. Dosing at 200 mg b.i.d. delivered a smaller but still significant 40% PASI score. In the 300 mg b.i.d cohort the clinical improvement in psoriasis persisted 2 weeks post dosing, an observation not noted with the 200 mg b.i.d dosing arm. Observed adverse events were mild in intensity and generally the compound was well tolerated. The overall results with sotrastaurin compare favorably to proof of concept studies with antibody agents directed towards TNFα and IL-12–IL-23, critical cytokines which are involved in the pathogenesis of psoriasis.

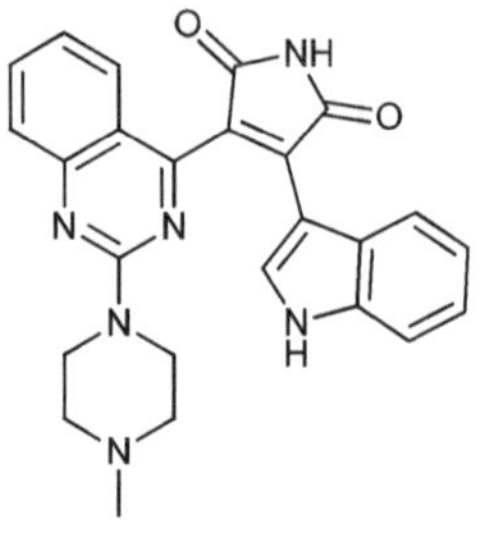

$$(7.11)$$

Figure 7.5 Structure of Sotrastaurin.

First clinical results in renal transplant recipients have been published.[47] *De novo* renal transplant patients received sotrastaurin 200 mg b.i.d + standard exposure tacrolimus (SET) or reduced exposure tacrolimus (RET). Mycophenolic acid (MPA) + SET formed the control regimen. After month 3, both sotrastaurin patient groups switched from tacrolimus to using MPA. Using composite efficacy failure as a primary endpoint, analysis of results preconversion showed both sotrastaurin + tacrolimus regimens (SET and RET) to be efficacious. In contrast inadequate efficacy, leading to premature termination of the study, was noted on post-conversion to sotrastaurin in combination with MPA. The results have triggered longer term evaluation of sotrastaurin + tacrolimus in renal transplantation, and an investigation in liver transplantation recipients.

The discovery and SAR of sotrastaurin has been communicated.[45] The compound is a potent single digit nM inhibitor on all classical and novel PKC isotypes assayed. Against a panel of 32 selected other kinases the compound proved highly selective, only inhibiting GSK3β with an IC_{50} just below 1 μM. Further results, in particular at the cellular level, have confirmed sotrastaurin as a selective inhibitor of T-cell activation with a different mode of action from currently used immunosuppressive agents, such as cyclosporine A and tacrolimus.[44]

7.6 Inhibitors of JAK

The binding of cytokines to cell surface receptors trigger activation of the Janus kinases, a family of cytoplasmic tyrosine kinases comprising of JAK1, JAK2, JAK3 and TYK2.[48,49] The activated JAKs play an integral role in the phosphorylation of STAT (signal transducers and activators of transcription) proteins, which subsequently combine as homo or heterodimers, translocate to the nucleus and induce gene transcription.[48,49] Built within this process is the alignment of specific cytokine signaling pathways to the engagement of specific JAKs in mediating the signaling events.[48] The γc chain containing cytokine receptors IL-2, IL-4, IL-7, IL-9, IL-15 and IL-21 signal through JAK1 and

JAK3. In contrast, receptors for the cytokines IL-10, IL-19, IL-20 and IL-22, as well as the common gp130 subunit of both the cytokine receptors IL-6 and IL-11, utilize primarily JAK1, but also associate with JAK2 and TYK2. Receptors binding hormones like cytokines, such as growth hormone, prolactin, Epo, thrombopoietin, and the IL-3 and GM-CSF receptors, which are central to hematopoietic cell development, use only JAK2. Finally, the p40 containing cytokines IL-12 and IL-23 employ JAK2 and TYK2, whereas IFNs rely on JAK1, but in combination with TYK2 for IFNα and JAK2 for IFNγ.[48] Given the dissection of the roles some cytokines play in driving certain autoimmune diseases, coupled with the learnings of the cell biology of JAKs and their intimate relationship with cytokine signaling, the discovery and development of JAK kinase inhibitors has attracted great attention. In the case of JAK3 this is further fueled through the observation of immune dysfunction in humans diagnosed with JAK3 mutated severe combined immunodeficiency (SCID), a rare inherited condition.[48,49]

In terms of expression JAK1, JAK2 and TYK2 are ubiquitously expressed, while JAK3 is restricted to only hematopoietic cells. Structurally the JAKs contain seven homology regions that span from the JH1 catalytic active kinase domain to the JH6–JH7 FERM domain. The role of the FERM domain is to anchor the JAK to the transmembrane cytokine receptor and regulate kinase activity. Adjacent to the JH1 domain is the JH2 pseudokinase domain, a unique feature of the JAKs within the protein tyrosine family, which provides critical regulatory functions despite lacking any intrinsic kinase activity.[48]

Amino acid alignment of the ATP binding site suggest only a handful of residues, which in nature represent a high level of conservation, differ across the four JAK isoforms. As a consequence achieving selectivity within the JAK family appears a formidable challenge.[50] However, the results of site directed mutational work, in which the ATP binding site of JAK2 was set to mimic JAK3, but retained wild type JAK2 characteristics in biochemical profiling against a panel of inhibitors, suggests a more complex picture, and the possibility that amino acids external to the ATP binding site, or domains outside the kinase domain, may influence specificity preferences for a given JAK isoform.[50] Of the reported inhibitors in the clinic several claim to display JAK isoform selectivity.

Pfizer's CP-690 550 (**7.12**) is currently the most advanced in development, with ongoing Phase III trials in RA and psoriasis. Although early reports described the inhibitor as JAK3 selective, re-examination of the human enzyme inhibitor potency by several groups has now classified the inhibitor as demonstrating a pan like profile (JAK1 K_i 0.68 nM, JAK2 K_i 0.99 nM, JAK3 K_i 0.24 nM, TYK2 K_i 4.39 nM).[51] A synopsis of the spectrum of inhibition of cytokine signaling by CP-690 550 in human whole blood has been published.[51] Results confirm the inhibitor can block multiple cytokine activation pathways associated with JAK1 and JAK3 signaling. Of note, however, is the observation of reduced potency in a human whole blood cell based assay dependent only on JAK2 signaling, suggesting that despite little separation of

JAK2 from JAK1–JAK3 at the *in vitro* enzyme level, increased selectivity windows (25–50-fold) may be at operation within a cellular context.[51] An impressive feature of CP-690 555 is its selectivity outside the JAK family, with lack of enzyme inhibition against > 300 other kinases tested.[50] The efficacy of CP-690 555 in various animal disease models has been reported, in particular with a focus on RA, where disease modifying effects are observed in concert with dose dependent decreases of pro-inflammatory cytokines, including IL-6.[51,52] The relevance of IL-6 is noteworthy given the recent approval of the IL-6 antibody (tocilizumab) to treat RA patients. Analysis of the PK–PD relationship from one RA study has confirmed a strong relationship with inhibition of both JAK1 and JAK3.[51]

The Phase III program for CP-690 550 in RA, known as ORAL, consists of multiple studies encompassing over 4000 RA patients. Results from one arm of this Phase III program have recently been released.[53] In ORAL solo (1045) moderately to severe RA patients, refractory at least to one DMARD, received 5 mg and 10 mg of CP-690 555 daily for six months. Analysis of efficacy measurements after three months confirmed the achievement of several primary endpoints, including a statistical reduction in the signs and symptoms of RA (*via* ACR scores) and improvement in physical function (*via* health assessment questionnaire disability index (HAQ-DI)). A third primary endpoint, focused on disease remission (*via* DAS28 scores), failed to reach statistical significance, but showed a numerically greater difference referenced to placebo.

To assess continued long-term safety and efficacy of CP-690 550 Pfizer have rolled over patients, who completed Phase II studies, into ORAL sequel (1024), an ongoing open label extension study, where moderately to severely active RA patients continue to receive CP-690 555, 5 mg and 10 mg twice daily, as either a monotherapy or on a methotrexate background.[53] ACR response rates, DAS28 and HAQ-DI scores at 24 months demonstrated efficacy had been maintained over an extended time period. The safety findings in ORAL solo remained consistent with those observed in the global Phase II RA program, with statistically significant decreases in neutrophil count and

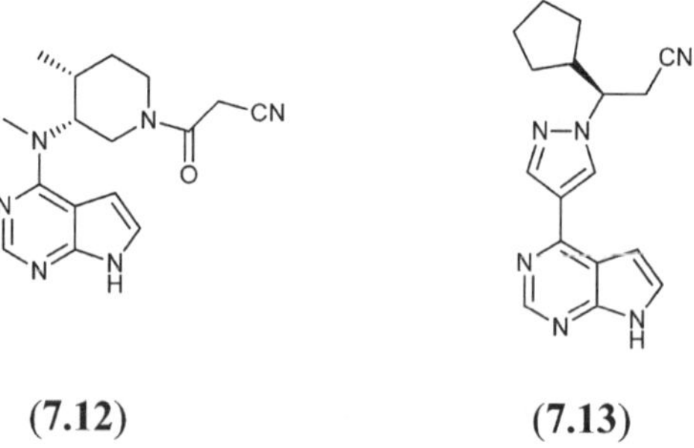

(7.12)   (7.13)

Figure 7.6   Structures of JAK inhibitors.

increases in both LDL and HDL cholesterol levels.[53] No new safety signals have emerged from the ORAL sequel extended study. Despite the findings of reduced neutrophil counts the incidence of infections remained relatively low. Mechanistically, the reduction in neutrophils could be explained by inhibition of JAK2 by CP-690 550. However, a number of recent studies focused specifically on PK–PD modeling aspects of the relationship between neutrophil reduction and JAK inhibition, in both RA patients and rodent models of RA, suggest the link to JAK2 inhibition at fully efficacious doses of CP-690 550 is unlikely to be the cause. The reduction in neutrophil count may in fact result as a consequence of the attenuation of disease inflammation.[51,54] A stronger safety concern is the clinical significance of increased LDL and HDL cholesterol levels and cardiovascular risk. The mechanism underlying this increase in lipid levels with CP-690 550 is currently unclear. Given the fact that RA patients treated with biologic agents also manifest this abnormal finding, it has been suggested that the effect is likely linked to the consequences of effective anti-inflammatory treatment on hepatic cholesterol production, rather than being related specifically to JAK.[55] Given the essential role JAK2 exerts on red blood cell homeostasis, monitoring of hemoglobin levels and anemia has become an important safety marker. In the ORAL solo study changes in hemoglobin levels were observed, but not deemed statistically significant.

Top level results from the latest Phase II study of the effects of oral CP-690 550 on patients with moderate to severe plaque psoriasis have also been recently released.[56] The primary endpoint of a statistically significant reduction in PASI 75 was achieved after 12 weeks dosing of CP-690 550 at 2, 5 and 15 mg twice daily. Further patient benefit was reported through significant increases in quality of life measurements using various clinically accepted dermatological, physical and mental scoring and indexing systems. As observed from RA trials dose dependent changes in neutrophil count, hemoglobin and LDL and HDL cholesterol levels were reported as potential safety flags, but not discussed in any detail. Nevertheless, the results have triggered Pfizer into initiation of a Phase III program called OPT (<u>O</u>ral <u>P</u>soriasis <u>T</u>reatment).

In Pfizer's current pipeline CP-690 550 is also listed in Phase II for transplant rejection and inflammatory bowel diseases. One 12-month Phase II study, involving comparative evaluation of the safety and efficacy in *de novo* renal allograft recipients of orally co-dosed CP-690 550 and mycophenolate mofetil (MMF) *versus* cyclosporine alone, has been completed but not yet reported. One may speculate that the results are encouraging, since patients from this study are being currently recruited to continue their medication in an open label extension trial to monitor long-term safety, tolerability and efficacy over a further five year period. A similar situation exists around renal transplant patients who responded well in a pilot study comparing orally co-dosed CP-690 550 and MMF with tacrolimus. In this case patient recruitment is complete, and the extension trial ongoing to monitor safety and efficacy of the drug dosing regimens through eight years post transplant, with a study completion date expected 2014. Pfizer have also completed, but not yet reported, Phase II safety and efficacy evaluations of twice daily oral dosing of CP-690 550 in patients with moderate and severe ulcerative colitis and Crohn's disease. Phase I data of CP-690 550 in transplant rejection and psoriasis, and the output of RA Phase II trials, has recently been reviewed.[52]

Dry eye disease is also an indication under investigation. To date a Phase I–II study comparing ophthalmic topical application of CP-690 550 to that of cyclosporine, delivered by the same route, has been completed but the results not disclosed. A follow on Phase II study is currently recruiting patients with dry eye disease, suggesting results from the initial trial merit further investigation and investment. The study design is centered on evaluating a dose response for the efficacy and safety of CP-690 550 in comparison to sodium hyaluronate, both drugs being applied to dry eye patients as eye drops.

Vertex Pharmaceuticals are currently conducting Phase II proof of concept studies with VX-509 (structure undisclosed) in moderate to severe RA patients. Interim data from the trial is expected in 2011. The compound is claimed as a potent JAK3 inhibitor, with 25–150 fold selectivity over the other JAK family members based on cellular readouts. Compared to non-JAK kinases selectivity is reported to be > 1000 fold, although the number and identity of kinases profiled has not been published. Dose ranging studies in healthy volunteers formed the basis of a 14 day Phase I trial. As part of the trial design biomarkers of JAK3 and JAK2 were measured. VX-509 demonstrated a dose dependent reduction in biomarkers for JAK3 whilst maintaining a high degree of selectivity over JAK2 biomarkers, in line with observations noted from cell based assays. Results also suggested VX-509 exerted a promising safety profile.

As an alternative to pan or selective JAK3 inhibitors, Incyte Corporation have focused on the inhibition of the JAK1–JAK2 axis, on the premise that this strategy can still deliver an effective treatment in autoimmune disorders through inhibition of specific JAK1 and JAK2 driven signaling of pathogenic cytokines, such as IL-6, IL-12 and IL-23. The recent approval of the IL-6 and IL-12–IL-23 antibody agents, tocilizumab and ustekinumab for RA and psoriasis respectively, add support to this concept.

Incyte have discovered and are investing in the clinical development of selective JAK1 and JAK2 inhibitors INCB18424 (**7.13**) and INCB28050 (structure undisclosed). INCB18424, with a JAK family IC_{50} biochemical enzyme profile of JAK1 = 2.7 nM, JAK2 = 4.5 nM, JAK3 = 322 nM and TYK2 = 19 nM, and a marked selectivity over other non-JAK kinases tested, is being pushed towards a psoriasis disease endpoint.[57] In support of this, top line clinical results from a 12-week Phase II study of topical INCB18424 in patients with mild to moderate psoriasis have been reported.[58] The primary endpoint of a statistical reduction in total lesion score (erythema + scaling + thickness) was achieved, in conjunction with a statistical significance in secondary endpoints connected with life assessment and PASI scores. From the three different dose strengths of INCB18424 cream formulation investigated no clinically significant effects were noted in hematology or other laboratory parameters. Plans to further progress the clinical development of INCB18424 are awaited with interest. Incyte have also used INCB18424 as a proof of concept compound in successful Phase II RA trials, but took the corporate decision to target and continue development of the compound against myeloproliferative disorders and cancer through a licensing agreement with Novartis. Psoriasis remains outside the agreement, leaving Incyte the rights to further develop and commercialize INCB18424 as a topical application for the disease.

INCB28050 replaced INCB18424 as the clinical development compound for evaluation in RA trials. INCB28050, like INCB18424, is selective over JAK3, with IC_{50} values of JAK1 = 5.9 nM, JAK2 = 5.7 nM, JAK3 = > 400 nM and TYK2 = 53 nM.[59] Selectivity against kinases outside the JAK family is of the order $\geqslant$ 100 fold over JAK1–JAK2. In cell based assays INCB28050 demonstrated inhibition of intracellular signaling of multiple pro-inflammatory cytokines, including IL-6 and IL-23 at concentrations < 50 nM.[59] The effects observed in cellular systems followed through to rodent disease models, where significant efficacy in a number of rat arthritis models, and a mouse delayed type hypersensitivity (DTH) model, was observed at oral doses of 10 mg kg^{-1} and below.[59] Further detailed analysis of the relationship between efficacy and JAK1–2 biomarkers concluded that complete and continuous pathway inhibition was not a prerequisite to exerting significant disease modifying effects in the arthritis models. Additional PK–PD modeling suggested concentrations of INCB28050 generated on oral dosing of 10 mg kg^{-1} would likely be insufficient to inhibit JAK3 signaling. To counter concerns over the potential hematological impact of JAK2 inhibition, analysis of terminal blood samples from one of the arthritis models revealed no significant differences in levels of hemoglobin, RBCs, total WBCs or absolute neutrophil or lymphocyte counts. Incyte hypothesize that the observations are consistent with a periodic incomplete inhibition paradigm established for efficacy.[59]

The latest Phase II results of oral INCB28050 dosed once daily at 4, 7 and 10 mg over a 24 week treatment period in patients with active RA has been recently reported at the 2010 American College of Rheumatology-Association of Rheumatology Health Professionals (ACR-ARHP) Annual Scientific Meeting.[60] Compared to placebo, ACR scores improved with numerically higher increases taken at week 24, compared to an interim analysis at week 12. Unfortunately, no statistics were presented with the data, the likely reason falling on insufficient patient numbers per treatment arm. DAS remission rates using CRP as a measure was also highlighted as evidence of encouraging efficacy. As with Pfizer RA clinical data increases in both HDL and LDL cholesterol were noted, up 25% depending on dosage. Although hemoglobin values declined in a dose dependent fashion, there appeared no dose related impairment of neutrophil counts. In addition, an unusual spike in platelet counts was recorded during the study for which reasons remain unclear at present.

Enthusiasm for INCB28050 as a potential new oral RA therapy, and use in other autoimmune diseases, is endorsed by the fact that in late 2009 Incyte and Eli Lilly and Company entered into an exclusive worldwide licensing and collaboration agreement concerning development and commercialization of INCB28050 and follow on compounds. The deal leveraged an upfront payment to Incyte of US $90 million, with up to a further US $665 million in potential milestones. At this moment in time a decision by Incyte and Lilly on whether to progress INCB28050 into Phase III RA trials on the back of the Phase II reported results is unknown.

7.7   Inhibitors of MAPKAPK5

Galapagos have developed a proprietary adenoviral short hairpin RNA (shRNA) based target discovery platform (SilenceSelect®) to probe for new drug targets within disease relevant cells. Starting from 12 000 shRNA knock-down constructs, application of the SilenceSelect® technology to RA synovial fibroblasts presented initially 323 hits which blocked cytokine mediated MMP1 expression. Of these only 152 inhibited collagen degradation. Through a process of further extensive target analysis and knock-in experiments the 152 hits yielded 14 validated RA targets, which included mitogen-activated protein kinase activated protein kinase 5 (MAPKAPK5).[61] Although MAPKAPK5 was initially cloned some 10 years ago its physiological role and downstream targets remain largely undefined, and its association with RA was not known prior to the Galapagos work. Using the Biofocus DPI SoftFocus® kinase collection a screening campaign against a functional MAPKAPK5 assay provided several hit series. Subsequent biochemical and oral ADME optimization delivered GLPG0259 (structure unknown), which demonstrated significant bone protection and reduced inflammation in an industry standard RA model on par with etanercept. On the back of successful Phase I clinical trials in healthy volunteers, including compatibility with methotrexate and confirmation of once a day capsule dosing, a Phase II study in 180 methotrexate insensitive RA patients has just begun, with the trial expected to be completed by the end of 2011. Under a collaboration agreement with Galapagos, Janssen Pharmaceuticals have the option to license the program on completion and success of this proof of concept clinical trial.

7.8   Multikinase Inhibitors

Pazopanib (**7.14**) is a multikinase inhibitor that dampens the intracellular signaling of a myriad of receptor tyrosine kinases, including VEGF receptors 1,2 and 3, PDGFR, c-Kit and fibroblast growth factor 1 (FGFR1). It has found utility in the treatment of choroidal neovascularization (CNV) associated with age related macular degeneration (AMD).[62] The compound is currently undergoing Phase II clinical trials as an ophthalmic formulation in AMD patients.

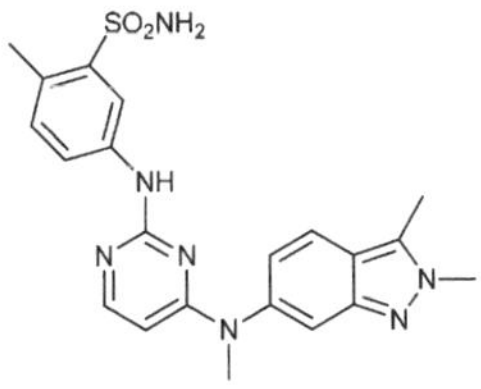

$$(7.14)$$

Figure 7.7   Structure of Pazopanib.

7.9   Inhibitors of Bcr-Abl, c-Kit and PDGFR – Imatinib (Gleevec®) and Nilotinib (Tasigna)

Pulmonary Arterial Hypertension (PAH) manifests itself when blood flow through pulmonary arteries is restricted, causing an increased strain on the right side of the heart to pump blood through to the lungs. In the early stages of the disease symptoms, such as breathlessness, chest tightness and fatigue can often be mistaken for other conditions, leading to a delay in diagnosis of PAH, often to the point where once the disease has been confirmed patients have progressed to a moderate-severe state. To aid clinicians in assessing the severity of PAH the World Health Organization (WHO) have created a 4-tier classification system based on the functional status of the patients, which correlates with prognosis of the disease. Patients diagnosed with WHO class I–II status who then start therapy, often prolong life expectancy by approximately five years. In contrast, the medium survival rate of patients belonging to the severe status class IV is cut to around six months. Treatment options are limited, opening up opportunities for new therapeutic approaches.

A suggestion that imatinib (**7.15**), an approved oncology drug, could provide a benefit for PAH came from the observation of improved clinical status in isolated patient settings.[63] To study the safety, tolerability and efficacy of oral imatinib in PAH under a more formal setting 59 patients were recruited to a controlled 24-week trial. Compared to placebo, imatinib-treated patients showed a significant decrease in pulmonary vascular resistance coupled with a significant increase in cardiac output.[64] However, no significant changes were observed in the mean 6-minute walking distance (6MWD), the primary endpoint of the study. Post hoc subgroup analysis of the same data did reveal a significant increase in the 6MWD associated with patients classified as more severe, suggesting imatinib may produce more profound effects against latter stages of PAH. An open label 2–3 year extension study of the treated patients showed functional stabilization to be maintained over this time period. As a result of these findings a Phase III study is currently recruiting severe PAH patients to evaluate the long-term safety, tolerability and efficacy of oral imatinib dosed at 200 mg and 400 mg. The study completion date is envisaged in 2013. Imatinib is a relatively selective kinase inhibitor, with notable inhibition only at Bcr-Abl, c-Kit and PDGFR. All three kinases can be implicated in the pathogenesis of PAH, and at this moment in time it is unclear whether the effects of imatinib are through inhibition of a single kinase pathway or through inhibition of more than one signaling pathway and exertion of polypharmacological effects.

Nilotinib (Tasigna) (**7.16**) is an FDA approved analogue of imatinib for the use in an oncology setting with patients who have become resistant to the actions of imatinib. Investigation of the efficacy of nilotinib in preclinical PAH models suggest it may elicit a greater payload than imatinib.[65] As a consequence a Phase II 24 week study to compare the efficacy, safety, PK and tolerability of nilotinib to imatinib is currently recruiting.

Imatinib is also being evaluated as a treatment for systemic sclerosis (SSc), an autoimmune disease that affects skin or an array of internal organs, including musculoskeletal, pulmonary, gastrointestinal and renal. Increased disability and mortality are experienced by patients with advanced SSc. Unfortunately, effective therapeutic options are currently lacking. In terms of pathophysiology, the disease is characterized by over production of collagen, *via* excessive activation of TGF-β and PDGF profibrotic pathways.[66] Within the TGF-β signaling pathway Bcr-Abl represents a key downstream mediator. As a consequence of the role Bcr-Abl and PDGF may play in SSc, attention has particularly focused on the evaluation of imatinib as a treatment option for the disease. Following demonstration of prevention of fibrosis and regression of established fibrosis in pre-clinical models of SSc, imatinib is undergoing a number of Phase II clinical trials to evaluate efficacy, safety and tolerability in a clinical setting.[66] To date interim results from a 1 year open label trial have reported encouraging data. Treatment of severe SSc patients with imatinib at 400 mg per day led to improvements in several endpoints, including skin scores and pulmonary function.[67] Plans are also ongoing to evaluate nilotinib in SSc, and patients are being actively recruited to support an open label small Phase IIa study.

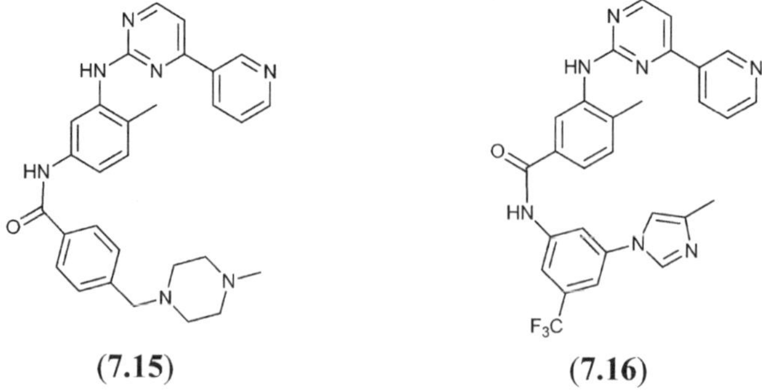

(7.15)   (7.16)

Figure 7.8 Structures of Imatinib and Nilotinib.

7.10 Conclusion

The first wave of kinase inhibitors to successfully reach the market centered on providing clinical benefit within the oncology franchise. Over the last decade or so with the advances in kinase knowledge, in particular in kinase selectivity profiling, kinase inhibitor design using structural and computational technologies, and the unraveling of kinase signaling networks, there has been a drive to discover and develop kinase inhibitors to treat non-oncology indications. This chapter has served to illustrate the leading kinase inhibitor players currently undergoing clinical evaluation towards this goal. Given several are entering Phase III status there is hope in the not too distant future that new kinase inhibitors will be approved that can sit along fasudil, as new treatments for diseases outside oncology.

References

1. D. R. Duckett and M. D. Cameron, *Expert Opin. Drug Metab. Toxicol.*, 2010, **6**, 1175.
2. Bioseeker Group AB. Protein Kinase Inhibitors in Oncology Drug Pipeline Update 2010. http://www.researchandmarkets.com/reports/1196697/.
3. O. Fedorov, S. Müller and S. Knapp, *Nat. Chem. Biol.*, 2010, **6**, 166.
4. A. J. Olaharski, N. Gonzaludo, H. Bitter, D. Goldstein, S. Kirchner, H. Uppal and K. Kolaja, *PLoS Comput. Biol.*, 2009, **5**, e1000446.
5. R. Morphy, *J. Med. Chem.*, 2010, **53**, 1413.
6. M. F. Olson, *Curr. Opin. Cell Biol.*, 2008, **20**, 242.
7. C. Hahmann and T. Schroeter, *Cell. Mol. Life Sci.*, 2010, **67**, 171.
8. J. Bain, L. Plater, M. Elliott, N. Shpiro, C. J. Hastie, H. McLauchlan, I. Klevernic, J. S. C. Arthur, D. R. Alessi and P. Cohen, *Biochem. J.*, 2007, **408**, 297.
9. M. Honjo, H. Tanihara, M. Inatani, N. Kido, T. Sawamura, B. Y. T. Yue, S. Narumiya and Y. Honda, *Invest. Ophthalmol. Visual Sci.*, 2001, **42**, 137.
10. Y. Delaney, T. E. Walshe and C. O'Brien, *Optom. Vis. Sci.*, 2006, **83**, 406.
11. Y. Kitaoka, Y. Kitaoka, T. Kumai, T. T. Lam, K. Kuribayashi, K. Isenoumi, Y. Munemasa, M. Motoki, S. Kobayashi and S. Ueno, *Brain Res.*, 2004, **1018**, 111.
12. M. Honjo, H. Tanihara, T. Kameda, T. Kawaji, N. Yoshimura and M. Araie, *Invest. Ophthalmol. Visual Sci.*, 2007, **48**, 5549.
13. W. M. Peterson, J. Lampe, T. Navratil, Y. Paran, Z. Kam, E. A. Hennes, B. T. Gabelt, B. Geiger, P. L. Kaufman and J. L. Vittitow, presented at the Association for Research in Vision and Ophthalmology (ARVO) Annual Meeting, Fort Lauderdale, Florida, 2008, Abstract 3816.
14. H. Tanihara, M. Inatani, M. Honjo, H. Tokushige, J. Azuma and M. Araie, *Arch. Ophthamol.*, 2008, **126**, 309.

15. D. M. Goldstein, A. Kuglstatter, Y. Lou and M. J. Soth, *J. Med. Chem.*, 2010, **53**, 2345
16. L. R. Coulthard, D. E. White, D. L. Jones, M. F. McDermott and S. A. Burchill, *Trends Mol. Med.*, 2009, **15**, 369.
17. S.-I. Choi and E. Brahn, *Autoimmunity*, 2010, **43**, 1.
18. P. Cohen, *Curr. Opin. Cell Biol.*, 2009, **21**, 317.
19. C. Liu, J. Lin, S. T. Wrobleski, S. Lin, J. Hynes, H. Wu, A. J. Dyckman, T. Li, J. Wityak, K. M. Gillooly, S. Pitt, D. R. Shen, R. F. Zhang, K. W. McIntyre, L. Salter-Cid, D. J. Shuster, H. Zhang, P. H. Marathe, A. M. Doweyko, J. S. Sack, S. E. Kiefer, K. F. Kish, J. A. Newitt, M. McKinnon, J. H. Dodd, J. C. Barrish, G. L. Schieven and K. Leftheris, *J. Med. Chem.*, 2010, **53**, 6629.
20. Z. Liu and W. Cao, *Endocr., Metab. Immune Disord.: Drug Targets*, 2009, **9**, 38.
21. P. Natarajan and C. P. Cannon, *Arterioscler., Thromb., Vasc. Biol.*, 2010, **30**, 2081.
22. D. R. H. Huntjens, M. Danhof and O. E. Della Pasqua, *Rheumatology*, 2005, **44**, 846.
23. L. Carter, K. Litwiler, J. Neale, S. Brown, N. Klopfenstein, S. Karan, S. Johnson, J. Vanderlugt, J. Yates and S. Rojas-Caro, presented at the American College of Rheumatology–Association of Rheumatology Health Professionals (ACR–ARHP) Annual Scientific Meeting, San Francisco, California, 2008, Abstract 359.
24. A. E. Remmers, C. Martinez, S. Daniels and J. Yates, presented at the American College of Rheumatology–Association of Rheumatology Health Professionals (ACR–ARHP) Annual Scientific Meeting, San Francisco, California, 2008, Abstract 357.
25. R.-R. Ji and M. R. Suter, *Mol. Pain*, 2007, **3**, 33.
26. T. Renda, S. Baraldo, G. Pelaia, E. Bazzan, G. Turato, A. Papi, P. Maestrelli, R. Maselli, A. Vatrella, L. M. Fabbri, R. Zuin, S. A. Marsico and M. Saetta, *Eur. Respir. J.*, 2008, **31**, 62.
27. S. Medicheria, M. F. Fitzgerald, D. Spicer, P. Woodman, J. Y. Ma, A. M. Kapoun, S. Chakravarty, S. Dugar, A. A. Protter and L. S. Higgins, *J. Pharmacol. Exp. Ther.*, 2008, **324**, 921.
28. W. MacNee, R. Allan, I. Jones, M. C. De Salvo and L. Tan, presented at the European Respiratory Society Annual Congress, Barcelona, Spain, 2010, Abstract P4000.
29. L. Xing, H. S. Shieh, S. R. Selness, R. V. Devraj, J. K. Walker, B. Devadas, H. R. Hope, R. P. Compton, J. F. Schindler, J. L. Hirsch, A. G. Benson, R. G. Kurumbail, R. A. Stegeman, J. M. Williams, R. M. Broadus, Z. Walden and J. B. Monahan, *Biochemistry*, 2009, **48**, 6402.
30. J. E. Clark, N. Sarafraz and M. S. Marber, *Pharmacol. Ther.*, 2007, **116**, 192.
31. N. M. Aston, P. Bamborough, J. B. Buckton, C. D. Edwards, D. S. Holmes, K. L. Jones, V. K. Patel, P. A. Smee, D. O. Somers, G. Vitulli and A. L. Walker, *J. Med. Chem.*, 2009, **52**, 6257.

32. R. N. Willette, M. E. Eybye, A. R. Olzinski, D. J. Behm, N. Aiyar, K. Maniscalco, R. G. Bentley, R. W. Coatney, S. F. Zhao, T. D. Westfall and C. P. Doe, *J. Pharmacol. Exp. Ther.*, 2009, **330**, 964.

33. S. P. Nash and R. M. Heuertz, *Int. Immunopharmacol.*, 2005, **5**, 1870.

34. S. Schnyder-Candrian, V. F. Quesniaux, F. Di Padova, I. Maillet, N. Noulin, I. Couillin, R. Moser, F. Erard, B. B. Vargaftig, B. Ryffel and B. Schnyder, *J. Immunol.*, 2005, **175**, 262.

35. D. Singe, L. Smyth, Z. Borrill, L. Sweeney and R. Tal-Singer, *J. Clin. Pharmacol.*, 2010, **50**, 94.

36. L. Sarov-Blat, J. M. Morgan, P. Fernandez, R. James, Z. Fang, M. R. Hurle, C. Baidoo, R. N. Willette, J. J. Lepore, S. E. Jensen and D. L. Sprecher, *Arterioscler., Thromb., Vasc. Biol.*, 2010, **30**, 2256.

37. M. Riccaboni, I. Bianchi and P. Petrillo, *Drug Discovery Today*, 2010, **15**, 517.

38. V. C. Kyttaris and G. C. Tsokos, *Clin. Immunol.*, 2007, **124**, 235.

39. P. Ruzza, B. Biondi and A. Calderan, *Expert Opin. Ther. Pat.*, 2009, **19**, 1361.

40. S. Matsubara, G. Li, K. Takeda, J. E. Loader, P. Pine, E. S. Masuda, N. Miyahara, S. Miyahara, J. J. Lucas, A. Dakhama and E. W. Gelfand, *Am. J. Respir. Crit. Care Med.*, 2006, **173**, 56.

41. S. Matsubara, T. Koya, K. Takeda, A. Joetham, N. Miyahra, P. Pine, E. S. Masuda, C. H. Swasey and E. W. Gelfand, *Am. J. Respir. Cell Mol. Biol.*, 2006, **34**, 426.

42. S. Braselmann, V. Taylor, H. Zhao, S. Wang, C. Sylvain, M. Baluom, K. Qu, E. Herlaar, A. Lau, C. Young, B. R. Wong, S. Lovell, T. Sun, G. Park, A. Argade, S. Jurcevic, P. Pine, R. Singh, E. B. Grossbard, D. G. Payan and E. S. Masuda, *J. Pharmacol. Exp. Ther.*, 2006, **319**, 998.

43. M. E. Weinblatt, A. Kavanaugh, M. C. Genovese, T. K. Musser, E. B. Grossbard and D. B. Magilavy, *N. Engl. J. Med.*, 2010, **363**, 1303.

44. J.-P. Evenou, J. Wagner, G. Zenke, V. Brinkmann, K. Wagner, J. Kovarik, K. A. Welzenbach, G. Weitz-Schmidt, C. Guntermann, H. Towbin, S. Cottens, S. Kaminski, T. Letschka, C. Lutz-Nicoladoni, T. Gruber, N. Hermann-Kleiter, N. Thuille and G. Baier, *J. Pharmacol. Exp. Ther.*, 2009, **330**, 792.

45. J. Wagner, P. von Matt, R. Sedrani, R. Albert, N. Cooke, C. Ehrhardt, M. Geiser, G. Rummel, W. Stark, A. Strauss, S. W. Cowan-Jacob, C. Beerli, G. Weckbecker, J. P. Evenou, G. Zenke and S. Cottens, *J. Med. Chem.*, 2009, **52**, 6193.

46. H. Skvara, M. Dawid, E. Kleyn, B. Wolff, J. G. Meingassner, H. Knight, T. Dumortier, T. Kopp, N. Fallahi, G. Stary, C. Burkhart, O. Grenet, J. Wagner, Y. Hijazi, R. E. Morris, C. McGeown, C. Rordorf, C. E. M. Griffiths, G. Stingl and T. Jung, *J. Clin. Invest.*, 2008, **118**, 3151.

47. K. Budde, C. Sommerer, T. Becker, A. Asderakis, F. Pietruck, J. M. Grinyo, P. Rigotti, J. Dantal, J. Ng, M. J. Barten and M. Weber, *Am. J. Transplant.*, 2010, **10**, 571.

48. K. Ghoreschi, A. Laurence and J. J. O'Shea, *Immunol. Rev.*, 2009, **228**, 273.

49. J. J. O'Shea, M. Pesu, D. C. Borie and P. S. Changelian, *Nat. Rev. Drug Discovery*, 2004, **3**, 555.

50. J. E. Chrencik, A. Patny, I. K. Leung, B. Korniski, T. L. Emmons, T. Hall, R. A. Weinberg, J. A. Gormley, J. M. Williams, J. E. Day, J. L. Hirsch, J. R. Kiefer, J. W. Leone, H. D. Fischer, C. D. Sommers, H. C. Huang, E. J. Jacobsen, R. E. Tenbrink, A. G. Tomasselli and T. E. Benson, *J. Mol. Biol.*, 2010, **400**, 413.

51. D. M. Meyer, M. I. Jesson, X. Li, M. M. Elrick, C. L. Funckes-Shippy, J. D. Warner, C. J. Gross, M. E. Dowty, S. K. Ramaiah, J. L. Hirsch, M. J. Saabye, J. L. Barks, N. Kishore and D. L. Morris, *J. Inflammation*, 2010, **7**, 41.

52. K. West, *Curr. Opin. Invest. Drugs*, 2009, **10**, 491.

53. http://www.pfizer.com/news/press_releases/pfizer_press_releases.jsp= (accessed Nov 7 2010).

54. P. Gupta, L. E. Friberg, M. O. Karlsson, S. Krishnaswami and J. French, *J. Clin. Pharmacol.*, 2010, **50**, 679.

55. J. M. Kremer, B. J. Bloom, F. C. Breedveld, J. H. Coombs, M. P. Fletcher, D. Gruben, S. Krishnaswami, R. Burgos-Vargas, B. Wilkinson, C. A. F. Zerbini and S. H. Zwillich, *Arthritis Rheum.*, 2009, **60**, 1895.

56. http://pfizer.mediaroom.com/index.php?s=5149&item=14387 (accessed Oct 7 2010).

57. J. Fridman, R. Nussenzveig, P. Liu, S. Shepard, J. Rodgers, T. Burn, P. Haley, P. Scherle, R. Newton, G. Hollis, S. Friedman, S. Verstovsek and K. Vaddi, presented at the American Society of Hematology, Atlanta, Georgia, 2007, Abstract 3538.

58. http://investor.incyte.com/phoenix.zhtml?c=69764&p=irol-newsArticle&ID=1333640&highlight= (accessed Sep 21 2009).

59. J. S. Fridman, P. A. Scherle, R. Collins, T. C. Burn, Y. Li, J. Li, M. B. Covington, B. Thomas, P. Collier, M. F. Favata, X. Wen, J. Shi, R. McGee, P. J. Haley, S. Shepard, J. D. Rodgers, S. Yeleswaram, G. Hollis, R. C. Newton, B. Metcalf, S. M. Friedman and K. Vaddi, *J. Immunol.*, 2010, **184**, 5298.

60. http://investor.incyte.com/phoenix.zhtml?c=69764&p=irol-newsArticle&ID=1494997&highlight= (accessed Nov 10 2010).

61. M. Andrew, R. Brys, N. Vandeghinste, P. Pujuguet, F. Namour, G. Lorenzon, M. Chambers, W. Schmidt, A. Clase, V. Birault and G. Dixon, presented at the Annual European Congress of Rheumatology, Copenhagen, Denmark, 2009, Abstract 146.

62. K. Takahashi, Y. Saishin, Y. Saishin, A. G. King, R. Levin and P. A. Campochiaro, *Arch. Ophthalmol.*, 2009, **127**, 494.

63. R. Souza, O. Sitbon, F. Parent, G. Simonneau and M. Humbert, *Thorax* 2006, **61**, 736.

64. H. A. Ghofrani, N. W. Morrell, M. M. Hoeper, H. Olschewski, A. J. Peacock, R. J. Barst, S. Shapiro, H. Golpon, M. Toshner. F. Grimminger and S. Pascoe, *Am. J., Respir. Crit. Care Med.*, 2010, **182**, 1171.
65. N. Duggan, O. Bonneau, M. Hussey, D. A. Quinn, P. Manley, C. Walker, J. Westwick and M. J. Thomas, *Am. J, Respir. Crit. Care Med.*, 2010, **181**, A6304.
66. J. K. Gordon and R. F. Spiera, *Curr. Opin. Rheumatol.*, 2010, **22**, 690.
67. J. Gordon, J. Mersten, S. Lyman, S. A. Kloiber, H. F. Wildman, M. K. Crow, K. A. Kirou and R. F. Spiera, *Arthritis Rheum.*, 2009, **60**, S10, 10, 606.

Allosteric Activators of Glucokinase (GK) for the Treatment of Type 2 Diabetes

KEVIN R. GUERTIN

Department of Discovery Chemistry, Roche Research Center, 340 Kingsland St., Nutley, NJ 07110, USA

8.1 Introduction

Kinases, also referred to as phosphotransferases, are key modulators of a variety of cellular and biochemical pathways and are known to phosphorylate a host of substrates including lipids, proteins, nucleotides, aminoacids and carbohydrates, thereby facilitating intracellular signaling. From a traditional standpoint, the inhibition of dysregulated kinases has received considerable focus in drug discovery to treat a number of diseases such as inflammation, cancer and diabetes.

Given that gain-of-function dysregulation of kinases is frequently observed in the oncology therapy area (*i.e.* MEK, CDK's *etc.*), kinase activation has typically not been an area of intense research focus in the drug discovery arena. However, the upregulation or activation of select and specific kinases such as AMP-activated protein kinase (AMPK), insulin receptor tyrosine kinase (IRTK), protein kinase C (PKC), and 3-phosphoinositide-dependant protein kinase 1 (PDK1) and glucokinase (GK) has recently received greater attention in the pharmaceutical industry. In contrast to targeting the orthosteric or ATP-binding site common in kinase inhibition strategies, the activation of a

RSC Drug Discovery Series No. 19
Kinase Drug Discovery
Edited by Richard A. Ward and Frederick Goldberg

Published by the Royal Society of Chemistry, www.rsc.org

kinase with a small molecule typically involves targeting an allosteric site on the catalytic domain of the kinase. Binding of the small molecule can induce a conformational change that translates to the active site and affects the reaction rate of the kinase, which may occur cooperatively with substrate. By targeting an allosteric site of a specific kinase, improved selectivity *versus* other kinases may be achieved.

The activation or upregulation of glucokinase (GK) has recently emerged as an important field of research specifically in the search for drugs targeting type 2 diabetes mellitus.[1] GK catalyses the phosphorylation of glucose to glucose 6-phosphate which is the first and the rate limiting reaction in glycolysis.[2] GK is mainly expressed in pancreatic β-cells and hepatocytes, both of which are known to play a pivotal role in whole-body glucose homeostasis.[3] Therefore, small molecule GK activators have considerable potential for the treatment of type 2 diabetes mellitus.[4,5]

8.2   Glucokinase Structure and Function

GK, also referred to as hexokinase IV (or D) is one of four members of the hexokinase family of sugar kinases. GK is a monomeric enzyme that contains both a large and small domain connected by a hinge region. In mammals, hexokinases I–III are $\sim$100 kDa monomeric proteins that exhibit a low K_m ($<$ 0.2 mM) for glucose. In contrast, GK is a $\sim$50 kDa monomeric enzyme that has a lower affinity for glucose (K_m = 8mM), and is not inhibited by glucose 6-phosphate.[6] Given that the K_m for glucose is in the normal physiological range of plasma glucose levels, GK acts as a molecular sensor and the flux of glucose through the GK pathway increases as the concentration of glucose in the blood rises from fasting (5 mM) to post-prandial (10–15 mM) levels following a carbohydrate based meal.[7]

GK displays positive kinetic cooperativity with glucose (Hill number [nH] $\sim$1.6) which is unusual for a monomeric enzyme bearing a single active site. With the aid of X-ray crystallography,[8] stopped-flow and equilibrium binding fluorescence spectrophotometry,[9,10] and molecular dynamic studies,[11] it was determined that the kinetic cooperativity apparently arises from slow glucose induced conformational changes between three distinct forms of GK (super-open, open and closed), each displaying different affinity and kinetic parameters for glucose. By comparing the X-ray structures of the apo and GK activator bound forms of GK,[8] Kamata and co-workers formulated a model that rationalizes the positive cooperativity of GK with respect to glucose (Figure 8.1). Low glucose concentrations favor the thermodynamically more stable "super-open", low affinity conformation resulting in catalysis through a slow catalytic cycle. In contrast, as glucose concentrations rise, the "open" and "closed" high affinity GK conformations are favored and the enzyme changes catalytic cycles employing the fast catalytic cycle to a greater extent. In principal, small molecule GK activators bind to and stabilize the high glucose affinity conformations and increase the affinity of GK for glucose

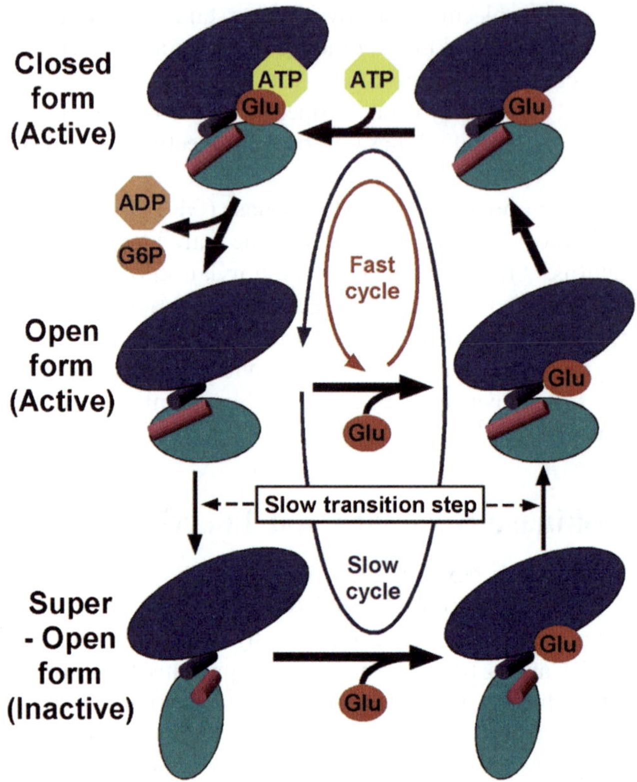

Figure 8.1 Kamata's model that rationalizes the positive cooperativity of GK with respect to glucose. Reproduced from K. Kamata, M. Mitsuya, T. Nishimura, J. Eiki and Y. Nagata, *Structure,* 2004, **12**, 429 with permission from Elsevier.

by binding to an allosteric site located in the hinge region at the interface of the large and small domains of the enzyme.[8] Specifically, the allosteric site is located 20 angstroms remote from and opposite to the glucose binding site. This site is not present in the "super-open" conformation, but is seen in both the open and closed conformations. With a small molecule GK activator bound, GK can only exist in the open or closed states effectively locking the enzyme into the fast catalytic cycle, even at low glucose concentrations. The X-ray crystal structure of a glucokinase activator 8.20 bound in the allosteric site of the active conformation of GK is illustrated in Figure 8.2,[8] with key residues from mutation studies (*vide infra*) highlighted. As illustrated in Figure 8.3, the loop encircling the GK activator 8.20 bound in the allosteric site of the active GK conformation collapses into the site in the "super-open" inactive conformation coupled with significant conformational changes in the lower domain of GK.[8]

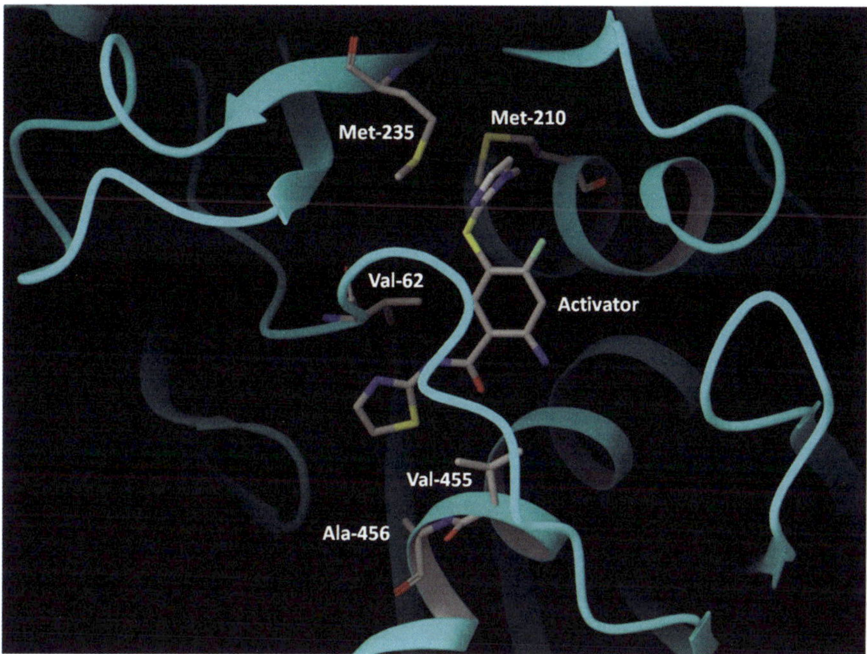

Figure 8.2   Co-crystal structure of the GK activator 8.20 bound in the allosteric site of glucokinase (pdb code 1V4S). Key residues from the mutation studies are highlighted.

A number of activating–deactivating mutations have been mapped directly to the allosteric small molecule binding site in the hinge region of GK. The natural inactivating mutations of residues V62M, M210K and M235V are known to cause the maturity onset diabetes of the young type 2 (MODY-2) phenotype.[12] In contrast, the V455M and A456V are natural activating mutations associated with persistent hyperinsulinemic hypoglycemia of infancy (PHHI).[13,14] Taken together, these observations would suggest that a small molecule GK activator may have considerable impact on glycemic control in patients with type 2 diabetes mellitus.

8.3   The Initial Discovery of Small Molecule GK Activators

The discovery of the first small molecule GK activator was reported in 2003 by researchers at Roche.[15] High throughput screening of a 120 000 member compound collection lead to the identification of the acyl urea screening hit **8.1**. A pharmacophore model **8.2** was developed and optimization of **8.1** was specifically carried out in three areas of the molecule, namely the R1, R2 and R3 regions.[16] Initially, systematic exploration of the R2 substituent revealed

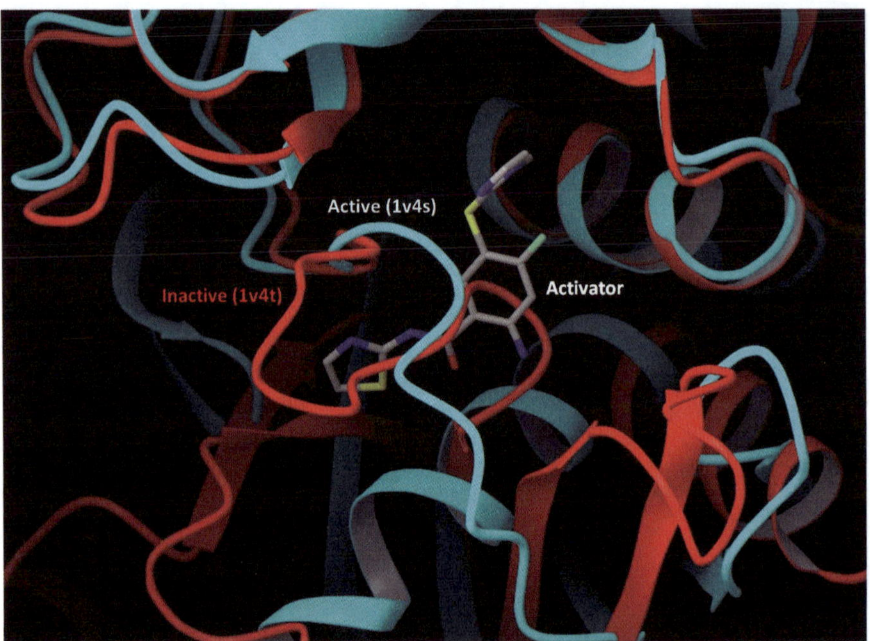

Figure 8.3   The crystal structures of the GK active conformation (pdb code 1V4S) in blue overlayed with the "super-open" inactive conformation (pdb code 1V4T) in red.

that a cyclopentyl group was preferred over isopropyl in this region and compound 8.3 was identified as an early lead structure.

8.1

8.2

8.3 R = H, $SC_{1.5}$ = 4.4 μM
8.4 R = Me, $SC_{1.5}$ = 1.0 μM

N-methylation of the terminal –NH of the urea in the R3 region provided GK activator **8.4** which displayed a 4-fold enhancement in potency relative to **8.3**. Racemic **8.4** was separated into optically pure *R*- and *S*- stereoisomers **8.5** and **8.6** and the absolute stereochemistry of **8.5** was confirmed by single-crystal X-ray structural analysis. The *R*-enantiomer **8.5** displayed potent GK activation ($SC_{1.5}$ = 0.41 µM), while the *S*-enantiomer **8.6** was inactive up to a concentration of 30 µM. In addition to confirming the absolute stereochemistry, the X-ray crystal structure of **8.5** also revealed that the acyl urea moiety adopts *cis*-amide geometry as indicated, due to an intramolecular hydrogen bonding arrangement between the carbonyl oxygen of the acyl group and the terminal –NH of the urea. Based on this key observation, it was proposed that heterocyclic ring systems bearing a hydrogen bond acceptor may function as a potential replacement for the terminal amide of the acyl urea moiety. As anticipated, it was found that the *N*-methyl acyl urea moiety of racemic derivative **8.4** could be effectively replaced by a 2-aminothiazole ring system to provide GK activator **8.7**. Optimization in the R1 region indicated that a variety of electron withdrawing groups were favored on the phenyl ring and the 4-methylsulfonyl substituted analog **8.8** with the *R*-configuration at the stereocenter was identified as a potent GK activator ($SC_{1.5}$ = 0.24 µM) with favorable ADME properties.

8.5 $SC_{1.5}$ = 0.41 µM

8.6 $SC_{1.5}$ > 30 µM

8.7 $SC_{1.5}$ = 0.94 µM

8.8 $SC_{1.5}$ = 0.24 µM

This prototype GK activator enhanced the catalytic activity of GK by increasing the maximum velocity (V_{max}) and decreasing the glucose concentration at 0.5 V_{max} or half-maximal velocity ($S_{0.5}$), consistent with a mixed-type non-essential activator. In an *in vivo* setting, oral administration of

8.8 to mice produced robust acute glucose lowering capacity coupled with a significant improvement in overall glucose tolerance. The oral bioavailability of **8.8** was determined to be 93%.[16] GK activator **8.8** (RO 28-1675) was evaluated in a clinical setting in healthy male volunteers.[17,18] At single doses of 200 mg and 400 mg, RO 28-1675 significantly blunted the glucose excursion in an oral glucose tolerance test (OGTT). These results provided a positive clinical proof of concept regarding the role of glucokinase in glucose homeostasis in humans and set the stage for the advancement of GK activators as a potential therapy for type 2 diabetes. Roche researchers also reported the identification of a second generation clinical candidate, Piragliatin **8.9** (RO4389620) that completed Phase II clinical trials.[19] However, further development of Piragliatin was halted for undisclosed reasons. Key structural differences between the initial candidate RO 28-1675 and Piragliatin include replacement of the R3 thiazole with pyrazine, incorporation of a carbonyl moiety onto the R2 cyclopentane group and introduction of a chlorine ortho to the methylsulfonyl group in the R1 region.

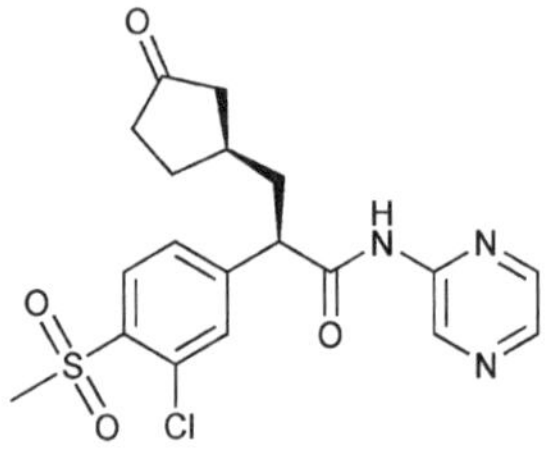

8.9 (Piragliatin)

8.4   Recent Advancements in the Identification of Small Molecule GK Activators

Since the initial disclosure of small molecule GK activators, there has been considerable interest and research activity in the field. A number of other research groups have reported the identification of potent small molecule GK activators. Researchers from OSI Prosidion and Tanabe reported the optimization of the screening hit **8.10** to the potent GK activator PSN-GK1 (**8.13**).[20] Interestingly, PSN-GK1 displays structural resemblance to the Roche GK activator RO28-1675. Key modifications including replacement of the R2 cyclopentyl group of RO28-1675 with pyran and replacement of the methylsulfonyl phenyl R1 group with cyclopropylsulfonyl phenyl. In addition, it was noted that the 2-aminothiazole moiety of **8.11** was susceptible to oxidative metabolism and the corresponding thiourea **8.12** was identified as a metabolite upon dosing of **8.11** in rats. In an attempt to block the formation of this potentially toxic metabolite, the researchers evaluated the effects of introducing substituents onto the thiazole ring. It was determined that a 5-fluorine substituent on the thiazole ring (*i.e.* PSN-GK1) provided a good balance between GK activation potency and oral exposure. Moreover, the putative thiourea metabolite **8.12** could not be detected by mass spectrometry in the plasma samples following oral administration of PSN-GK1 to rats. In an *in vivo* setting, PSN-GK1 displayed potent glucose lowering capacity in rodent models of type 2 diabetes.[21] The same researchers also reported research efforts leading to the identification of the urea based GK activator **8.14** (reported EC_{50} = 6.6 μM).[22] Although significantly less potent than PSN-GK1, compound **8.14** was orally active and displayed significant glucose lowering capacity in mice albeit at a dose of 100 mg kg^{-1} PO.

8.10

8.11
Reported EC50 = 0.57 μM

8.12

8.13 (PSN-GK1)
Reported EC50 = 0.13 μM

8.14
Reported EC50 = 6.6 μM

Researchers at Eli Lilly reported the identification of GK activator **8.15** (LY-2121260) bearing a central cyclopropyl ring.[23,24] From a structural standpoint, LY-2121260 retains some of the structural features of the Roche GK activator RO28-1675 particularly in the R1 and R3 regions. Following dosing in Wistar rats, LY-2121260 showed an improvement in glucose tolerance following an OGTT.

8.15 (LY-2121260)

Astrazeneca researchers have reported a number of GK activators based on a 1,3,5-trisubstituted benzamide core. An early report was centered around GK activators **8.16** and **8.17**.[25] These two compounds provided a significant increase in affinity of GK for glucose (by 4- and 11-fold, respectively) at a concentration of 10 μM. Unlike the Roche GK activator RO28-1675, compounds **8.16** and **8.17** did not have any significant impact on the V_{max} of GK, however, compound **8.16** produced a significant and dose dependent reduction in glucose levels in mice.[26] Further optimization of **8.16** and **8.17** led to the identification of **8.18** and **8.19**, also referred to as GKA-22 and GKA-50, respectively.[27,28] Despite having no apparent affect on the V_{max} of GK, GKA-50 was found to be particularly potent and robust glucose lowering effects were observed in an OGTT in high fat diet fed Zucker rats at oral doses as low as 1 mg kg^{-1}. Furthermore, GKA-50 has been demonstrated to raise intracellular calcium (Ca^{2+}) levels and stimulate insulin secretion from human and rodent islets, thus functioning as a Ca^{2+} mediated glucose-dependent insulin secretagogue.[29] The AstraZeneca group has reportedly generated two GK activator clinical development compounds AZD1656 and AZD6370 (both of undisclosed structure) which are currently in Phase II trials.

8.16

8.17

8.18 (GKA-22)

8.19 (GKA-50)

Recently, researchers at Merck–Banyu have reported GK activators bearing a 2-amino benzamide,[30] or picolinamide,[31] central core structure. By solving the X-ray co-crystal structure of the benzamide analog **8.20** bound to the allosteric site of GK, the researchers formulated and proposed a model which rationalizes the cooperativity of GK with respect to glucose (Figure 8.1).[8] The close analog **8.21** was also found to bind to the allosteric site,[30] and both **8.20** and **8.21** activate GK by increasing the V_{max} and lowering the $S_{0.5}$, similar to the Roche GK activator RO28-1675 (**8.8**).[32] Efforts to replace the undesirable and potentially toxic aniline moiety of **8.20** and **8.21** culminated in the identification of the picolinamide **8.22**,[31] and the 3,5-disubstituted benzamide **8.23** bearing a substituted phenoxy group.[33] Interestingly, GK activator **8.23** displayed improved aqueous solubility and significant blood glucose lowering capacity in KKAy mice maintained on a high fat diet.

8.20

8.21

8.22

8.23

All of the GK activators shown contain a common amide bond in the central core region of the molecule. Researchers at Array Biopharma have specifically targeted the replacement of the amide bond and have reported the identification of a series of non-amide derivatives of representative structure **8.24**. Their optimization efforts culminated in the identification of ARRY-588, a potent GK activator with a reported EC_{50} of 42 nM. From a pharmacokinetic standpoint, ARRY-588 displayed low IV clearance, high permeability and was readily absorbed, with a reported oral bioavailability of 59% in the CD-1 mouse.[34] In an *in vivo* setting, ARRY-588 produced a significant reduction in the post-prandial glucose excursion in an OGTT in C57Bl–6J mice. Similar results were obtained in an ob/ob mouse study where the animals were dosed with ARRY-588 sub-chronically for 14 days. In this sub-chronic study, the researchers also noted a significant reduction in insulin and cholesterol on day 14 with no significant changes in body weight.[34]

8.24

Although ARRY-588 was not advanced for further development, an apparently improved GK activator ARRY-403 emerged from subsequent optimization of ARRY-588. It was recently disclosed that ARRY-403 was advanced to a single ascending dose Phase I clinical study in patients with type 2 diabetes. The outcome of this important study was recently publicized.[35] Overall, ARRY-403 was well tolerated, displayed linear increases in systemic exposure up to 400 mg and a pharmacokinetic profile consistent with once-daily dosing. Furthermore, ARRY-403 produced a significant dose-dependent reduction in post-prandial glucose levels. At doses of 200 mg or higher, a significant reduction in 24 hour fasting blood glucose level was achieved. The study outcome also revealed a significant increase in insulin and C-peptide, consistent with the mechanism of action expected of a GK activator. Taken together, these clinical results indicate that ARRY-403 holds considerable promise as a potential new therapy for type 2 diabetes and further clinical evaluation is warranted.

8.4 Selected Novel Glucokinase Activator Structures from Recent Patent Literature

Several pharmaceutical companies have continued to actively pursue GK activators and several patent applications have been filed on a diverse set of

Table 8.1   Selected representative and novel GK structures from recent patent literature.

Representative Structure	Class	Patent No. and Publication Date	Company
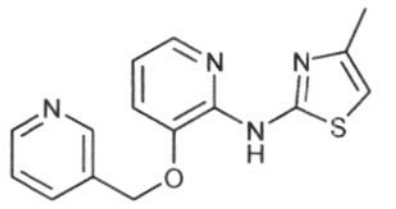	Non-amide or 2-aminopyridine	WO2007053345 May 10, 2007	Array Biopharma
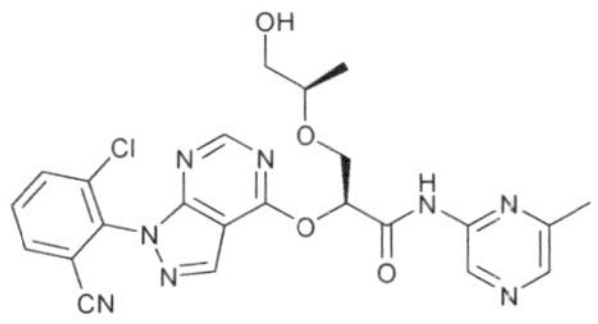	Acetamide	WO2010116176 Oct. 14, 2010	Astrazeneca
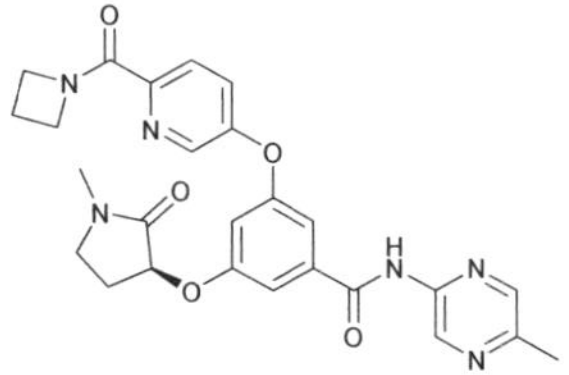	1,3,5-trisubstituted Benzamide	WO2008050101 May 2, 2008	Astrazeneca
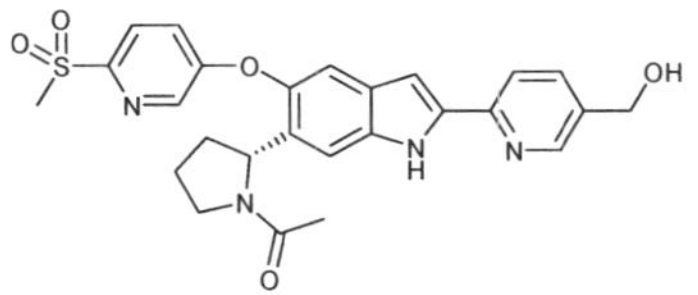	Substituted Indole	WO2010018800 Feb. 18, 2010	Banyu

Table 8.1 (*Continued*)

Representative Structure	Class	Patent No. and Publication Date	Company
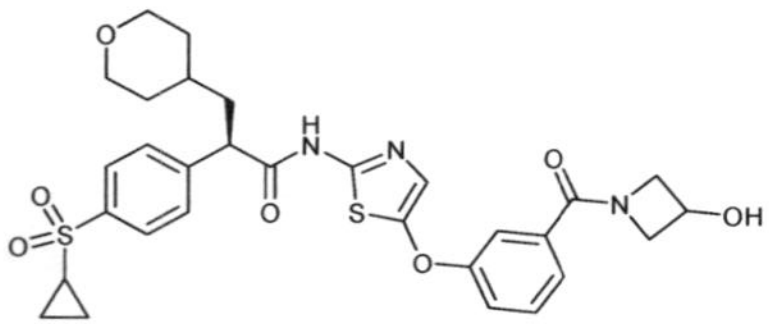	Phenylacetamide	WO2008005914 Jan. 10, 2008	Bristol Myers Squibb
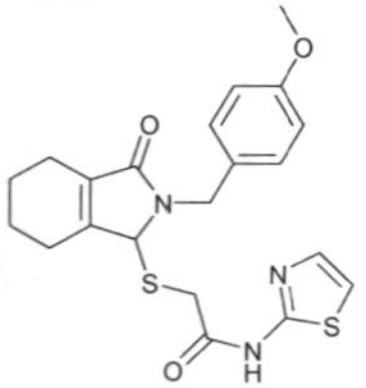	Indolinone	US2007–0117808 May 24, 2007	Johnson & Johnson
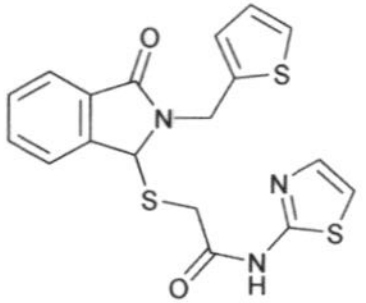	Indolinone	US2007–0099936 May 3, 2007	Johnson & Johnson

Table 8.1   (Continued)

Representative Structure	Class	Patent No. and Publication Date	Company
	Phenylacetamide	US2008–0103167 May 1, 2008	Novartis
	Urea	US2008–0319028 Dec. 25, 2008	Novo Nordisk
	Pyrazole Fused Benzamide	WO2010103438 Sept. 16, 2010	Pfizer
	1,3,5-trisubstituted Benzamide	WO2007122482 Nov. 1, 2007	Pfizer

Table 8.1 (*Continued*)

Representative Structure	Class	Patent No. and Publication Date	Company
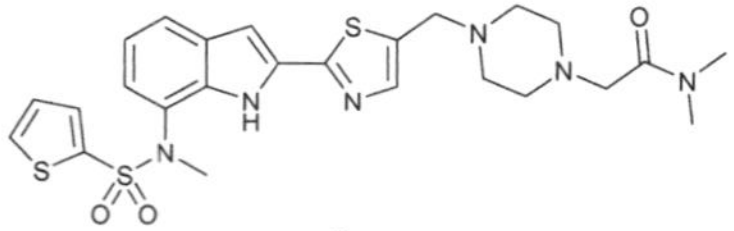	Substituted Indole	US2008–096877 April 24, 2008	Takeda
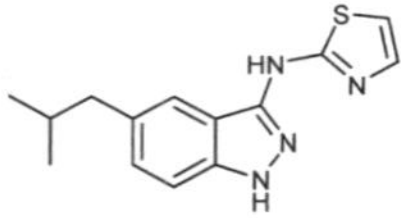	Indazole	US2007–0244169 Oct. 18, 2007	Takeda
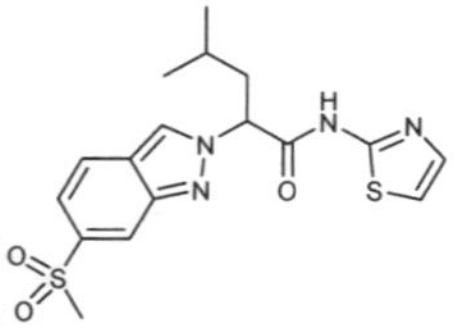	Aminoacid Derivative	US2007–0281942 Dec. 6, 2007	Takeda
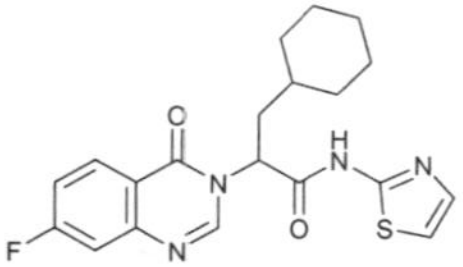	Aminoacid Derivative	US2007–0213349 Sept. 13, 2007	Takeda
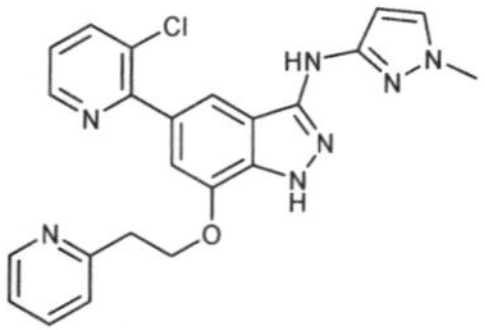	Indazole	WO2008156757 Dec. 24, 2008	Takeda

structurally novel compounds. Selected representative structures appearing in the recent patent literature are listed in Table 8.1.

8.5 Conclusion

There is currently a large unmet medical need for oral antidiabetic drugs with a superior safety and efficacy profile for the treatment of type 2 diabetes. In contrast to older therapies with a single mode of action such as sulfonylureas, GK activators have considerable therapeutic potential as they are anticipated to improve overall glucose homeostasis *via* a dual mode of action involving the liver and pancreas. In several animal models of type 2 diabetes, GK activators have demonstrated robust glucose lowering capacity with favorable safety characteristics. As a result, a number of GK activators have progressed to human clinical trials and some have demonstrated successful outcomes. Results from longer-term clinical trials will be required in order to determine if GK activators will provide the next generation of oral antidiabetic therapies. For this reason, it is anticipated that the allosteric activation of GK with small molecules will likely continue to be an important field of research in the pharmaceutical industry for many years to come.

Acknowledgements

We would like to thank Dr Richard A. Ward for kindly producing Figures 8.2 and 8.3.

References

1. F. M. Matschinsky, *Nat. Rev.* 2009, **8**, 399.
2. M. D. Meglasson and F. M. Matschinsky, *Diabetes/Metab. Rev.*, 1986, **2**, 163.
3. S. R. Chipkin, K. L. Kelly and N. B. Ruderman in *Joslin's Diabetes*, ed. C. R. Khan and G. C.Weir, Lea and Febiger, Philadelphia, PA, 1994, p 97.
4. K. R. Guertin and J. Grimsby, *Curr. Med. Chem.*, 2006, **13**, 1839.
5. M. Pal, *Curr. Med. Chem.*, 2009, **16**, 3858.
6. E. Van Schaftingen, *Encycl. Biol. Chem.*, 2004, **2**, 372.
7. R. G. Printz, M. A. Magnusson and D. K. Oranner in *Annual Reviews of Nutrition*, ed. R. E. Olson, D. M. Bier, and D. B. McCormick, Annual Reviews Inc., Palo Alto, CA, 1993, vol. 13, p 463.
8. K. Kamata, M. Mitsuya, T. Nishimura, J. Eiki and Y. Nagata, *Structure*, 2004, **12**, 429.
9. V. V. Heredia, J. Thomson, D. Nettleton and S. Sun, *Biochemistry*, 2006, **45**, 7553.
10. Y. B. Kim, S. S. Kalinowski and J. Marcinkeviciene, *Biochemistry*, 2007, **46**, 1423.

11. J. Zhang, C. Li, K. Chen, W. Zhu, X. Shen, and H. Jiang, *Proc. Natl. Acad. Sci. U.S.A.*, 2006, **103**, 13368.

12. S. S. Fajans, G. I. Bell and K. S. Polonsky, *N. Engl. J. Med.*, 2001, **345**, 971.

13. B. Glaser, P. Kesavan, M. Heyman, E. Davis, A. Cuesta and A. Buchs, *N. Engl. J. Med.*, 1998, **338**, 226.

14. H. B. Christesen, B. B. Jacobsen, S. Odili, C. Buettger, A. Cuesta-Munoz, T. Hansen, K. Brusgaard, O. Massa, M. A. Magnuson, C. Shiota, F. M. Matschinsky and F. Barbetti, *Diabetes,* 2002, **51**, 1240.

15. J. Grimsby, R. Sarabu, W. L. Corbett, N. E. Haynes, F. T. Bizzarro, J. W. Coffey, K. R. Guertin, D. W. Hilliard, R. F. Kester, P. E. Mahaney, L. Marcus, L. Qi, C. L. Spence, J. Tengi, M. A. Magnuson, C. A. Chu, M. T. Dvorozniak, F. M. Matschinsky and J. F. Grippo, *Science,* 2003, **301**, 370.

16. N. E Haynes, W. L. Corbett, F. T. Bizzarro, K. R. Guertin, D. W. Hilliard, G. W. Holland, R. F. Kester, P. E. Mahaney, L. Qi, C. L. Spence, J. Tengi, M. T. Dvorozniak, A. Railkar, F. M. Matschinsky, J. F. Grippo, J. Grimsby and R. Sarabu, *J. Med. Chem.,* 2010, **53(9)**, 3618.

17. J. Grimsby, Keystone Symp. Diabetes Mellitus Insulin Action Resist. Jan. 22 – 27, Breckinridge, Co. 2008, Abstract # 151.

18. R. F. Kester, R. Sarabu, W. L. Corbett, N. E. Haynes, F. T. Bizzarro, K. R. Guertin, D. W. Hilliard, P. E. Mahaney, L. Qi, J. Tengi, G. W. Holland, A. Focella, J. F. Grippo, J. Grimsby, J. W. Coffey, L. Marcus, C. L. Spence and M. T. Dvorozniak, 237th ACS National Meeting, March 22–26, Salt Lake City, UT, 2009, Abstract MEDI 162.

19. J. Zhi, S. Zhai, M.-E. Mulligan, J. Grimsby, C. Arbet-Engels, M. Boldrin and R. Balena, in *Diabetologia*, ed. E. Gale, Springer, Amsterdam, 2008, **51(Suppl. 1)**, Abstract 42.

20. L. S. Bertram, D. Black, P. H. Briner, R. Chatfield, A. Cooke, M. C. T. Fyfe, P. J. Murray, F. Naud, M. Nawano, M. J. Procter, G. Rakipovski, C. M. Rasamison, C. Reynet, K. L. Scholfield, V. K. Shah, F. Spindler, A. Taylor, R. Turton, G. A. Williams, P. Wong-Kai-In and K. Yasuda, *J. Med. Chem.,* 2008, **51**, 4340.

21. M. C. T. Fyfe, J. White, A. Taylor, R. Chatfield, E. Wargent, R. L. Printz, T. Sulpice, J. G. McCormack, M. J. Procter, C. Reynet, P. S. Widdowson and P. Wong-Kai-In, *Diabetologia,* 2007, **50**, 1277.

22. A. L. Castelhano, H. Dong, M. C. T. Fyfe, L. S. Gardner, Y. Kamikozawa, S. Kurabayashi, M. Nawano, R. Ohashi, M. J. Procter, L. Qiu, C. M. Rasamison, K. L. Scholfield, V. K. Shah, K. Ueta, G. M. Williams, D. Witter and K. Yasuda, *Bioorg. Med. Chem. Lett.,* 2005, **15**, 1501.

23. S. Heuser, D. G. Barrett, M. Berg, B. Bonnier, A. Kahl, M. L. De La Puente, N. Oram, R. Riedl, U. Roettig, S. S. Gil, E. Seger, D. J. Steggles, J. Wanner and A. G. Weichert, *Tetrahedron Lett.,* 2006, **47(16)**, 2675.

24. A. M. Efanov, D. G. Barrett, M. B. Brenner, S. L. Briggs, A. Delaunois, J. D. Durbin, U. Giese, H. Guo, M. Radloff, G. S. Gil, S. Sewing, Y. Wang,

A. Weichert, A. Zaliani and J. Gromada, *Endocrinology,* 2005, **146(9)**, 3696.

25. K. J. Brocklehurst, V. A. Payne, R. A. Davies, D. Carroll, H. L. Vertigan, H. J. Wightman, S. Aiston, I. D. Waddell, B. Leighton, M. P. Coghlan and L. Agius, *Diabetes,* 2004, **53**, 535.

26. B. Leighton, A. Atkinson and M. P. Coghlan, *Biochem. Soc. Trans.,* 2005, **33**, 371.

27. D. McKerrecher, J. V. Allen, S. S. Bowker, S. Boyd, P. W. Caulkett, G. S. Currie, C. D. Davies, M. L. Fenwick, H. Gaskin, E. Grange, R. B. Hargreaves, B. R. Hayter, R. James, K. M. Johnson, C. Johnstone, C. D. Jones, S. Lackie, J. W. Rayner and R. P. Walker, *Bioorg. Med. Chem. Lett.,* 2005, **15**, 2103.

28. D. McKerrecher, J. V. Allen, P. W. Caulkett, C. S. Donald, M. L. Fenwick, E. Grange, K. M. Johnson, C. Johnstone, C. D. Jones, K. G. Pike, J. W. Rayner and R. P. Walker, *Bioorg. Med. Chem. Lett.,* 2006, **16**, 2705.

29. D. Johnson, R. M. Shepherd, D. Gill, T. Gorman, D. M. Smith and M. J. Dunne, *Diabetes,* 2007, **56**, 1694.

30. T. Nishimura, T. Iino, M. Mitsuya, M. Bamba, H. Watanabe, D. Tsukahara, K. Kamata, K. Sasaki, S. Ohyama, H. Hosaka, M. Futamura, Y. Nagata and J.-I. Eiki, *Bioorg. Med. Chem. Lett.,* 2009, **19(5)**, 1357.

31. M. Mitsuya, K. Kamata, M. Bamba, H. Watanabe, Y. Sasakai, K. Sasaki, S. Ohyama, H. Hosaka, M. Futamura, Y. Nagata, J.-I. Eiki and T. Nishimura, *Bioorg. Med. Chem. Lett.,* 2009, **19(10)**, 2718.

32. M. Futamura, H. Hosaka, A. Kadotani, H. Shimazaki, K. Sasaki, S. Ohyama, T. Nishimura, J. Eiki and Y. Nagata, *J. Biol. Chem.,* 2006, **281**, 37668.

33. T. Iino, N. Hashimoto, K. Sasaki, S. Ohyama, R. Yoshimoto, H. Hosaka, T. Hasegawa, M. Chiba, Y. Nagata, J-I. Eiki and T. Nishimura, *Bioorg. Med. Chem.,* 2009, **17(11)**, 3800.

34. http://www.arraybiopharma.com/_documents/Publication/ PubAttachment288.pdf

35. http://www.arraybiopharma.com/Documents/PDF/Slides.pdf

Drug Discovery and Non-Human Kinomes

ANDREW F. WILKS[a,b] AND ISABELLE LUCET[c]

[a] SYN|thesis Med Chem, PO Box 450, South Yarra, VIC 3020, AUSTRALIA;
[b] Monash Institute of Pharmaceutical Sciences, 399 Royal Parade, Parkville,
VIC 3052, AUSTRALIA; [c] Department of Biochemistry and Molecular
Biology, School of Biomedical Sciences, Monash University, Wellington Rd,
Clayton, VIC 3800, Australia

9.1 The Burden of Human Parasitic Diseases

Whilst we in the broader pharmaceutical industry are largely preoccupied with
the development of new pharmaceuticals for the treatment of cancer, cardiovascular disease and inflammatory diseases, it is an "inconvenient truth" that
more than 1 billion people, fully one sixth of the world's population, are
suffering from one (or more) of a range of parasitic diseases. These diseases,
collectively known as the "Neglected Tropical Diseases" (NTDs), include
Leishmaniasis, Shistosomiasis, Chagas Disease, Sleeping Sickness and Malaria
amongst others, and collectively account for more than a million deaths per
year. The adjective "Neglected" is not applied to these diseases without
justification. Whilst they collectively account for perhaps 10% of the human
disease burden worldwide, they attract less than 0.2% of the global investment
in medical research. Although things have improved substantially in recent
years, through the agency of the public private partnerships such as Drugs for
Neglected Disease Initiative (http://www.dndi.org/), Medicines for Malaria
Venture (http://www.mmv.org/) and the Gates Foundation, it is pertinent to

RSC Drug Discovery Series No. 19
Kinase Drug Discovery
Edited by Richard A. Ward and Frederick Goldberg
© Royal Society of Chemistry 2012
Published by the Royal Society of Chemistry, www.rsc.org

note that of 1393 new drugs developed between 1975–1999, only 4 (0.3%) were antimalarials. Clearly new targets and new approaches are required, and some careful thinking with respect to prioritisation of targets and approach needs to be done. With the intense investment of the pharmaceutical industry into kinase-focussed drug discovery in the last twenty years, much has been learnt about the most effective strategies for delivering potent and specific molecules with appropriate pharmaceutical properties. Armed with this knowledge, and the conviction that a significant proportion of the kinases in the genome of these parasites will be critical to their survival at some stage of their life cycle, it is logical to pursue the development of potent and specific inhibitors against key parasite kinases with a view to providing new and mechanistically different drugs for the treatment of these diseases.

9.2 Non-Human Kinomes

The eukaryotic Protein Kinases (ePK) are a highly diverse branch of a much broader family of proteins found in prokaryotes and eukaryotes, the Protein Kinase-like, or PKL Family.[1] The breadth and diversity of this mega-family is astonishing, with more than 45 000 members identified so far across all species investigated. There are some 20 distinct sub-families of PKL, of which the ePK family accounts for nearly 25 000 members. Interspecies comparison of members of the ePK family underscores the plasticity of this family, with gains and losses of entire classes of ePK family elaborated in the kinomes of particular organisms. The major families of ePK predate the eukaryotic radiation, and have been elaborated in particular ways in different Kingdoms, Phyla and Classes of organism; for example, the loss of the Tyrosine Kinase-like (TKL) family in yeast, and the expansion of receptor kinases in plants is typical of such elaboration.

Alongside the macro changes in kinase repertoire, there are subtle alterations in the constellation of amino acids that make up the essential elements of the kinase fold. The kinase fold is constructed of ten subdomains,[2] the core of which (namely subdomains I, II, III, IV, VIa, VIb, VII, and IX) is built around ten essential, and largely immutable, amino acids. These conserved amino acids are the key signature of the ePK family, as well as the other nineteen PKL-related families defined in broad sequencing exercises such as the GOS (Global Ocean Sampling) expedition.[3] What is interesting, and most relevant here, is that these two evolutionary drives lead to the development of an exceedingly high degree of diversity within the kinomes of different species. For those of us who have been spending our working days seeking to design molecules that can distinguish between two closely related human kinase targets (Human JAK2 and JAK3 have only nine differences within the ATP binding cavity and yet these are each critical "off targets" for the development of inhibitors against the other kinase), the huge evolutionary distance between humans and parasites offers significant diversity of overall shape and composition of the catalytic sites of members of their kinomes. The kinomes of pathogens such as *Plasmodium falciparum* (Malaria) and *Trypanosoma brucei* (Sleeping Sickness) *Trypanosoma cruzei* (Chagas Disease) and *Leishmania* sp. (Leishmaniasis) thus offer a host of species-specific kinase families, as well as interesting points of diversity within the catalytic domains of targets held in common between humans and the parasite.

With this in mind, it seems entirely plausible that a concerted attempt to develop kinase inhibitors that target kinases from parasitic pathogens could offer fertile ground for the development of novel treatments for the appalling diseases that they cause.

The protein kinase family is one of the largest gene families in the genomes of Eukaryotes. The number of kinases account for somewhere between 2% and 4% of all genes in the Eukaryotes genome. In humans there are 538 different members of the protein kinase family (G. Manning, Salk Institute: personal communication) which can be divided up on the basis of similarity into 9 defined families, plus one catch-all category ("Other–Orphan"). These are: Tyrosine kinases (TK) (including both receptor kinases and non-receptor kinases), Tyrosine Kinase-like (TKL) (which are similar to, but distinct from the Tyrosine kinases), AGC (PKA-, PKG-, PKC-like), CAMK (Calcium–Calmodulin regulated kinases), CK1 (Casein Kinase 1 Group), CMGC (CDK-, MAPK-, GSK3- and DDK-like), STE (MAP Kinase cascade kinases; homologs of yeast Ste7 (MAP2K), Ste11 (MAP3K) and Ste20 (MAP4K) kinases), RGC (Receptor Guanylate Cyclase-related), Atypical (including families of typical kinases whose kinase domains do not have significant sequence similarity to eukaryotic protein kinases) and Other/Orphan (Kinases that are not members of a larger family). The perception that the human kinome is a smorgasbord for drug discovery is one that has driven the investment of many billions of dollars into the activity of kinase-focussed drug discovery and whilst the current return on investment appears to be relatively poor,[4,5] our understanding of the design of new kinase targeted chemotypes has evolved enormously to the point where for a new kinase the repertoire of tools available to the kinase focussed medicinal chemist is now considerable.

9.3   Kinetoplastids

The kinetoplastids are a group of unicellular protozoa that includes the trypanosomatids, a family of parasites responsible for a number of debilitating human diseases. *Leishmania major* (cutaneous leishmaniasis), *Trypanosoma brucei* (African sleeping sickness) and *Trypanosoma cruzi* (Chagas disease) are amongst the most debilitating diseases of humankind. These parasites have evolved a complex heteroxenous, requiring more than one obligatory host to complete the life cycle, often involving an insect borne sexual phase and a blood borne asexual phase. With *T. brucei*, the insect vector is the Tsetse fly, with *T. Cruzi* it is the Reduviid bug, and with *Leishmania,* the sand fly.

9.3.1 The Kinomes of Trypanosoma

The kinomes of three species of trypanosomatid, *L. major, T. brucei* and *T. cruzi* have been explored and compared.[6] Approximately 2% of the genes in the genomes of these parasites are kinases, with *L. Major (179 ePK), T. brucei (156 ePK)* and *T. cruzi (171 ePK)* showing a similar distribution of kinase family types. Using *L. Major* as an example, an overview of the trypansomal kinome shows intriguing features. Whilst most of the major kinase families are represented in this kinome, there is a complete absence of kinases that fall into the Tyrosine kinase (TK) and Tyrosine kinase-like (TKL) families, nor are there any examples of the RGC-family. Examples of the AGC-family (11 members), CAMK-family (16 members), CK1-family (6 members), CMGC-family (45 members), STE-family (34 members), are all represented. Interestingly, the CAMK-family and the AGC-family are under-represented *versus* the complement of these families in the Human kinome, whilst the GMGC- and STE families appear to be over-represented *versus* the Human kinome. An additional 40 "Other" kinases can be found with orthologues within the Human kinome (including AUR, Aurora; PLK, polo-like kinases; and Wee1). From this group, the highly expanded NIMA-related (Nek) family (22 members, *versus* 15 in the Human kinome) is particularly intriguing. Looking further into the familial relationships within this family of Nek kinases, in cladistic analysis several members appear to cluster with yeast and metazoan Nek family members, whilst twelve of these kinases form their own cluster of Trypanosome specific Nek kinases.[6] The remaining 26 kinases appear to be unique to the Trypanosome kinome.

9.4 *Apicomplexa*

The taxon Apicomplexa includes members of a number of significant human parasites, including Babesiosis (*Babesia*), Toxoplasmosis (*Toxoplasma*) and Malaria (*Plasmodium*). Whilst each parasite is an important vector for human disease, we will focus on *Plasmodium*, and particularly *P. falciparum* in this review.

9.4.1 Malaria

Malaria is a devastating human disease that causes more than 850 000 deaths each year. The disease is caused by a protozoan parasite of the genus *Plasmodium,* which is obligate intracellular protozoan parasites of humans and animals, and the symptoms of the disease are largely a consequence of the asexual multiplication of these parasites within erythrocytes in the host. The Human disease is caused by the Plasmodium species *P. falciparum, P. vivax, P. ovale* and *P. malariae.* The life cycle of the Malaria parasite is extremely complex involving distinct cellular morphologies for infection of the host organism (human or other animal) and the infectious vector, the mosquito

(*Anopheles*). The parasites must therefore encounter and overcome a range of different physiological, and indeed pharmacological, assaults in order to continue to grow and divide and complete its lifecycle. Since the late 1990s there has been a significant rise in the number of parasite strains with reduced sensitivity to even the newest drugs, and new approaches to this parasite are desperately needed.

Infection in humans begins with the bite of an infected female Anopheles mosquito. Sporozoites released from the salivary glands of the mosquito quickly invade liver cells (hepatocytes) and undergo schizogony producing several thousand merozoites able to invade its host erythrocytes. This life cycle is shown diagrammatically in Figure 9.1. Malaria pathogenesis is a consequence of the asexual multiplication of parasites within erythrocytes in the host. Some merozoites arrest their cell cycle within the host red blood cell and differentiate into male or female gametocytes. These sexual cells do not contribute to the pathology of the disease, but are essential for the transmission of parasite to the mosquito vector where the sexual cycle is completed with the generation of infective sporozoites that accumulate in the insect salivary glands. Notionally, all phases of the life cycle present interesting points of potential pharmacological intervention, including those that occur within the insect vector.

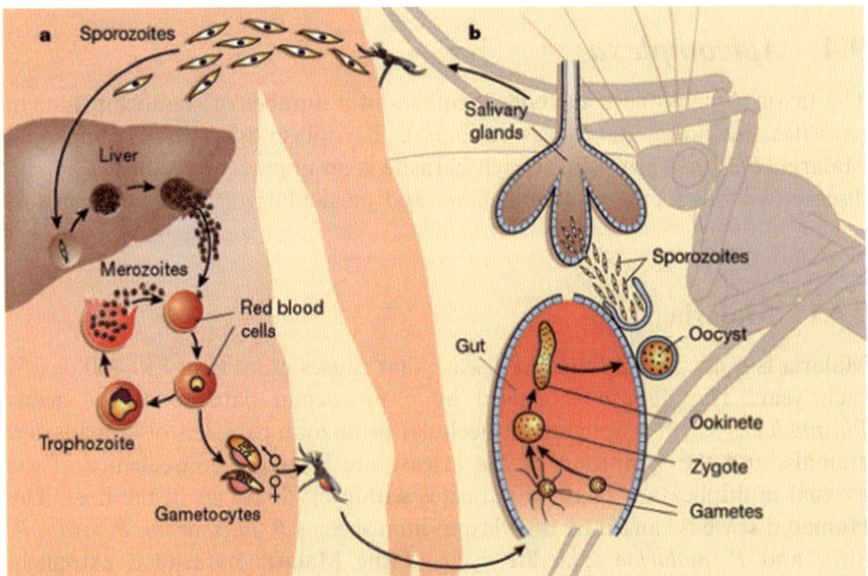

Figure 9.1   Life cycle of the malarial parasite *Plasmodium falciparum.*

9.4.2 The *Plasmodium* Kinome

Early exploration of the *P. falciparum* kinome was undertaken by Doerig and co-workers.[7,8] The field of Parasitology was late to enter the "Genomics" *era*, although significant progress has been made in the last decade. Genome-wide sequencing efforts have yielded the complete sequence of the Human parasite *P. falciparum,*[9] and the rodent parasite *Plasmodium yoelii,*[10] (both delivered in 2002), with other human parasites (*Plasmodium vivax*),[11] primate (*Plasmodium knowlesi*),[12] and rodent (*Plasmodium berghei, Plasmodium chabaudi*) now available.[13]

The Malarial parasite *P. falciparum* is a protist and is therefore unrelated either to plants, fungi or animals. It is intriguing therefore to note that the *P. falciparum* kinome has a number of specific kinases that are held in common with all other classes of living organisms, which might imply that these kinases play a fundamental role in the cellular metabolism of all Eukaryotic cells. Parenthetically it is interesting that the *P. falciparum* kinome does not possess any of the kinases of the three component MAPK pathways.[14] The *Plasmodium* kinome is significantly more compact than the Human kinome, with less than 70 members in any *Plasmodium* species. *P. falciparum* kinome possesses at least one member of six of the nine sub-families of kinases defined in the human kinome. There are no members of the protein tyrosine kinase family, nor the STE family, in the kinome of *P. falciparum*, indeed there is a relative scarcity of obvious orthologues of most mammalian ePKs. Even those *Plasmodium* enzymes that clearly cluster within an established ePK group are generally difficult to align with a human or yeast orthologue. The *Plasmodium* kinome is shown in Figure 9.2. Kinase families identified to date include:

9.4.2.1 The AGC Group

There are six *P. falciparum* kinases that belong to AGC group; namely PfPKA, PfPKB, PfPKG, PF11_0227, PFC0385, and cPFI1285w. Each member of this family appears to demonstrate essentiality in the rodent parasite *P. Berghei* knock-out survey.[15] PfPKA is the only known cAMP effector kinase known in *P. falciparum*, and its *P. Berghei* homologue is purportedly essential to Erythrocytic schizogony. PKG appears to be critical to *Plasmodium* gametogenesis, demonstrating both genetic,[16] and chemical genetic necessity,[17,18] for the activation of asexual proliferation. Interestingly, whilst PfPKB bears a strong resemblance to mammalian PKB the Malarial parasite enzyme lacks a PH domain and appears to be activated by calcium-bound Calmodulin through a Calmodulin binding domain present at the end terminus of the protein.[19-21] It therefore seems likely that this enzyme is not regulated by phosphoinositides as in the mammalian system but rather through calcium and Calmodulin.

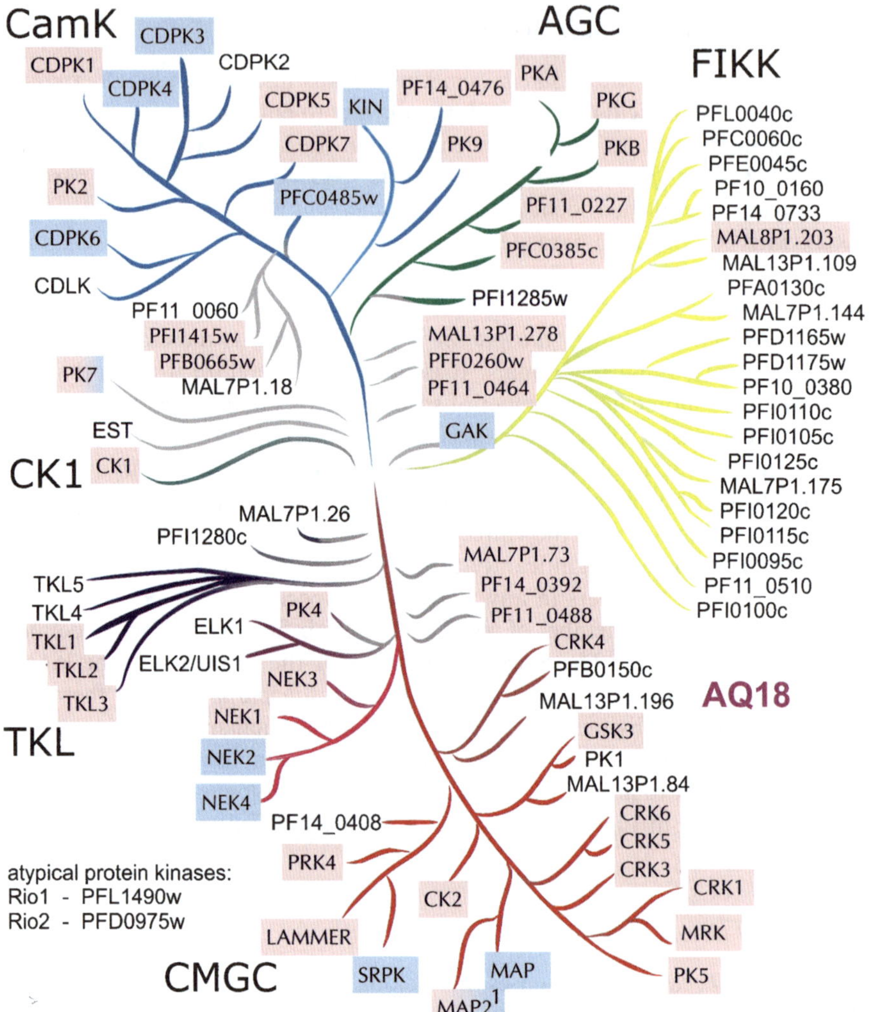

Figure 9.2 The *Plasmodium falciparum* kinome. A representation of the *P. falciparum* kinome has been sketched to show the broad familial relationships between the individual members of the kinase family. Where paralogues are present in the Human kinome, the *Plasmodium* equivalent is usually named after this kinase. Where no paralogue has been identified, the gene name (*e.g.* PF14_0476) is usually used. The colour coding employed to highlight each kinase relates to the potentially essential role demonstrated by individual kinases in the genetic screen undertaken by Tewari, Billker and co-workers (Ref. 15) in the rodent parasite, *P Berghei*. Kinases essential for growth of the parasite in the erythrocytic phase of parasite growth are shown in red (*e.g.* CDPK1), whereas those shown with a blue background are the kinases demonstrated to be essential for extra-erythrocytic and sexual phases of the parasite life cycle (*e.g.* NEK2). Kinases that have been shown to be non-essential for either the erythrocytic or extra-erythrocytic phases of growth are not coloured. (Original kinane image courtesy of Billker and Tewari.)

9.4.2.2 CAMK (Calcium–Calmodulin Modulated Kinases)

Calcium is a particularly important second messenger in parasites, and appears to control a number of essential pathways, including activation of motility, cell invasion and egress.[22] The Human Kinome possesses 74 members of the CAMK family, which can be classified into 17 families of related kinases. Interestingly, none of the *P. falciparum* CAMK family members fall easily into any of the human clusters, but rather in the parasite kinome this class of kinases is dominated by a family of plant-related calcium-dependent kinases, the Calcium-dependent kinase family (CDPKs).[14] A defining characteristic of this family is that they possess a Calmodulin-related domain that is required to control the activity of the kinase domain. There are six (*P. Berghei*) or seven (*P. falciparum*) members of this family in *Plasmodium* kinomes. PfCDPK1 is involved with the regulation of myosin function in the process of parasite motility,[23] and there is strong genetic evidence from the *P. berghei* functional genomics study that CDPK3 and CDPK4 are involved in ookinete gliding motility and sexual reproduction in the midgut.[24,25] The CDPK5 gene is transcribed in mature blood stage schizonts and invasive merozoites.[26] In *P. falciparum*, the use of the destabilising domain (DD) system to regulate the level of PfCDPK5 expression in *P. falciparum* led to similar conclusions.[26] The role(s) of CDPK2 is at present not known. The gene is not present in the *P. berghei* kinome.[15]

9.4.2.3 The Casein Kinase Group

This family of kinases has been greatly expanded in a number of different organisms (*e.g.* 10 in *Drosophila melanogaster*, 12 in Humans; and 87 (!) in *C. elegans*). However, the *P. falciparum* parasite only possesses a single member of this group (PfCK1). This kinase was defined as "possibly essential" in the functional genomics screen.[15]

9.4.2.4 The CMGC Group

Defined operationally as those kinases related to the CDK, MAPK, GSK3 and CLK families, this is the largest family of kinases in the *Plasmodium* kinome with 26 representatives in *P. falciparum*. With recognisable representatives of each sub-family of the greater CMGC family, (namely CDK [5 members], GSK3 [3 members], MAPK [2 members], CLK [4 members]), the pivotal role played by members of this family is underscored by the demonstration of apparent essentiality for most of these kinases in the *P. berghei* knock out study.[15]

9.4.2.5 The NIMA Group

The NIMA related kinases (Nek) family has four members in the *P. falciparum* kinome (Pfnek-1, Pfnek-2, Pfnek-3 and Pfnek-4). In other Eukaryotic systems these enzymes appear to play important roles in cell division, including regulating centrosome replication.[27] Interestingly, Pfnek-1 is expressed in both asexual and sexual blood stages whereas the other three members of the family are predominantly expressed in gametocytes.[28] The *P. berghei* orthologue of Pfnek-1 is present in male gametocytes and not in female,[29] a finding consistent with the plausible role, revealed by knock-out studies for this enzyme in male gametogenesis. In the *P.berghei* studies, Pfnek-3 does not appear to be essential for either erythrocytic shizogony nor gametogenesis.[15] Nek-2 and nek-4 are each essential for the period of DNA replication that precedes meiosis in the zygote. Knock-out of the genes encoding both of these NIMA family members results in defective ookinete maturation.[30,31]

9.4.2.6 Tyrosine Kinase-Like (TKL)

There are four members of the *P. falciparum* TKL family, whereas in *P. berghei*, the TKL group has an additional member. TKL1, TKL3, TKL4, TKL6 appear to be essential in the *P. berghei* genetic screen. In *P. falciparum*, PfTKL3 has recently been shown to be essential for completion of the erythrocytic asexual cycle.[15] Little has been done biochemically with these kinases to date. PfTKL1 and PfTKL3 share an overall domain organisation with MORN and SAM motifs at the N-terminal to the kinase domain. This organisation is not seen in any mammalian TKLs.[32]

9.4.2.7 Other–Orphan

Most of the *Plasmodium* kinome falls into the "Other–Orphan" category. Whilst many of these kinases bear only a nondescript gene tag for a name, the biology and genetics of a number have been explored in more depth, with interesting outcomes. PfPK7 for example, is an orphan kinase, with an N-terminal region with robust similarity to fungal PKAs, combined with a C-terminal region more closely related to MEK3–6.[33] Genetic studies, including knock-outs carried out in both *P. falciparum,*[34] and *P. Berghei,*[15] have lead to the conclusion that this kinase is required for intraerythrocytic replication and production of daughter merozoites by shizonts. Structural studies of this enzyme have been recently completed. The kinase bears two small inserts in the small lobe of the kinase domain, each of which appear to be well ordered in the structure, and may play a role in the regulation of this enzyme.

PfPK9 is expressed in the later stages of the erythrocytic phase of life cycle and is found in the parasitophorous vacuolar membrane as well as the cytosol. *In vitro* PfPK9 phosphorylates the ubiquitin-conjugating enzyme PfUBC13 which in turn regulates that enzyme's function of ubiquitin E2 conjugation activity.[35] UBC13 is a highly conserved enzyme in eukaryotes and its activity controls various cellular processes including immune receptor signalling, DNA repair and mitotic progression. PfPK9 phosphorylation of PfUBC13 strongly suppressed its ubiquitin E2 activity suggesting that PfPK9 is an important regulator of the parasite, involved in a range of cellular processes, such as DNA repair and cell cycling.

9.4.2.8 Plasmodium *sp. Specific Kinase Families – FIKKs*

Clearly the plasticity of kinome evolution is evident in the kinome of *P. falciparum*. A fascinating example of gene family expansion is exemplified in the FIKK family.[7,36] The family is named after an eponymous sequence element Phe-Ile-Lys-Lys, found in the N-terminal region of this family. Whilst only a single copy is found in most *Plasmodium* species, there has been an extraordinary expansion of their number in two *Plasmodium* species, with six members in *Plasmodium reichenow* and 20 (!) in *P. falciparum*. Whilst the function of these kinases remains somewhat obscure, it is interesting to note that the particular virulence of *P. falciparum* is due in large part to its ability to remodel the host cell by exporting its own proteins into the cytoplasm and plasma membrane of the infected erythrocyte. It may be no coincidence that the FIKK paralogues are located in the highly variable subtelomeric regions of 11 different chromosomes, close to other genes (the var genes) that are differentially expressed and play a role in the remodelling of the IE membrane by modifying its antigenic and functional properties. In support of this notion, a recent study demonstrated that some members of the FIKK family are indeed exported to the infected erythrocyte, localised to the intra-erythrocytic structures where they associate with the RBC membrane skeleton.[37,38] Some of these, notably FIKK7.1 and FIKK12, interacted specifically with as yet unidentified protein components of the RBC membrane skeleton and had major quantifiable effects on the mechanical properties of iRBCs that are well known to be associated with severe pathological consequences of malaria infection. These data suggest that FIKK proteins are therefore likely to be involved in specific host–parasite interaction and may be crucial in the trafficking of signalling pathway components.

Furthermore, and quite unexpectedly, recent studies (Cooke, personal communication) now indicate that a subset of FIKK genes are refractory to genetic deletion and are therefore likely to be directly involved in essential metabolic or other processes that are vital for parasite survival.

9.5 Drugability: Prospects for New Drugs

9.5.1 Will P. falciparum Kinase Inhibitors Work as Anti-Malarial Drugs?

A review of the Target Product Profile for an optimal anti-malarial drug emphasises oral dosing; excellent safety profile, and speed of efficacy (one dose). Can an inhibitor for one or several *P. falciparum* kinases deliver against such high expectations? The key question is: "will the inhibition of one or several key *P. falciparum* kinases result in a rapid and irreversible parasitocidal effect?" There are already strong data suggesting that certain *P. falciparum* kinases are required for particular phases of the parasite growth cycle, however, are there mechanisms that the parasite might adopt to overcome the effects of drugs such as these?

Whilst there is evidence of functional redundancy from research on mammalian kinases (the SRC family of kinases amongst others), there is also ample evidence from both vertebrates and non-vertebrate systems that strong phenotypes (including death) can result from the loss or inhibition of many individual kinases. It is important to note that the redundancy or otherwise of any given kinase may be hard to assess without sophisticated genetic approaches, and that the selection of one or even a few particular kinases from the *P. falciparum* kinome may miss the real opportunities offered by this family of signalling molecules.

9.5.2 Specificity & Safety

The safety profile of any drug is in significant measure a function of both its on-target and off-target biological activity. For example, is the target needed for a function other than that which causes the pathology? Clearly, if the *P. falciparum* kinase target is not present in the human kinome, there is no chance of on-target toxicity for such a drug. More difficult to assess is the prospect of off-target toxicity; particularly given the anticipated degree of similarity of all kinase targets be they human or *P. falciparum*. We propose three responses suggesting this challenge can indeed be met. Firstly, whilst the first generation kinase inhibitors were designed approximately ten years ago, our current understanding of how to generate potent and specific kinase inhibitors has greatly improved. The palette of techniques available to kinase professionals is now sufficiently broad that specificity for any particular target can be anticipated, if not yet expected. The evolutionary divergence of many *P. falciparum*-specific kinases *versus* the entire human kinome greatly improves the prospects of generating specific inhibitors with minimal off-target activity. Secondly, the target product profile for an anti-malarial kinase inhibitor requires the period of drug exposure to be up to three days (at most), whereas much of the kinase-related toxicity seen with candidate kinase drugs in humans is related to the prolonged drug exposure required for the treatment of cancers and inflammatory disease. Thirdly, any limiting adverse effects resulting from short-term exposure to the kinase drugs will be observed in preclinical toxicity studies or early clinical safety studies. Not quite "case closed", but the notion that the *Plasmodium* kinome might offer new class of important anti-malaria drug targets is a compelling one.

9.5.3 Prioritisation of Targets

9.5.3.1 Current Best Practice and the Eradication Agenda

The symptoms of malaria are entirely due to the proliferation of *Plasmodium* parasites within the erythrocytes of infected individuals. The treatment of this is currently accomplished through the use of drugs such as Chloroquine. Whilst the erythrocytic phase of the life cycle is an obvious opportunity, a growing wave of enthusiasm has built over the last five years for the development of drugs and–or drug combinations that would diminish the capacity of the parasite to complete its life cycle, and block the transmission of the parasite into new Human hosts. The eradication agenda therefore seeks to identify drug targets that could hobble the capacity of the parasite to reproduce sexually. One example of this type of drug is Artemesinin, which is capable of exerting profound effects on the growth and proliferation of erythrocytic Shizonts, and also efficiently kills early stage gametocytes. The search for novel kinase inhibitors that would target key *Plasmodium* kinases offers an interesting opportunity to target proteins that are key to the growth and proliferation of the malarial parasite in both sexual and asexual growth phase. Whilst the notion underpins the current enthusiasm for the overall approach, there remains a great deal of work to do to identify these key kinase targets and validate them by both genetic and chemical genetic means.

A key step in the identification of these key kinase targets has recently been achieved through the application of functional genomics in the mouse malarial vector system, namely *P. berghei.*

9.5.3.2 Functional Genomics

The non-availability of reverse genetic tools and the inability to generate sufficient materials for proteomic analysis has rather limited the progress towards the genetic and biochemical validation of these targets. More recently however it has become possible to undertake worthwhile proteomic studies on this organism,[29] and to undertake quantitative proteomic approaches (SILAC),[39] plus the improvement in sensitivity in mass spectrometers will have a significant impact on research capacity to undertake these studies.

The completed life cycle of the malarial organism means that a systematic approach to reverse genetics and chemical genetics has been difficult to achieve. Pioneering work of Janse, Waters and co-workers,[40] appears to have formalised and greatly enhanced the capacity of the knockout experiments to be completed successfully. It still remains difficult to interpret the data from such experiments since they rely largely on the inability to generate a knockout phenotype to ascribe essentiality to the genes that have been knocked out. Whilst the *P. falciparum* kinome is relatively small in comparison to the human kinome these technical difficulties ensures that it remains a significant barrier to the generation of a fuller understanding of which kinase targets will be appropriate to undertake in depth drug discovery campaigns.

A remarkable genetic study of the function of all of the members of the *Plasmodium* kinome has recently been completed by Tewari, Billker and co-workers.[15] Whilst such a study, if carried out in a human *Plasmodium* species (*e.g. P. falciparum* or *P. vivax*) would be limited by being applicable to the erythrocytic phase of the life cycle, Tewari and co-workers used the rodent species, *P. berghei*, which permitted an analysis of the effects of each kinase knock-out in both asexual and sexual stages of the life cycle. As noted elsewhere, the *P. berghei* and *P. falciparum* kinomes exhibit a very high degree of similarity (72 orthologous kinase pairs, from 73 *P. berghei* kinase-like genes). Important differences are nonetheless found between these two kinomes; the FIKK family is but a single member in *P. berghei*, but has been amplified into a family of 21 members in *P. falciparum*. In contrast, *P. berghei* possesses six members for the TKL family (including TKL6), whereas the *P. falciparum* kinome has five. Otherwise, orthologues of each of the kinases in the *P. falciparum* kinome are to be found in *P. Berghei*.

Systematic gene deletion of all of the kinases in the *P. berghei* kinome placed each PK into one of three groups: (i) deletable with no detectable developmental phenotype (defined as "non-essential"); (ii) deletable with a developmental phenotype in sexual development; and (iii) non-deletable, defined as "*possibly* essential" in this study.[15] We have included this as a colour-coding in Figure 9.2. The third category contains 43 genes that are refractory to deletion, indicating a possible essentiality for asexual proliferation in the erythrocyte. The authors freely indicate that their failure to generate KO alleles of these genes is not watertight proof of their essentiality, (technical failure could easily account for the outcome of the experiment). Nonetheless, they make a persuasive case that the majority of these experiments are likely to reflect the essentiality of the kinase(s) in question, though for any given kinase, the case is somewhat weakened through the lack of an ability to cross reference through the use of other functional genetic approaches, such as RNAi *etc.*

Importantly, a further 23 kinases appear to be non-essential for asexual proliferation in the host erythrocyte. Eleven knockouts from this group appear to be able to complete a full life cycle (from rodent host, through mosquito phase, and back again into mice). The remaining 12 kinases appear to have an important role in the sexual phase of the parasite's life cycle, exhibiting a number of interesting defects in gametogenesis (for example PbCDPK4 and PbMap2), oocyte maturation (for example, Pbnek-4 and Pbnek-2) and ookinete migration (for example PbCDPK3). These studies provide an important underpinning to a rational selection of those *Plasmodium* kinases that might be most effectively targeted to treat malaria.

9.5.3.3 *Host Kinases*

There is strong evidence for the involvement of host (*i.e.* Human) cell kinases in the facilitation of the life cycle of the parasite. This is not surprising; the parasite has evolved and adapted to its complicated life cycle and has presumably made use of the most efficient mechanism to drive its own proliferation, evade the host immune system, and preserve and protect those cells it has elected to grow within. Two recent studies have underscored this notion, and presented a number of host candidate kinases. The first study makes use of a collection of human kinome-related siRNAs to test their potential for reducing the proliferation of malaria parasites in human hepatocytes.[41] Five host cell kinases were identified as being required for proliferation of the parasite namely, PKCzeta (which regulates NF-kappaB), MET (a receptor for Hepatocyte Growth Factor), PRKWNK1 (involved in osmotic control), SGK2 (involved in osmotic control), and STK35 (a regulator of actin–myosin cytoskeleton). It is interesting to note that each of these kinases has a plausible role in regulating a process that would have a direct bearing upon the 'health' of intrahepatocytic parasites.

9.6 Technical Challenges

A number of technical barriers present themselves in the development of *P. falciparum* kinase focused drug discovery. Firstly, it must be said that the expression of *P. falciparum* proteins has proved to be challenging in many reported cases. The AT content of the coding sequences within the *P. falciparum* genome means that the generation of a suitable construct for expression in another organism such as yeast, bacteria or insect cells may not be straightforward. Additionally, there are examples of large insertions within the catalytic domain within a number of *P. falciparum* kinases. Some of these insertions are of low complexity and may represent sequences that are either not translated into protein, or excised soon after production of the protein. Others are apparently additional domains that are located within the kinase domain. It seems likely that this is merely additional corroboration of the general theme of kinase regulation that is apparent in certain Human kinases, including the so-called "Insertion Element" located between the large and small lobes of the kinase domains of class III receptor tyrosine kinases. Many of these additional domains could be regulatory in function, or indeed may modify or modulate the ultimate role proteins bearing them. The work of Doerig *et al.*, and now more recently ourselves (Lucet *et al.*, unpublished data) suggests that a significant number of the kinase complement of the malarial genome can be expressed in significantly large quantities, suggesting that inhibitor screening ought to be possible. Nonetheless, selection of those kinases that are most likely to produce parasitical activity from their inhibitors remains difficult. As described above, a number of genetic studies have been underway for the last several years and reports are now starting to emerge about a number of allegedly essential kinases for *P. falciparum* viability. Additionally, a number of structural biology studies have been undertaken with specific kinases allowing methods such as *in silico* screening and–or fragment based design to be carried out.

9.6.1 Structural Biology

The integration of X-ray crystallography in kinase-focussed drug discovery programmes can provide a powerful rationale for chemical modification by allowing a unique glimpse of the bound inhibitor to its target allowing rapid modification of the compound to lead compound. With the kinome of *P. falciparum* now defined and the large number of *P. falciparum* kinases found to be widely divergent from human PKs, structural characterisation of *P. falciparum* kinases should provide a powerful tool for the design of plasmodium specific inhibitors. A number of images of *Plasmodium* kinase structures appear in Figure 9.3. The first crystal structure of a *P. falciparum* kinase to be reported was of PfPK5,[42] a cyclin-dependent kinase (CDK).

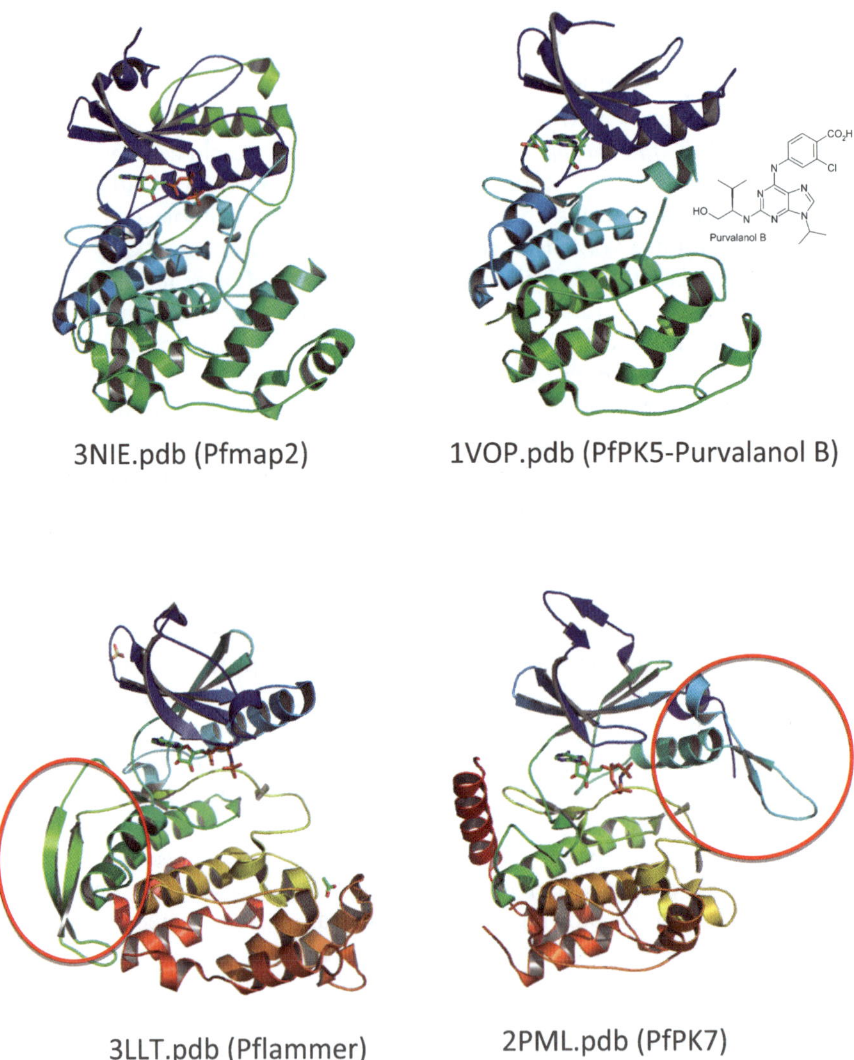

3NIE.pdb (Pfmap2)   1VOP.pdb (PfPK5-Purvalanol B)

3LLT.pdb (Pflammer)   2PML.pdb (PfPK7)

Figure 9.3   Examples of *Plasmodium falciparum* kinase structures. Ribbon diagrams of a number of *Plasmodium* kinases are displayed with their PDB number. The red rings show structured elements that are unique to their *Plasmodium* kinases, and may play an important role in the signalling choreography of these kinases.

PfPK5 has about 60% sequence identity with the human cyclin dependent kinases 1 and 5. By analogy, PfPK5 is involved in cell cycle regulation making it an attractive target. While the activation pathway of PfPK5 *in vivo* remains unclear, the crystal structure of PfPK5 bound to indirubin-5-sulfonate revealed a close similarity to activated human CDK2 (PDB 1V0O.pdb).

More recently the crystal structure of PfPK7 (2PML.pdb; 2PMN.pdb), a composite orphan kinase with the N-terminal lobe homologous to fungal PKA and the C-terminal lobe homologous to the MAPKKs (MEK) kinases has highlighted considerable divergence compared to any ePKs, suggesting that the design of specific PfPK7 inhibitors is achievable.[43]

Today, more *Plasmodium* kinase structures are being released as part of the Structural Genomics Consortium initiative. The crystal structure of Pfmap2 (PDB 3NIE.pdb) has provided an excited starting point for a structure-based drug discovery effort established by The Scientists Against Malaria (SAM) consortium (http://scientistsagainstmalaria.net/).

Recent research has highlighted the role several CDPKs play in key parasitic life cycles.[23,25,44] With no homologues in host species, they are marked as attractive targets for drug design. Recently, several crystal structures of CDPKs from apicomplexan organisms have been published throwing light on their mechanism of action and providing an exciting starting point for drug design.[45,46]

9.7   Chemical Biology and Drug Discovery Programmes

In the last two decades we have learnt a great deal about the ways in which small molecular drugs can be generated against specific kinase targets. We have also learnt much about the potential pitfalls; poor specificity profiles, poor ADME-T profiles, and a wide range of resistance mechanisms are problems that have dogged our early efforts in delivering kinase inhibitors as drugs. In targeting the *Plasmodium* kinome, we must still face many of these challenges; indeed, some, such as the potential range of mechanisms by which the parasite may become immune to the effects of a particular drug, may be considered to be even more formidable challenges, given the propensity for this organism to evade attack. Nonetheless, the challenge with generating kinase inhibitor drugs for a disease such as Malaria is that of hitting only the parasite kinases and leaving the host kinases untouched. Whilst some inhibitor design motifs may require structural elements that are held in common by many members of the broader kinase family (*e.g.* hinge binding or type 2 binding mode) much has now been learnt about the ways in which specificity, as well as potency, can be generated from particular chemotypes. Encouragingly, the diverse nature of many members of the *Plasmodium* kinome suggests that it should be well within our collective capabilities to design inhibitors against key *Plasmodium* kinases, and have these molecules sufficiently clean of unwanted off-target activity that their administration to otherwise healthy individuals would offer therapeutic benefit, without unwanted side effects.

At present, there are two streams of drug discovery activity that could lead to the successful development of a kinase inhibitor as a drug for *Plasmodium* Malaria, namely: *in vitro* screening activities around validated kinase targets and deconvolution of key targets revealed from the screening of various "Malaria Boxes" on various stages of the life cycle of the parasite.

9.7.1   In Vitro Screening of Validated *Plasmodium* Kinases

The development of novel selective inhibitors against *Plasmodium* kinases is still in its infancy. A number of compounds have been reported with varying levels of potency on particular *Plasmodium* enzymes, though very little counter-screening, either against other *Plasmodium* enzymes, or perhaps more importantly against a panel of human kinases, has been done. Thus PfPKG,[18] PfPK5,[47] PfPK7,[18,48] PfCDPK1,[49] and PfGSK3,[50,51] have each begun to be screened, with varying degrees of success. A selection of compounds identified as inhibitors for each of these targets is shown in Table 9.1. Whilst they represent a modest degree of success, it is unlikely that these compounds will be easy to elaborate towards lead and candidate stage. The tri-sibstituted pyrrole (compound **1**), was originally uncovered as an inhibitor of a paralogous PKG enzyme from *Toxoplasma gondii*. It was demonstrated to have potent *in vitro* activity (2 nM) against the *P. falciparum* PfPKG,[18] and is active *in vitro* against blood stages of the parasite lifecycle in the micromolar range. Interestingly, in a *P. berghei* mouse model of infection, compound **1** delays the onset of parasitaemia but does not cure the parasite infection. There are no data presented in this study on the activity of this compound against human targets (or *Plasmodium* targets), and whilst these studies present the intriguing possibility that the inhibition of certain kinases may have a profound impact on the growth and proliferation of Malarial parasites, there remains a great deal to do to turn such hits into safe and useful new drugs.

Two studies have targeted the calcium dependent kinase pfCDPK1. In Lemercier *et al.*,[49] a collaboration between a pharmaceutical company (Merck-Serono) and an academic group (from University of Heidelberg) was able to deliver very potent (compound **3**, 9 nM) inhibitors of PfCDPK1. The series was identified from an initial screen of a diverse screening set of 54 000 compounds, and from the elaboration of the lead into a small library of imidazopyridazines. No whole parasite data were shown, nor any broader screening on this compound. Kato *et al.*,[23] screened PfCDPK1 against a library of 20 000 compounds and identified a series of structurally related 2,6,9-trisubstituted purines as promising hits. The lead compound, renamed Purfalcamine **2**, as well as potent activity against the enzyme *in vitro*, exhibited reasonable levels of paracitocidal activity during late schizongony in whole parasite assays. Whilst no Human kinase counter-screening was undertaken with these compounds, a purported 32-fold window appeared to be available between the activity exhibited on parasites and that determined on mammalian cells.

Table 9.1 Examples of compounds with inhibitory activity on *Plasmodium* kinases.

Compound	Structure	Targets	Potency	PDBs for target	References
Compound 1		PfPKG	~2 nM	N/A	Ref. 18, 54
Purfalcamine 2		CDPK1	IC_{50} = 17 nM EC_{50} = 230 nM	TG homologue	Ref. 23
Compound 3		CDPK1	9 nM (K_i)	N/A	Ref. 49
Compound 4		PfPK7	131 nM	2PMO 2PMN 2PML	Ref. 48
Purported Kinase Inhibitors from GSK TCAMS Library, each active at < 2 μM on whole parasites					
TCMDC-133561		Unknown	IC_{80} < 2 μM		Ref. 52
TCMDC-141334		Unknown	IC_{80} < 2 μM		Ref. 52
TCMDC-133396		Unknown	IC_{80} < 2 μM		Ref. 52

Finally, the imidazopyridazine series elaborated as inhibitors of PfPK7 in Bouloc *et al.*,[48] was progressed from an initial 11 μM hit to the final compound (compound **4**) of 131 nM. The most potent compounds exhibited some capacity to inhibit the proliferation of live parasites in erythrocytes, albeit at levels an order of magnitude higher than those exhibited on the enzyme. The major problem, however, was that counterscreening them on a panel of 80 Human kinases showed that they are promiscuous inhibitors of many potentially undesirable off-target kinases.

9.7.2 "Malaria Boxes"

One strand of drug discovery targeting the erythrocytic phase of the life cycle of the malarial parasite is the screening of diverse chemical libraries against a whole parasite screen. In this case, there is no weighting placed upon the validation, genetic or otherwise, of the target; if a compound is paracitocidal on the protein(s) it hits, it is *ipso facto* a good target for drug discovery. By extension, the compounds that evoke a suitably cytotoxic effect on the malarial parasite could be deemed to be a suitable starting point from which to design a drug. The missing piece of the jigsaw puzzle here, then, is linking the compound with a molecular mechanism of action. Defining the targets of the hits in such a screen is likely to be a significant effort, and involve proteomic and genetic studies, but the net effect will be a pairing of the compound that "works" with a target that is druggable. Collections of hits from screens such as these, (collectively and colloquially known as "Malaria Boxes") may therefore form the starting points for more focussed drug discovery programmes going forward. One remarkable example of such screening efforts was reported by members of the GSK team at Tres Cantos in Spain,[52] in 2010. In this screen, 1 986 056 unique compounds, selected at the scaffold level from GSK's compound collection to provide diverse, comprehensive coverage of bioactive space, were screened against *P. falciparum* strain 3D7 at a concentration of 2 µM. Of these, 13 533 compounds showed sufficient activity to merit further examination. The majority (greater than 80%) of these compounds had never been tested against malarial screens before. Previously, 4205 compounds had been tested against human drug targets, and the team was able to compare the *Plasmodium* data with the annotations regarding their activity in biochemical assays against human (3435 compounds) or microbial targets (770 compounds). Overall, activities against 146 potential targets were noted in this set, with kinases making up nearly 50% of these targets. It seems highly likely, therefore, that this set of compounds has within it high quality lead compounds against a number of *Plasmodium* kinases.

The structures of these compounds were made available alongside the assay data in an effort to encourage further work towards a lead identification effort and further research in the disease. The availability of this data set combined with the structure of the compounds presents a landmark platform of purported kinase inhibitors that are demonstrably active against the erythrocytic form of this parasite. What is required now is to define the kinase target (or targets) of each of these molecules and to determine which of among the 69 (89 including the expanded FIKK family) kinases in the malarial kinome are most likely to yield potent parasitocidal compounds in hit to lead expansion programmes.

Similar screening exercises from other compound collections have successfully revealed additional target classes,[53] though in these studies the more difficult path of *in vitro* testing combined with the absence of a broad kinase screen for the *Plasmodium* kinome means that this class of targets is underrepresented in these analyses. It is highly likely that these additional screens will also reveal many potential kinase inhibitors for further study also.

9.8 Summary

It is clearly early days in the exploitation of parasite kinases as drug targets. In Malaria, for example, we have only recently begun to arrive at an understanding as to which kinases are essential for the growth and development of the parasite in the various stages of its complex life cycle. Screening too has been rather haphazard, and there is presently little progress towards the elaboration of a parasite-specific kinase inhibitor that kills the Malarial parasite. Nonetheless, there have been intriguing developments that foreshadow the true promise of targeting these enzymes. The most promising is the elaboration of a number of sets of compounds derived from the curated compound collections of various pharmaceutical companies, which are potent inhibitors of whole parasite growth (the "Malaria Boxes"). Most exciting is the prospect that within these Malaria Boxes are compounds that are targeting one or more of the members of the *Plasmodium* kinome. Unlocking the true targets of these potent hits should allow the field to take a giant leap forward, towards turning these compounds into clinical proof of concept molecules that could finally demonstrate the validity of this family of targets for the treatment of parasitic diseases.

References

1. N. Kannan, S. S. Taylor, Y. Zhai, J. C. Venter and G. Manning, *PLoS Biol.*, 2007, **5**, e17.
2. S. K. Hanks, A. M. Quinn and T. Hunter, *Science*, 1988, **241**, 42–52.
3. S. Yooseph, G. Sutton, D. B. Rusch, A. L. Halpern, S. J. Williamson, K. Remington, J. A. Eisen, K. B. Heidelberg, G. Manning, W. Li, L. Jaroszewski, P. Cieplak, C. S. Miller, H. Li, S. T. Mashiyama, M. P. Joachimiak, C. van Belle, J. M. Chandonia, D. A. Soergel, Y. Zhai, K. Natarajan, S. Lee, B. J. Raphael, V. Bafna, R. Friedman, S. E. Brenner, A. Godzik, D. Eisenberg, J. E. Dixon, S. S. Taylor, R. L. Strausberg, M. Frazier and J. C. Venter, *PLoS Biol.*, 2007, **5**, e16.
4. B. D. Marsden and S. Knapp, *Curr. Opin. Chem. Biol.*, 2008, **12**, 40–45.
5. J. Eswaran and S. Knapp, *Biochim. Biophys. Acta*, **1804**, 429–432.
6. M. Parsons, E. A. Worthey, P. N. Ward and J. C. Mottram, *BMC Genomics*, 2005, **6**, 127.
7. P. Ward, L. Equinet, J. Packer and C. Doerig, *BMC Genomics*, 2004, **5**, 79.
8. Anamika, N. Srinivasan and A. Krupa, *Proteins*, 2005, **58**, 180–189.
9. M. J. Gardner, N. Hall, E. Fung, O. White, M. Berriman, R. W. Hyman, J. M. Carlton, A. Pain, K. E. Nelson, S. Bowman, I. T. Paulsen, K. James, J. A. Eisen, K. Rutherford, S. L. Salzberg, A. Craig, S. Kyes, M. S. Chan,

V. Nene, S. J. Shallom, B. Suh, J. Peterson, S. Angiuoli, M. Pertea, J. Allen, J. Selengut, D. Haft, M. W. Mather, A. B. Vaidya, D. M. Martin, A. H. Fairlamb, M. J. Fraunholz, D. S. Roos, S. A. Ralph, G. I. McFadden, L. M. Cummings, G. M. Subramanian, C. Mungall, J. C. Venter, D. J. Carucci, S. L. Hoffman, C. Newbold, R. W. Davis, C. M. Fraser and B. Barrell, *Nature*, 2002, **419**, 498–511.

10. J. M. Carlton, S. V. Angiuoli, B. B. Suh, T. W. Kooij, M. Pertea, J. C. Silva, M. D. Ermolaeva, J. E. Allen, J. D. Selengut, H. L. Koo, J. D. Peterson, M. Pop, D. S. Kosack, M. F. Shumway, S. L. Bidwell, S. J. Shallom, S. E. van Aken, S. B. Riedmuller, T. V. Feldblyum, J. K. Cho, J. Quackenbush, M. Sedegah, A. Shoaibi, L. M. Cummings, L. Florens, J. R. Yates, J. D. Raine, R. E. Sinden, M. A. Harris, D. A. Cunningham, P. R. Preiser, L. W. Bergman, A. B. Vaidya, L. H. van Lin, C. J. Janse, A. P. Waters, H. O. Smith, O. R. White, S. L. Salzberg, J. C. Venter, C. M. Fraser, S. L. Hoffman, M. J. Gardner and D. J. Carucci, *Nature*, 2002, **419**, 512–519.

11. J. M. Carlton, J. H. Adams, J. C. Silva, S. L. Bidwell, H. Lorenzi, E. Caler, J. Crabtree, S. V. Angiuoli, E. F. Merino, P. Amedeo, Q. Cheng, R. M. Coulson, B. S. Crabb, H. A. Del Portillo, K. Essien, T. V. Feldblyum, C. Fernandez-Becerra, P. R. Gilson, A. H. Gueye, X. Guo, S. Kang'a, T. W. Kooij, M. Korsinczky, E. V. Meyer, V. Nene, I. Paulsen, O. White, S. A. Ralph, Q. Ren, T. J. Sargeant, S. L. Salzberg, C. J. Stoeckert, S. A. Sullivan, M. M. Yamamoto, S. L. Hoffman, J. R. Wortman, M. J. Gardner, M. R. Galinski, J. W. Barnwell and C. M. Fraser-Liggett, *Nature*, 2008, **455**, 757–763.

12. A. Pain, U. Bohme, A. E. Berry, K. Mungall, R. D. Finn, A. P. Jackson, T. Mourier, J. Mistry, E. M. Pasini, M. A. Aslett, S. Balasubrammaniam, K. Borgwardt, K. Brooks, C. Carret, T. J. Carver, I. Cherevach, T. Chillingworth, T. G. Clark, M. R. Galinski, N. Hall, D. Harper, D. Harris, H. Hauser, A. Ivens, C. S. Janssen, T. Keane, N. Larke, S. Lapp, M. Marti, S. Moule, I. M. Meyer, D. Ormond, N. Peters, M. Sanders, S. Sanders, T. J. Sargeant, M. Simmonds, F. Smith, R. Squares, S. Thurston, A. R. Tivey, D. Walker, B. White, E. Zuiderwijk, C. Churcher, M. A. Quail, A. F. Cowman, C. M. Turner, M. A. Rajandream, C. H. Kocken, A. W. Thomas, C. I. Newbold, B. G. Barrell and M. Berriman, *Nature*, 2008, **455**, 799–803.

13. N. Hall, M. Karras, J. D. Raine, J. M. Carlton, T. W. Kooij, M. Berriman, L. Florens, C. S. Janssen, A. Pain, G. K. Christophides, K. James, K. Rutherford, B. Harris, D. Harris, C. Churcher, M. A. Quail, D. Ormond, J. Doggett, H. E. Trueman, J. Mendoza, S. L. Bidwell, M. A. Rajandream, D. J. Carucci, J. R. Yates, 3rd, F. C. Kafatos, C. J. Janse, B. Barrell, C. M. Turner, A. P. Waters and R. E. Sinden, *Science*, 2005, **307**, 82–86.

14. D. Leroy and C. Doerig, *Trends Pharmacol. Sci.*, 2008, **29**, 241–249.

15. R. Tewari, U. Straschil, A. Bateman, U. Bohme, I. Cherevach, P. Gong, A. Pain and O. Billker, *Cell Host Microbe*, **8**, 377–387.
16. W. Deng and D. A. Baker, *Mol. Microbiol.*, 2002, **44**, 1141–1151.
17. H. M. Taylor, L. McRobert, M. Grainger, A. Sicard, A. R. Dluzewski, C. S. Hopp, A. A. Holder and D. A. Baker, *Eukaryotic Cell*, **9**, 37–45.
18. C. A. Diaz, J. Allocco, M. A. Powles, L. Yeung, R. G. Donald, J. W. Anderson and P. A. Liberator, *Mol. Biochem. Parasitol.*, 2006, **146**, 78–88.
19. A. Vaid, D. C. Thomas and P.Sharma, *J. Biol. Chem.*, 2008, **283**, 5589–5597.
20. A. Vaid and P. Sharma, *J. Biol. Chem.*, 2006, **281**, 27126–27133.
21. A. Kumar, A. Vaid, C. Syin and P. Sharma, *J. Biol. Chem.*, 2004, **279**, 24255–24264.
22. A. K. Wernimont, M. Amani, W. Qiu, J. C. Pizarro, J. D. Artz, Y. H. Lin, J. Lew, A. Hutchinson and R. Hui, *Proteins*, **79**, 803–820.
23. N. Kato, T. Sakata, G. Breton, K. G. Le Roch, A. Nagle, C. Andersen, B. Bursulaya, K. Henson, J. Johnson, K. A. Kumar, F. Marr, D. Mason, C. McNamara, D. Plouffe, V. Ramachandran, M. Spooner, T. Tuntland, Y. Zhou, E. C. Peters, A. Chatterjee, P. G. Schultz, G. E. Ward, N. Gray, J. Harper and E. A. Winzeler, *Nat. Chem. Biol.*, 2008, **4**, 347–356.
24. I.Siden-Kiamos, A. Ecker, S. Nyback, C. Louis, R. E. Sinden and O. Billker, *Mol. Microbiol.*, 2006, **60**, 1355–1363.
25. R. Ranjan, A. Ahmed, S. Gourinath and P. Sharma, *J. Biol. Chem.*, 2009, **284**, 15267–15276.
26. J. D. Dvorin, D. C. Martyn, S. D. Patel, J. S. Grimley, C. R. Collins, C. S. Hopp, A. T. Bright, S. Westenberger, E. Winzeler, M. J. Blackman, D. A. Baker, T. J. Wandless and M. T. Duraisingh, *Science*, **328**, 910–912.
27. L. O'Regan, J. Blot and A. M. Fry, *Cell Div.*, 2007, **2**, 25.
28. D. Dorin, K. Le Roch, P. Sallicandro, P. Alano, D. Parzy, P. Poullet, L. Meijer and C. Doerig, *Eur. J. Biochem.*, 2001, **268**, 2600–2608.
29. S. M. Khan, B. Franke-Fayard, G. R. Mair, E. Lasonder, C. J. Janse, M. Mann and A. P. Waters, *Cell*, 2005, **121**, 675–687.
30. L. Reininger, R. Tewari, C. Fennell, Z. Holland, D. Goldring, L. Ranford-Cartwright, O. Billker and C. Doerig, *J. Biol. Chem.*, 2009, **284**, 20858–20868.
31. L. Reininger, O. Billker, R. Tewari, A. Mukhopadhyay, C. Fennell, D. Dorin-Semblat, C. Doerig, D. Goldring, L. Harmse, L. Ranford-Cartwright and J. Packer, *J. Biol. Chem.*, 2005, **280**, 31957–31964.
32. A. Abdi, S. Eschenlauer, L. Reininger and C. Doerig, *Cell. Mol. Life Sci.*, **67**, 3355–3369.
33. D. Dorin, J. P. Semblat, P. Poullet, P. Alano, J. P. Goldring, C. Whittle, S. Patterson, D. Chakrabarti and C. Doerig, *Mol. Microbiol.*, 2005, **55**, 184–196.
34. D. Dorin-Semblat, A. Sicard, C. Doerig and L. Ranford-Cartwright, *Eukaryotic Cell*, 2008, **7**, 279–285.

35. N. Philip and T. A. Haystead, *Proc. Natl. Acad. Sci. U. S. A.*, 2007, **104**, 7845–7850.

36. A. G. Schneider and O. Mercereau-Puijalon, *BMC Genomics*, 2005, **6**, 30.

37. M. C. Nunes, J. P. Goldring, C. Doerig and A. Scherf, *Mol. Microbiol.*, 2007, **63**, 391–403.

38. M. C. Nunes, M. Okada, C. Scheidig-Benatar, B. M. Cooke and A. Scherf, *PLoS One*, **5**, e11747.

39. S. Hanke, H. Besir, D. Oesterhelt and M. Mann, *J. Proteome Res.*, 2008, **7**, 1118–1130.

40. C. J. Janse, J. Ramesar and A. P. Waters, C. Martin, E. Real, L. A. Goncalves, C. Carret, R. Dorkin, I. Rohl, K. Jahn-Hoffmann, A. J. Luty, R. Sauerwein, C. J. Echeverri and M. M. Mota, *Nat. Protoc.*, 2006, **1**, 346–356.

41. M. Prudencio, C. D. Rodrigues, M. Hannus, C. Martin, E. Real, L. A. Goncalves, C. Carret, R. Dorkin, I. Rohl, K. Jahn-Hoffmann, A. J. Luty, R. Sauerwein, C. J. Echeverri and M. M. Mota, *PLoS Pathog.*, 2008, **4**, e1000201.

42. S. Holton, A. Merckx, D. Burgess, C. Doerig, M. Noble and J. Endicott, *Structure*, 2003, **11**, 1329–1337.

43. A. Merckx, A. Echalier, K. Langford, A. Sicard, G. Langsley, J. Joore, C. Doerig, M. Noble and J. Endicott, *Structure*, 2008, **16**, 228–238.

44. O. Billker, S. Lourido and L. D. Sibley, *Cell Host Microbe*, 2009, **5**, 612–622.

45. A. K. Wernimont, M. Amani, W. Qiu, J. C. Pizarro, J. D. Artz, Y. H. Lin, J. Lew, A. Hutchinson and R. Hui, Structures of parasitic CDPK domains point to a common mechanism of activation *Proteins*.

46. A. K. Wernimont, J. D. Artz, P. Finerty, Jr., Y. H. Lin, M. Amani, A. Allali-Hassani, G. Senisterra, M. Vedadi, W. Tempel, F. Mackenzie, I. Chau, S. Lourido, L. D. Sibley and R. Hui, *Nat. Struct. Mol. Biol.*, **17**, 596–601.

47. S. M. Keenan and W. J. Welsh, *J. Mol. Graphics Modell.*, 2004, **22**, 241–247.

48. N. Bouloc, J. M. Large, E. Smiljanic, D. Whalley, K. H. Ansell, C. D. Edlin and J. Bryans, *Bioorg. Med. Chem. Lett.*, 2008, **18**, 5294–5298.

49. G. Lemercier, A. Fernandez-Montalvan, J. P. Shaw, D. Kugelstadt, J. Bomke, M. Domostoj, M. K. Schwarz, A. Scheer, B. Kappes and D. Leroy, *Biochemistry*, 2009, **48**, 6379–6389.

50. I. Garcia, Y. Fall and G. Gomez, *Curr. Pharm. Des.,* **16**, 2666–2675.

51. I. Garcia, Y. Fall and G. Gomez, *Molecules*, **15**, 5408–5422.

52. F. J. Gamo, L. M. Sanz, J. Vidal, C. de Cozar, E. Alvarez, J. L. Lavandera, D. E. Vanderwall, D. V. Green, V. Kumar, S. Hasan, J. R. Brown, C. E. Peishoff, L. R. Cardon and J. F. Garcia-Bustos, *Nature*, 465, 305–310.

53. W. A. Guiguemde, A. A. Shelat, D. Bouck, S. Duffy, G. J. Crowther, P. H. Davis, D. C. Smithson, M. Connelly, J. Clark, F. Zhu, M. B. Jimenez-

Diaz, M. S. Martinez, E. B. Wilson, A. K. Tripathi, J. Gut, E. R. Sharlow, I. Bathurst, F. El Mazouni, J. W. Fowble, I. Forquer, P. L. McGinley, S. Castro, I. Angulo-Barturen, S. Ferrer, P. J. Rosenthal, J. L. Derisi, D. J. Sullivan, J. S. Lazo, D. S. Roos, M. K. Riscoe, M. A. Phillips, P. K. Rathod, W. C. Van Voorhis, V. M. Avery and R. K. Guy, *Nature*, 465, 311–315.

54. L. McRobert, C. J. Taylor, W. Deng, Q. L. Fivelman, R. M. Cummings, S. D. Polley, O. Billker and D. A. Baker, *PLoS Biol.*, 2008, **6**, e139.

55. L. Meijer, A. M. Thunnissen, A. W. White, M. Garnier, M. Nikolic, L.-H. Tsai, J. Walter, K. E. Cleverley, P. C. Salinas, Y.-Z. Wu, J. Biernat, E.-M. Mandelkow, S.-H. Kim and G. R. Pettit, *Chem. Biol.*, 2000, **7**, 51–63.

56. M. Klein, P. Diner, D. Dorin-Semblat, C. Doerig and M. Grotli, *Org. Biomol. Chem.*, 2009, **7**, 3421–3429.

The Future of Kinase Drug Discovery

CARLOS GARCÍA-ECHEVERRÍA

Oncology Drug Discovery and Preclinical Research, Sanofi Oncology, Vitry-sur-Seine, France

10.1 Introduction

As extensively illustrated in several chapters of this book, protein and lipid kinases have been an important source of therapeutic targets over the past few decades. However, and despite numerous drug discovery and development projects and considerable research work and investments, only a limited number of low-molecular mass kinase modulators have received marketing approval from the health authorities, and significantly improved survival and quality of life of cancer patients as well as individuals suffering from other malignancies. Extrapolation from current practice invites speculation about the future of kinase drug discovery in several areas of particular interest. To this end, three topics have been selected to illustrate current challenges, opportunities and emerging strategies in the identification and development of low-molecular mass kinase modulators.

10.2 Overcoming resistance

One of the recurrent clinical problems of molecularly targeted kinase inhibitors in oncology has been the inevitable occurrence of resistance on account of point mutations that impair drug binding and hence effective enzymatic

---

RSC Drug Discovery Series No. 19
Kinase Drug Discovery
Edited by Richard A. Ward and Frederick Goldberg
© Royal Society of Chemistry 2012
Published by the Royal Society of Chemistry, www.rsc.org

blockade. In this context, critical lessons learned from the development of antibacterial drugs could be taken into consideration for the future development of kinase inhibitors in oncology. As in the case of bacterial diseases, tumor cell burden needs to be diminished as efficiently and early as possible to prevent the emergence of drug resistant clones. To this end, several approaches to maximize efficacy while minimizing toxicity and delay the onset of resistance mechanisms are briefly reviewed in this section.

10.2.1 Irreversible kinase inhibitors

Irreversible inhibition has been a proven mechanism of action for different therapeutic targets and indications, but often this was not the intended medicinal chemistry strategy and the covalent inhibition of the target was discovered in hindsight.[1,2] Despite the number of marketed drugs that operate by irreversible mechanisms, concerns about non-selective covalent binding to macromolecules and the potential adverse effects associated to the non-reversible nature of these modifications (*e.g.* idiosyncratic drug toxicity) have prompted academic and pharmaceutical drug discovery laboratories to avoid or deprioritize this approach. In the kinase drug discovery field, the potential reactivity of the thiol side-chain of cysteine residues in the ATP-binding cleft of a limited number of protein kinases (around 20% of the total kinome contains a cysteine in the ATP-binding pocket) have been exploited to design potent and selective irreversible kinase modulators for relevant oncology therapeutic targets (for representative recent examples, see Epidermal Growth Factor Receptor (EGFR) T790M,[3] and Fibroblast Growth Factor Receptor (FGFR)).[4] Proof-of-concept for this strategy in preclinical settings was achieved in the early 1990s by showing that thioadenosine covalently inactivated EGFR by forming a disulfide bond with the side-chain of Cys-773, a residue located on the extended coil stretch at the bottom of the ATP-blinding cleft of this receptor tyrosine kinase. These pan-EGFR modulators contained a Michael acceptor-type substituent and the covalent interaction with the side-chain of cysteine provided impressive selectivity (ratio of nearly 10^5-fold) against other receptor or intracellular protein kinases. Following this approach, irreversible kinase inhibitors have emerged from structure-based design optimization of reversible modulators and refinement of warhead placement. Three irreversible EGFR inhibitors, BIBW2992 (afatinib, compound **10.1**, Figure 10.1), HKI-272 (neratinib, compound **10.2**, Figure 10.1) and PF-00299804 (PF-299, structure not disclosed), and two Bruton's tyrosine kinase (Btk) inhibitors PCI-32765 (compound **10.3**, Figure 10.1) and AVL-292 (structure not disclosed), which contain electrophilic functionalities based on acrylamides or derivatives thereof, are currently undergoing late- and early-stage clinical testing, respectively. The identification and clinical development of these compounds have illustrated the possibility to minimize promiscuous binding and achieve a high level of selectivity over related pharmacological targets, in particular those that contain a cysteine residue at the same or

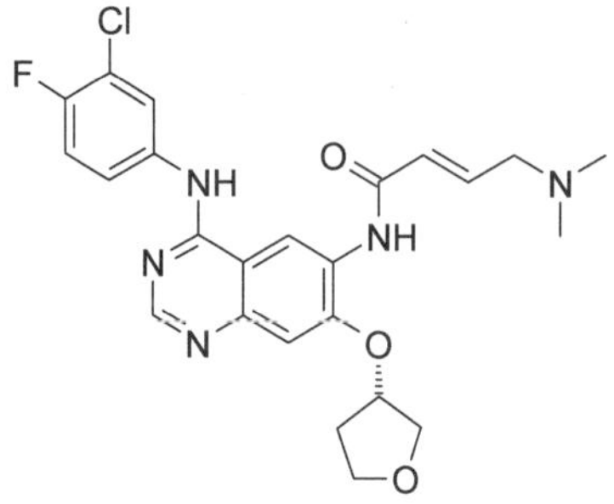

10.1

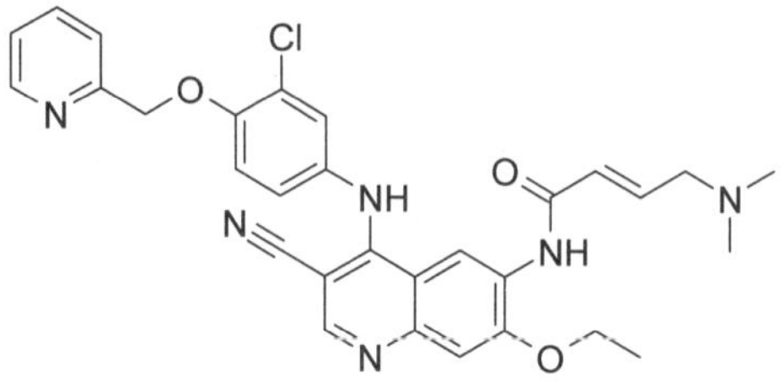

10.2

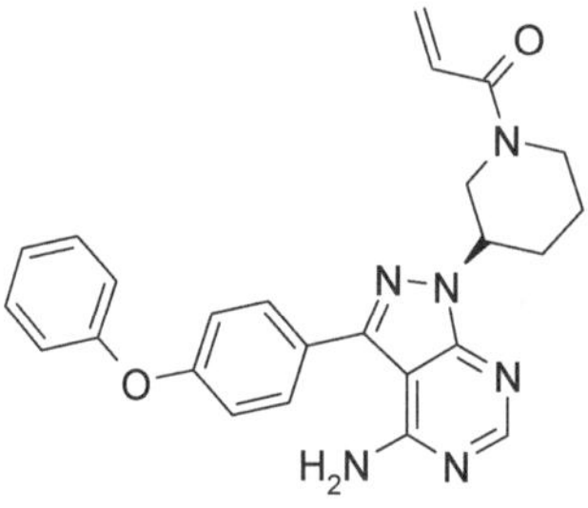

10.3

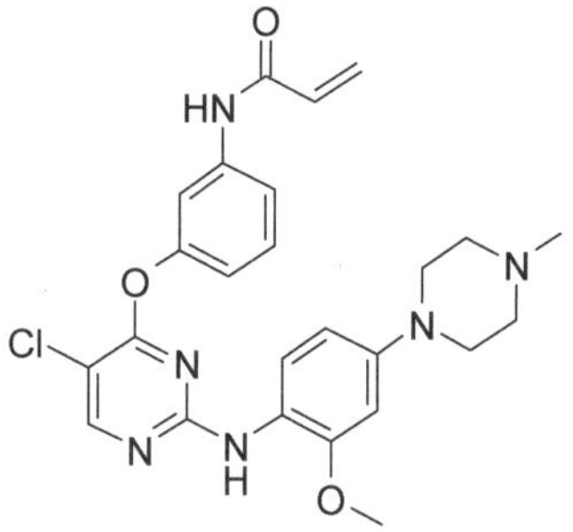

10.4

Figure 10.1 Representative examples of irreversible kinase inhibitors for the treatment of solid tumors and hematological malignancies.

close-by position in the ATP-binding cleft (*e.g.* ten kinases share a cysteine that structurally corresponds to Cys-773 in EGFR). In the case of the EGFR modulators, preclinical studies have shown that these molecules can effectively inhibit growth of tumors harboring the EGFR T790M gatekeeper mutant, despite the increased ATP binding affinity conferred by this secondary mutation and subsequent lack of sensitivity to reversible and commercially approved EGFR inhibitors.[5] Further, durable antitumor activity has been observed in clinical settings,[6] and the most advanced compounds are currently undergoing pivotal Phase III trials for non-small cell lung cancer (NSCLC).

The available data shows that plasma drug concentrations of these inhibitors need only to be available for a period of time long enough to achieve significant and sustained target coverage. As expected, and contrary to the non-covalent kinase inhibitors, the pharmacodynamic activity of these covalent agents is driven by target occupancy and is dependent on the half-life of the target rather than compound exposure in the media. To guide their clinical development and determine early-on proof-of-target inhibition in clinical settings, fluorescent probes to readily quantify target occupancy in tumor or surrogate tissues have been developed. For example, durable (pre-dose *versus* 4 h and 24 h) and dose-dependent pharmacodynamic effects (% of target occupancy in peripheral blood cells) were recently reported in a dose-escalation, multicenter, open label Phase I study of PCI-32765 in recurrent B cell lymphoma patients (for a presentation reporting this data, visit www.pharmacyclics.com.).

The possibilities of achieving high and sustained target coverage and minimizing the occurrence of allele resistant mutants are considered to be potential advantages of this mechanism of action. The current clinical irreversible kinase inhibitors should offer insight into this drug discovery strategy and serve to validate the use of covalent modifiers in oncology or other settings that require sustained target modulation (*e.g.* antibacterial drugs). If these compounds achieve clinical benefit at well-tolerated doses and eventually delay the appearance of allele resistant mutants or other forms of resistance, this will probably spur the development of other covalent kinase inhibitors and stimulate interest in broader applications of this approach. To this end, pharmaceutical companies will have to set-up more systematic strategies to exploit covalent modification as a more general orthogonal approach in kinase drug discovery. Additional medicinal chemistry efforts will be required to further fine tune the electrophilic reactivity of the selected warheads,[7] and expand the possibility to form covalent interactions with the side-chain of other amino acids (*e.g.* lysine, threonine, serine or tyrosine) in the ATP-binding cleft or suitable allosteric pockets. The availability of fragments sets for screening, such as the one recently described to positively or negatively modulate the kinase activity of 3′-phosphoinositide-dependent protein kinase 1 (PDK1) by binding to a specific allosteric site,[8] will be important in providing suitable starting points and exploring different strategies (*e.g.* ATP-compctitive *versus* non-competitive kinase modulators).

However, and in spite of the advantages of this mechanism of action, the potential exacerbation of the adverse effects associated with the on-target activity of irreversible kinase inhibitors should be taken into consideration in the general implementation of this approach in drug discovery. This point has been recently illustrated in the identification of mutant-selective irreversible EGFR kinase inhibitors.[3] The concurrent inhibition of wild-type EGFR in lung cancer patients treated with the current irreversible EGFR inhibitors may result in a higher incidence and–or severity of on-target adverse effects—in particular, skin rash and diarrhea, that may limit the possibility to achieve compound concentrations sufficient to block EGFR T790M kinase activity in cancer cells. To overcome this limitation, novel mutant-selective inhibitors against EGFR T790M and other mutant forms that spare activity against wild-type were pursued by screening an irreversible kinase inhibitor library. This effort resulted in the identification of covalent pyrimidine derivatives that are 30- to 100-fold more potent against EGFR T790M, and up to 100-fold less potent against wild-type EGFR, than current quinazoline-based EGFR clinical candidates (*e.g.* WZ4002, compound **10.4**, Figure 10.1). The possibility to design inhibitors that target exclusively kinase mutant forms is of high interest, but further studies will be needed to determine the clinical effectiveness of this approach and its general applicability to other proteins and allele resistant oncokinase mutants.

10.2.2 Optimizing target modulation and pharmacological properties

As with many other therapeutic targets, medicinal chemistry efforts to optimize kinase modulators have often generated highly hydrophobic and conformationally constrained drugs that have limited drug-like properties and a thermodynamic signature that is characterized by an entropically driven binding affinity. Taking into consideration these features, it is not a surprise that poor human pharmacological properties and limited *in vivo* target inhibition and efficacy continue to be major sources of clinical attrition in oncology (the clinical attrition rate of kinase inhibitors in oncology from Phase I to registration is 53%),[9] and a potential source of resistance. The human pharmacokinetic data obtained from Phase I clinical trials with kinase inhibitors in cancer patients invariably shows that these compounds tend to have high intra- and inter-patient variability in c_{max} and plasma drug exposure. Moreover, high doses are often needed in order to achieve significant and sustained abrogation of kinase activity. Elevated doses can compromise patient compliance (*e.g.* number or size of tablets or capsules), and high variability in c_{max} or plasma Area Under the Curve (AUC) values could translate in an erratic target inhibition, providing an ideal scenario for the emergence of resistance clones or an improper determination of biologically relevant clinical dosages or regimes (*e.g.* early onset of Dose Limiting Toxicities (DLTs) due to outliers in c_{max} or plasma exposure values within the same dose cohort.).

The BRaf inhibitor PLX4032,[10] is a perfect case in point to illustrate the development challenges encountered with kinase modulators in oncology clinical settings. Although proof-of-concept has been achieved with this compound in BRafV600E mutant melanoma cancer patients, significant and sustained target modulation ($\geqslant$ 80% inhibition of phospho-extracellular signal-regulated kinase (pERK) levels in tumor tissue) requires oral doses of 960 mg b.i.d per day to achieve plasma $AUC_{0-24\,h}$ values of 1741 µM h. As a reference to these values, it is worth noting that the compound inhibits BRafV600E in preclinical biochemical and cellular settings with IC_{50} values in the low nM range.[11] The durability of the response to PLX4032 is still under evaluation, but tumor re-growth has already occurred in treated patients and several mechanisms of resistance have been elucidated.[12–14] The precise molecular understanding of resistance in BRafV600E melanoma patients should rapidly point to the optimal combination of targeted agents (see next section) or the identification of compounds with alternative mechanism of action or biological profiles. How can the previously mentioned deficiencies be properly addressed?

The need to achieve a significant and sustained target modulation in cancer cells with kinase inhibitors may require improving the lifetime of the drug-target complex. To this end, measurements of drug-target residence time should be systematically incorporated into drug discovery and be an integral part of the lead optimization process. The importance of residence time in controlling the on-target pharmacodynamic effects of drug action and its potential impact on the optimal development of lead compounds has been reviewed in recent publications.[15,16] It is fair to say that improvements in residence time are often accomplished serendipitously and that the rational modulation of this property remains a much greater challenge than optimizing the affinity of drug-target bindings. The available data indicates that residence time of a drug on its target results from the difference in free energy between the relevant ground and transition states on the reaction coordinate (for a recent review on this topic, see Ref. 17). Thus, an increase in residence time will occur by a decrease in off-rates, and minimizing the dissociation rate of the drug-target complex is the most practical way to approach this medicinal chemistry objective. As in the case of irreversible inhibitors, maximizing the dissociative half-life can result in complete blockade of the enzymatic activity of the target to the point that recovery of biological function could only happen as the result of new target protein bio-synthesis by the organism. The compounds that present this mechanism of action have been termed the 'ultimate physiological inhibitors' to illustrate that this is the maximum efficacy that can be attained by a therapeutic agent.[18]

It is important to recognize that treatment of non-oncology malignancies with kinase inhibitors may require a short and transient drug-target interaction to achieve the expected therapeutic responses at well-tolerated doses. However, and independently of the intended therapeutic area and disease application, understanding the drug-target binding kinetics that are required to achieve efficacy at tolerated doses should be an integral part of the drug discovery activities for kinase targets.

As previously discussed, kinase inhibitors in oncology tend to show high intra- and inter-patient variability in cancer patients, and a greater attention to the optimization of the Absorption, Distribution, Metabolism, and Excretion (ADME) properties should help achieve reproducible and homogenous target inhibition upon repeat dosing. A shift from affinity–potency-focused strategy to a more holistic and informed preclinical medicinal chemistry optimization approach to obtain the desirable or balanced activity and pharmacological profile of kinase modulators have begun to be implemented throughout the different phases of drug discovery. To this end, efforts should also be directed to understand if efficacy and on–off-target toxicity for the intended therapeutic kinase target are c_{max} or exposure (AUC) driven. As it has been shown recently for dasatinib and imatinib,[19] transient potent inhibition is equivalent to prolonged target coverage. The possibility to achieve "oncogenic shock" by high-dose pulse therapy may provide interesting opportunities to increase the tolerability and efficacy of targeted anticancer agents and, at the same time, delay the onset of resistance mechanisms by inducing cell death across pre-existing heterogenic cell cancer clones. A broad range of pharmacological properties (*e.g.* plasma half-lives or tissue distribution) should be considered to systematically investigate different doses and schedules as single agent or in suitable combinations in preclinical settings. Further, the available clinical findings obtained with kinase inhibitors in oncology advocate the need to monitor plasma exposure levels more closely and, eventually, to adjust dosing and scheduling for individual patients in order to ensure maximum antitumor activity without exacerbating the development of adverse effects. The implementation of the personalized medicine concept may require, in addition to administration of the right drug to biomarker-defined cancer patients,[20] considering the possibility of adjusting the dose and schedule of the targeted agent to maximize individual responses while minimizing treatment-associated adverse effects.

10.2.3 Polypharmacological inhibitors and combinations thereof

An alternative approach to overcome resistance mechanisms may require targeting tumor cells at multiple nodes, through compounds that bind to multiple protein or lipid kinases in one or more interconnected biological pathways, or by a combination of highly selective kinase modulators. Identification and development of polypharmacological inhibitors (compounds that target exclusively a set of protein kinases to give a desired phenotype) is expected to be even more challenging than for the most selective inhibitors. Medicinal chemists would have to carry out optimization for complex multidimensional target profiles while also improving or maintaining adequate pharmacological and drug-like properties. Although it is often unclear if polypharmacological inhibition was part of the original optimization strategy or if it was found retrospectively, there is already evidence that this is possible for certain target combinations.[21,22] Thus, several examples of compounds that are biologically active against phosphoinosite-3-kinase (PI3K) and mammalian target of rapamycin (mTOR) (*e.g.* NVP-BEZ235, compound **10.5**,[23] and GSK2126458, compound **10.6**,[24] Figure 10.2), which are key nodes in the PI3K–mTOR pathway, are currently undergoing clinical trials in cancer patients.[22] Although it is too early to determine if PI3K drug-resistance mutations will emerge in cancer patients as is the case for protein kinase modulators, preclinical studies have shown that the dual pan-PI3K–mTOR inhibitors may be less susceptible to PI3K drug resistance than selective PI3K-targeted agents owing to their preserved activity against mTOR.[25] This early preclinical finding provides evidence that the simultaneous engagement of multiple targets may represent a suitable strategy to delay the appearance of resistance mechanisms within a pathway.

From the analysis of kinase selectivity profiles obtained with a broad range of inhibitor scaffolds, clusters of kinases that could be simultaneously inhibited

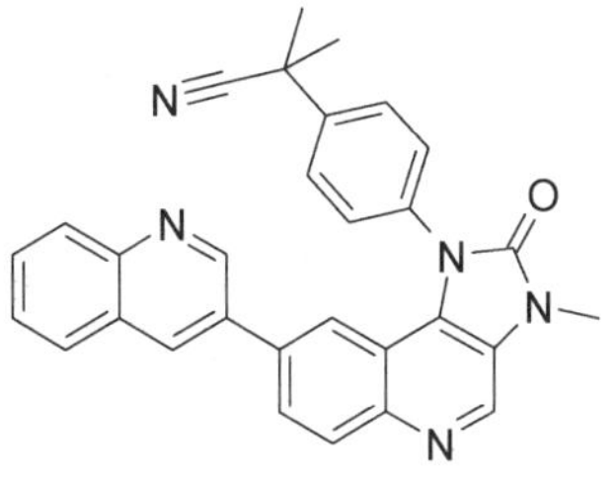

10.5

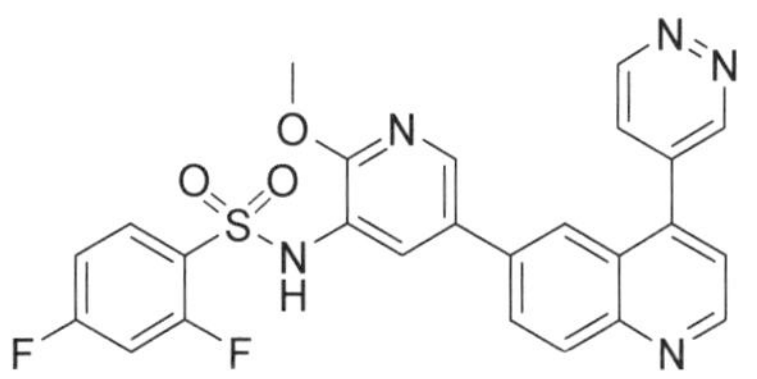

10.6

Figure 10.2 Representative examples of polypharmacological inhibitors.

start to emerge,[26] providing good starting points for a more systematic approach to identify targeted polypharmacological inhibitors with new and interesting biological activities. In this context, the availability of well annotated compound kinase activity fingerprints will be fundamental in determining the tractability of individual kinases or of a specific set thereof.[27]

Additional development challenges will also be evident for this type of approach in the clinic, in particular for a compound that demonstrates significantly different potencies for each relevant therapeutic target. Under these circumstances, and assuming that the polypharmacological compound has a good selectivity profile against other proteins, it is reasonable to assume that the maximum tolerated dose is going to be dictated by the blockade of the physiologically more essential target for the functional activity of non-tumor cells, which may or may not be identical for the therapeutic target most pivotal to tumor growth, survival and maintenance. If this is the case, combinations of selective kinase inhibitors should provide more flexibility to match the dose and schedule to the oncogenic driver lesion(s) presented in tumor cells.[28] The experience gained over the past few years with molecularly targeted kinase cancer therapeutics suggests that inter-tumoral heterogenicity,[29] induction–activation of cross-pathways or feedback loops,[30,31] and other resistance mechanisms,[13,32] are likely to require that kinase inhibitors will need to be administered in combination with other targeted or standard-of-care cancer agents. To this end, combinations of experimental stage kinase inhibitors in the clinic is a growing trend and could become commonplace in the near future, for example, in June 2009, AstraZeneca and Merck & Co. entered into an agreement to conduct early stage Phase I clinical trials of their mitogen-activated protein kinase (MEK) and protein kinase B (PKB) inhibitors, respectively.

The combination of agents that inhibit the same target through different mechanisms of action has also been proposed as a suitable approach to suppress or delay the emergence of resistance due to the acquisition of secondary mutations. This concept has been elegantly demonstrated in preclinical settings by combining ATP-competitive inhibitors with allosteric modulators of the same target. Thus, the allosteric breakpoint cluster region–abelson tyrosine

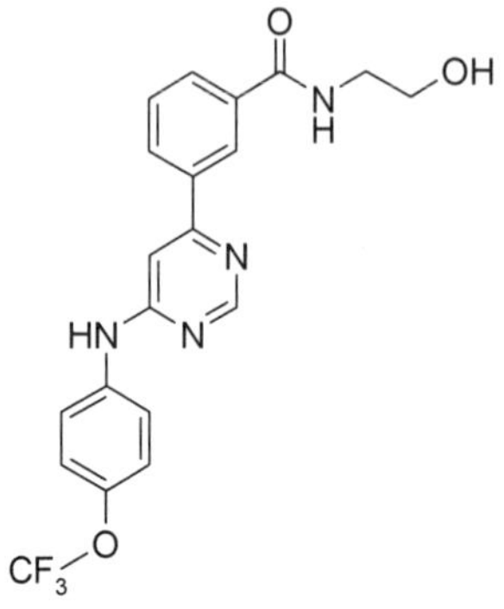

10.7

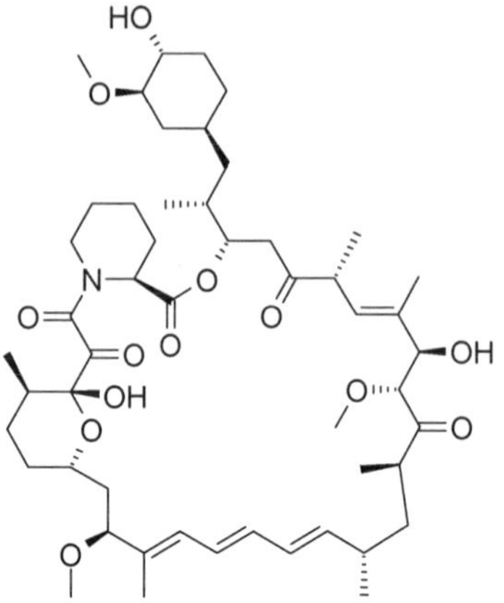

10.8

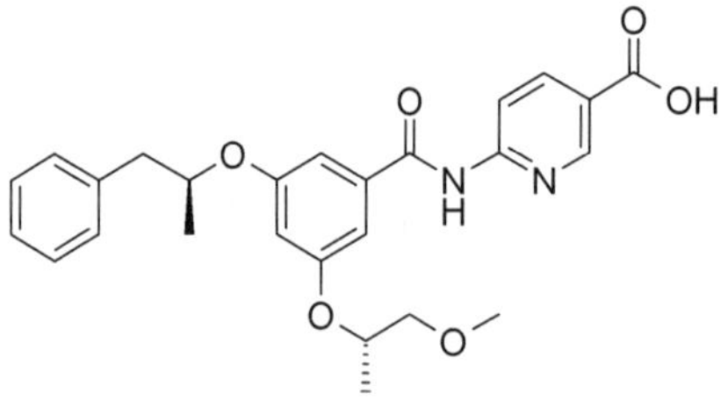

10.9

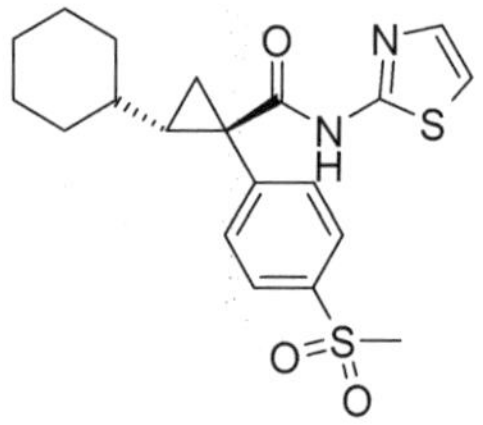

10.10

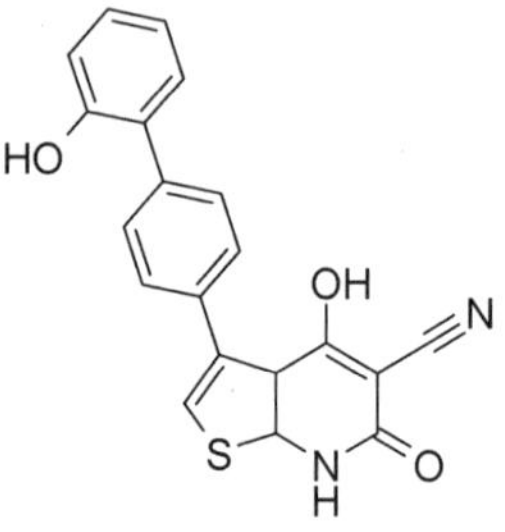

10.11

Figure 10.3 Representative examples of allosteric kinase modulators.

kinase (bcr-abl) inhibitor GNF-5 (compound **10.7**, Figure 10.3),[33] binds to the myristate-binding pocket of abl and demonstrates an additive effect when combined with imatinib or nilotonib.[34,35] By exploiting fundamentally different molecular interactions with the target, allosteric kinase inhibitors in combination with "classical" ATP-competitive modulators are likely to further decrease the chances for development of allele resistant mutations.[36]

The results obtained with bcr-abl also provide early evidence that targeting pockets outside the traditional ATP-binding clef can provide important and alternative pharmacological agents for kinases in cancer and other therapeutic indications. This topic is briefly reviewed in the next section.

10.3 Magic bullets and new chemical space

Biochemical and cell-based assays have been established to assess the specificity of kinase inhibitors against a large number of proteins or signal transduction pathways. Early *in vitro* efforts to determine compound kinase profiles have shown that specificity varies widely and is not strongly correlated with the chemical class or the primary sequence of the intended target(s).[27] From these extensive screening efforts, novel and unexpected interactions have been identified for previously described highly selective inhibitors.[37,38] Although few drugs are truly selective for a single target, the broad kinase inhibitory activity of some drugs, which were not designed to have the observed multitarget profile, has been successfully exploited in clinical settings. For example, imatinib has shown significant clinical benefit against malignancies dependent on bcr-abl (chronic myelogenous leukemia), c-Kit (gastrointestinal stromal tumors) and platelet-derived growth factor receptor (PDGFR) (dermatofibro sarcoma protuberans). Similarly, sunitinib, sorafenib, dasatinib and nilotinib have also demonstrated that clinical response in the treatment of solid tumors or hematological malignancies with manageable adverse effects is possible with broad-acting kinase modulators. Although these non-selective kinase inhibitors have had a positive impact on the treatment of cancer, the risk to benefit ratio has to be appropriate for the disease being treated, and for most indications (in particular chronic, non-oncology diseases) this means that cleaner selectivity will be required.[39]

In spite of the information that is available today (only in the private sector) there is still a high unmet need to understand the biological consequences of inhibiting individual or particular combinations of protein and lipid kinases. Integration of the results obtained by genetic approaches and extensive biochemical characterization with results from cell-based or *in vivo* preclinical tolerability studies, and ultimately with clinical observations, should enable a more complete understanding of the biological consequences of inhibiting individual or particular combinations of kinases. If we take into consideration the different biological or pharmacological properties across species (often the sequence identity for a target across different organisms is not high or compounds have species differences in drug metabolic fate or tissue distribution), adequate determination of molecular interaction profiles for chemically diverse compounds in biological systems will help identify kinases whose inhibition leads, or could lead to adverse effects, and the combination of kinases whose inhibition has or could have a synergistic beneficial effect in particular diseases states.

P. Ehrlich's faith in the 'magic bullet'—his term-concept and his ideal of treating malignancies with highly efficacious and selective drugs "aiming precisely" at the intended target is going to require identification and optimization of kinase inhibitors with mechanisms of action independent of the "classical" ATP binding site competitive approach. Furthermore, competition in the identification and design of chemical templates that could mimic the key binding features of ATP—the binding modes of most of the current kinase inhibitors can be easily categorized in terms of the ligand scaffold and a few simple empirical rules—has led to a crowded intellectual property chemical space. Knowledge-based kinase drug discovery should leverage the profiles of drugs directed to other targets or serendipitous findings. In fact, one of the most incidentally selective and the first kinase inhibitor to obtain marketing approval—rapamycin (compound **10.8**, sirolimus, Figure 10.3) approved as an immunosuppressant by the FDA in September 1999 is an allosteric, selective modulator of the mTORC1 complex.[40] These "non-classical kinase modulators" could offer the possibility of greater selectivity because they can establish interactions with pockets unique in sequence and structure. Alternative mechanisms of kinase modulation (*e.g.* compounds that modulate intracellular localization of the kinase) could also provide opportunities to differentially regulate subtypes of kinases within a closely related sub-family. To this end, novel technologies that take advantage of the cumulative understanding of kinase function and its catalytic regulation, including the structural elements that are involved in the process, start to emerge, and could foster a new generation of tool compounds or drugs (for recent reviews on this topic and representative new methods to identify this type of inhibitors see Ref. 41–44).

10.4   Beyond kinase inhibitors

10.4.1   Kinase activation—a new paradigm in drug discovery

Due to the pathological role that certain up-regulated or gain-of-function kinases play in disease states, particularly in cancer, it is logical that most of the drug discovery activities around this family of therapeutic targets were, or are, directed to the inhibition of their enzymatic activity. However, these proteins also constitute nodes in key signal-transduction cascades that may need to be transiently activated to lead to favorable disease outcomes suggesting the potential therapeutic use of kinase activators. In addition to the medicinal chemistry challenges associated with the identification of compounds with this mechanism of action, the need to achieve a transient effect in order to avoid a potential pro-oncogenic constitutive activation of certain pathways or targets represents a major hurdle to this approach; consequently, the paucity of concrete examples in the public domain on this area of research is unsurprising (for a recent review, see Ref. 45). However, and although only a relative small number of kinase activators with biological activities in the low micromolar range are known in the literature (*e.g.* adenosine monophosphate-activated protein kinase (AMPK) or glucokinase (GK)), potential design principles that could be applied to other targets in the near future start to emerge. From an in depth understanding of the mechanism(s) of kinase regulation, different and complementary approaches have been considered. The possibility to target an allosteric site or induce a conformational change in the intended target seems to have been successful for structurally different targets. For example, GK can be activated to reduce blood glucose by targeting an allosteric pocket 20 Å away from the catalytic site of this enzyme. Interestingly, this area coincides with the region where the activating clinical mutations of GK are clustered.[46] Confirmation of binding has been obtained by solving the X-ray structure of the ligand-bound complex and has enabled the optimization of different compound classes (*e.g.* compounds **10.9** and **10.10**, Figure 10.3) (see review, Ref. 47). Modest activation of PDK1 has also been achieved,[48] by targeting the so-called hydrophobic motif (HM)–PDK1-interacting fragment (PIF) pocket, which is a key regulatory motif that is highly conserved across the AGC Ser–Thr protein kinase family. As an alternative to targeting remote binding sites, activators that cause a relief of auto-intra or -inter protein-inhibition by inhibiting protein-protein interactions or causing conformational changes have also been described and, in some cases such us A-769662, (compound **10.11**, Figure 10.3) have reached clinical trials.[49]

The pursuit of kinase activators is still in its infancy, but could be an interesting area for drug discovery in the near future. Undoubtedly, the direct activation of kinases is a challenging drug discovery endeavor that could only be justified for targets with a strong scientific rationale supporting the favorable disease outcome *via* this mechanism of action, *e.g.*, blocking progression in Alzheimer's and Parkinson's disease or treating traumatic brain injury by abrogating neurodegeneration trough activation of the receptor kinase tropomyosin-related (Trk) receptor B (TrkB).[50] If this is the case, more systematic approaches to achieve specific activation of the intended target are eagerly needed in academia and pharmaceutical industry.

10.4.2 Pseudokinases—new kids on the block

Pseudokinases are a protein family that constitute approximately 10% of the human kinome (for reviews on this topic, see Ref. 51–53). These proteins are characterized by the presence of a kinase-homology domain predicted to lack enzymatic activity due to the absence of at least one of the three conserved critical catalytic motifs: (1) the Val-Ala-Ile-Lys (VAIK) motif in subdomain II, in which the side-chain of Lys interacts with the α and β phosphates of ATP; (2) the His-Arg-Asp (HRD) motif in subdomain VIb, in which the aspartic acid is the catalytic residue; and (3) the Asp-Phe-Gly (DFG) motif in subdomain VII, in which the carboxylic moiety of aspartic acid binds the Mg^{II} ion that coordinates the β and γ phosphates of ATP. Owing to their lack of intrinsic phosphoryl-transfer catalytic activity, pseudokinase domain-containing proteins have long been thought of as bystanders in intracellular processes rather than active participants and therefore have not been the subject of intense drug discovery approaches. One notorious exception is erbB3 for which several monoclonal antibodies that target its extracellular domain are currently undergoing clinical trials in cancer patients.[54] In preclinical studies erbB3 has been shown to play a part in both ligand-dependent and independent oncogenic signaling. Amplification of c-Met confers constitutive erbB3 signaling and contributes to resistance to EGFR kinase inhibitors in NSCL tumors. In breast cancer cells that overexpress erbB2, increased levels of membrane-bound erbB3 drives continued oncogenic signaling and contributes to resistance of erbB2 targeted therapies. In addition to the compelling data obtained so far on the involvement of erbB3 in the development and progression of cancer, recent articles have provided scientific evidence in support of the view that other members of the pseudokinase family may play crucial roles in signal transduction, principally *via* the allosteric regulation of their kinase-competent counterparts, or have enzymatic activity even if they lack essential catalytic residues (*e.g.*, WNK1 and Haspin missing the VAIK and DFG motifs, respectively).[51] Although few of the 48 human pseudokinase domains have been functionally or structurally characterized, and consequently their roles in cell signaling remain to be discovered, the molecular, structural and functional details underlying pseudokinase modulation of signal transduction and protein kinase activity in normal and cancer cells is starting to emerge. Of critical importance, mutations affecting pseudokinase domains underlie the dysfunctional regulation of several clinically important kinases by their partner pseudokinase regulators. For example, the V617F mutation in the Janus kinase (Jak) homology 2 (JH2) pseudokinase domain of Janus kinase 2 (Jak2) leads to constitutive activation of the tyrosine kinase activity of this enzyme with phosphorylation and abnormal upregulation of signal transducers and activators of transcription 5 (STAT5) and other downstream effectors. The discovery of the Jak2 V617F mutation in DNA clinical samples,[55,56] initiated a search for Jak2 modulators to treat patients with myeloproliferative neoplasms (*e.g.* polycythemia vera, essential thrombocythemia and primary myelofibrosis). The identification of additional links between pseudokinase-mediated dysregulation or control of signal networks and human diseases should provide interesting opportunities for new targets and renew interest in inhibiting (or activating) kinases or pseudokinases by exploiting alternative mechanisms to the ones pursued to date. In this context, structural insights, like the recent report on the molecular mechanism underlying the activation of LKB1 by the pseudokinase STRAD,[57] should provide interesting points for medicinal chemistry approaches to control kinase fold and enzymatic activity, and contribute to the search for new therapeutic agents across different indications.

10.5   Outlook

In spite of the progress made over the past few years, the field of kinase drug discovery and development is still on the steepest part of the learning curve.

The relatively limited number of kinase inhibitors, notable in oncology alone, that have received marketing approval so far, in comparison to the number of compounds that have entered clinical trials, requires an in depth evaluation of the way drug discovery has been conducted for this family of therapeutic targets. We need to move away from a 'one-size-fits-all' ATP-competitive inhibitory approach to a tailored strategy in which the mechanism of action, including the activity and selectivity profile, of the drug are adjusted to the target to be modulated and disease to be treated. As shown in the preceding sections, strategies for intentionally designing kinase modulators with different mechanisms of action (*e.g.* ATP-competitive *versus* allosteric modulators) or selectivity profiles (*e.g.* selective *versus* polypharmacological inhibitors) are emerging and being implemented. Furthermore, challenges in target validation have led to a focus on a relatively narrow number of kinases as targets for therapeutic intervention while the function of many other kinases (around 25% of the kinome),[9] and pseudokinases is still unknown. To explore the untargeted therapeutic kinome, including nearby family members like pseudokinases, will require significant research efforts in chemistry and structural biology to identify chemical probes to functionally annotate these new enzymes and determine their role in human malignancies. Our ability to validate new kinase targets in different disease applications and deliver a novel generation of effective and safe kinase inhibitors will ensure that this incredible class of therapeutic targets has a future.

Acknowledgements

I thank my present and former colleagues for their outstanding work and dedication to identify and develop kinase inhibitors, bringing hope to cancer patients and their families.

References

1. J. Singh, R. C. Petter and A. F. Kluge, *Curr. Opin. Chem. Biol.*, 2010, **14**, 475.
2. M. H. Potashman and M. E. Duggan, *J. Med. Chem.*, 2009, **52**, 1231.
3. W. Zhou, D. Ercan, L. Chen, C.-H. Yun, D. Li, M. Capelletti, A. B. Cortot, L. Chirieac, R. E. Iacob, R. Padera, J. R. Engen, K.-K. Wong, M. J. Eck, N. S. Gray and P. A. Janne, *Nature*, 2009, **462**, 1070.
4. W. Zhou, W. Hur, U. McDemott, A. Dutt, W. Xian, S. B. Ficarro, J. Zhang, S. V. Sharma, J. Brugge, M. Meyerson, J. Scttlcman and N. S. Gray, *Chem. Biol.*, 2010, **17**, 285.
5. C. P. Belani, *Cancer Invest.*, 2010, **28**, 413.
6. T. A. Yap, L. Vidal, J. Adam, P. Stephens, J. Spicer, H. Shaw, J. Ang, G. Temple, S. Bell, M. Shahidi, M. Uttenreuther-Fishcer, P. Stopfer,

A. Futreal, H. Calvert, J. S. de Bono and R. Plummer, *J. Clin. Oncol.*, 2010, **28**, 3965.

7. C. Carmi, A. Cavazzibi, S. Vezzosi, F. Bordi, F. Vacondio, C. Silva, S. Rivara, A. Lodola, R. R. Alfieri, S. La Monica, M. Galetti, A. Ardizzoni, P. G. Petronini and M. Mor, *J. Med. Chem.*, 2010, **11**, 5.

8. J. D. Sadowsky and J. A. Wells, *239th ACS National Meeting*, 2010, Abstract BIOL 274.

9. O. Fedorov, S. Müller and S. Knapp, *Nat. Chem. Biol.*, 2010, **6**, 166.

10. G. Bollag, P. Hirth, J. Tsai, J. Zhang, P. N. Ibrahim, H. Cho, W. Spevak, C. Zhang, Y. Zhang, G. Habets, E. A. Burton, B. Wong, G. Tsang, B. L. West, B. Powell, R. Shellooe, A. Marimuthu, H. Nguyen, K. Y. J. Zhang, D. R. Artis, J. Schlessinger, F. Su, B. Higgins, R. Iyer, K. D'Andrea, A. Koehler, M. Stumm, P. S. Lin, R. J. Lee, J. Grippo, I. Puzanov, K. B. Kim, A. Ribas, G. A. McArthur, J. A. Sosman, P. B. Chapman, K. T. Flaherty, X. Xu, K. L. Nathanson and K. Nolop, *Nature*, 2010, **467**, 596.

11. E. W. Joseph, C. A. Pratilas, P. I. Poulikakos, M. Tadi, W. Wang, B. S. Taylor, E. Halilovic, Y. Persaud, F. Xing, A. Viale, J. Tsai, P. B. Chapman, G. Bollag, D. B. Solit and N. Rosen, *Proc. Natl. Acad. Sci. U. S. A.*, 2011, **107**, 14903.

12. C. M. Johannessen, J. S. Boehm, S. Y. Kim, S. R. Thomas, L. Wardwell, L. A. Johnson, C. M. Emery, N. Stransky, A. P. Cogdill, J. Barretina, G. Caponigro, H. Hieronymus, R. R. Murray, K. Salehi-Ashtiani, D. E. Hill, M. Vidal, J. J. Zhao, X. Yang, O. Alkan, S. Kim, J. L. Harris, C. J. Wilson, V. E. Myer, P. M. Finan, D. E. Root, T. M. Roberts, T. Golub, K. T. Flaherty, R. Dummer, B. L. Weber, W. R. Sellers, R. Schlegel, J. A. Wargo, W. C. Hahn and L. A. Garraway, *Nature*, 2010, **468,** 968.

13. R. Nazarian, H. Shi, Q. Wang, X. Kong, R. C. Koya, H. Lee, Z. Chen, M.-K. Lee, N. Attar, H. Sazegar, T. Chodon, S. F. Nelson, G. McArthur, J. A. Sosman, A. Ribas and R. S. Lo, *Nature*, 2010, **468**, 973.

14. D. Solit and C. L. Sawyers, *Nature*, 2010, **468**, 902.

15. R. A. Copeland, D. L. Pompliano and T. D. Meek, *Nat. Rev. Drug Discov.*, 2006, **5**, 730.

16. J. Gabrielsson, H. Dolgos, P.-G. Gillberg, U. Bredberg, B. Benthem and G. Duker, *Drug Discovery Today*, 2009, **7/8**, 358.

17. H. Lu and P. J. Tonge, *Curr. Opin. Chem. Biol.*, 2010, **14**, 467.

18. A. Lewandowicz, P. C. Tyler, G. B. Evans, R. H. Furneaux and V. L. Schramm, *J. Biol. Chem.*, 2003, **278**, 31465.

19. N. P. Shah, C. Kasep, C. Weier, M. Balbes, J. M. Nicoll, E. Bleickardt, C. Nicaise, and C. Sawyers, *Cancer Cell*, 2008, **14**, 485.

20. R. L. Schilsky, *Nat. Rev. Drug Discovery*, 2010, **9**, 366.

21. B. Apsel, J. A. Blair, B. Gonzalez, T. M. Nazif, M. E. Feldman, B. Aizenstein, R. Hoffman, R. L. Williams, K. M. Shokat and Z. A. Knight, *Nat. Chem. Biol.*, 2008, **4**, 691.

22. C. Garcia-Echeverria and W. R. Sellers, *Oncogene*, 2008, **27**, 5511.

23. S.-M. Maira, F. Stauffer, J. Brueggen, P. Furet, C. Schnell, C. Fritsch, S. Brachmann, P. Chène, A. De Pover, K. Schoemaker, D. Fabbro, D. Gabriel, M. Simonen, L. Murphy, P. Finan, W. Sellers and C. García-Echeverría, *Mol. Cancer Ther.*, 2008, **7**, 1851.

24. S. D. Knight, N. D. Adams, J. L. Burgess, A. M. Chaudhari, M. G. Darcy, C. A. Donatelli, J. I. Luengo, K. A. Newlander, C. A. Parrish, L. H. Ridgers, M. A. Sarpong, S. J. Schmidt, G. S. Van Aller, J. D. Carson, M. A. Diamond, P. A. Elkins, C. M. Gardiner, E. Garver, S. A. Gilbert, R. R. Gontarek, J. R. Jackson, K. L. Kershner, L. Luo, K. Raha, C. S. Sherk, C.-M. Sung, D. Sutton, P. T. Tummino, R. J. Wegrzyn, K. R. Auger and D. Dhanak, *ACS Med. Chem. Lett.*, 2010, **1**, 39.

25. E. R. Zunder, Z. A. Knight, B. T. Houseman, B. Apsel and K. M. Shokat, *Cancer Cell*, 2008, **14**, 107.

26. Z. A. Knight, H. Lini and K. M. Shokat, *Nat. Rev. Cancer*, 2010, **10**, 130.

27. P. Bamborough, D. Dewry, G. Harper, G. K. Smith and K. Schneider, *J. Med. Chem.*, 2008, **5**, 7898.

28. F. Stegmeier, M. Warmuth, W. R. Sellers and M. Dorsch, *Clin. Pharmacol. Ther.*, 2010, **87**, 543.

29. J. E. Visvader, *Nature*, 2011, **469**, 314.

30. K. E. O'Reilly, F. Rojo, Q. B. She, D. Solit, G. B. Mills, D. Smith, H. Lane, F. Hofmann, D. J. Hicklin, D. L. Ludwig, J. Baselga and N. Rosen, *Cancer Res.*, 2006, **66**, 1500.

31. A. Carracedo, L. Ma, J. Teruya-Feldstein, F. Rojo, L. Salmera, A. Alimonti, A. Egia, A. T. Sasaki, G. Thomas, S. C. Kozma, A. Papa, C. Nardelia, L. C. Cantley, J. Baselga and P. P. Pandolfi, *J. Clin. Invest.*, 2008, **118**, 3065.

32. R Barouch-Bentov and K Sauer, *Expert Opin. Invest. Drugs*, 2011, **20**, 153

33. X. Deng, B. Okram, Q. Ding, J. Zhang, Y. Choi, F. J. Adrian, A. Wojciechowski, G. Zhang, J. Che, B. Bursulaya, S. W. Cpwan-Jacob, G. Rummel, T. Sim and N. S. Gray, *J. Med. Chem.*, 2010, **53**, 6934.

34. J. Zhang, F. J. Adrián, W. Jahnke, S. W. Cowan-Jacob, A. G. Li, R. E. Iacob, T. Sim, J. Powers, C. Dierks, F. Sun, G.-R. Guo, Q. Ding, B. Okram, Y. Choi, A. Wojciechowski, X. Deng, G. Liu, G. Fendrich, A. Strauss, N. Vajpai, S. Grzesiek, T. Tuntland, Y. Liu, B. Bursulaya, M. Azam, P. W. Manley, J. R. Engen, G. Q. Daley, M. Warmuth and N. S. Gray, *Nature*, 2010, **463**, 501.

35. R. E. Iacob, J. Zhang, N. S. Gray and J. R. Engen, *PLoS One*, 2011, **6**, e15929.

36. R. Eglen and T. Reisine, *Pharmacol. Ther.*, 2011, **130**, 144.

37. M. W. Karaman, S. Herrgard, D. K. Treiber, P. Gallant, C. E. Atteridge, B. T. Campbell, K. W. Chan, P. Cicieri, M. I. Davis, P. T. Edeen, R. Faraoni, M. Floyd, J. P. Hunt, D. J. Lockhart, Z. V. Milanov, M. J. Morrison, G. Pallares, H. K. Patel, S. Pritchard, L. M. Wodicka and P. P. Zarrinkar, *Nat. Biotechnol.*, 2008, **26**, 127.

38. M. A. Fabian, W. H. Biggs III, D. K. Treiber, C. E. Atteridge, M. D. Azimiora, M. G. Benedetti, T. A. Carter, P. Civeri, P. T. Edeen, M. Floyd, J. M. Fors, M. Galvin, J. L. Gerlach, R. M. Grotzfeld, S. Herrgard, D. E. Insko, M. A. Insko, A. G. Lai, J.-M. Léilas, S. A. Mehta, Z. V. Milanov, A. M. Velasco, L. M. Wodicka, H. K. Patel, P. P. Zarrinkar and D. J. Lockhart, *Nat. Biotechnol.*, 2005, **3**, 336.
39. R. Morphy, *J. Med. Chem.*, 2010, **53**, 1413.
40. C. Garcia-Echeverria, *Bioorg. Med. Chem. Lett.*, 2010, **20**, 4308.
41. F. Zuccotto, E. Ardini, E. Casale and M. Angiolini, *J. Med. Chem.*, 2010, **53**, 2681.
42. F. J. Moy, A. Lee, L. K. Gavrin, Z. B. Xu, A. Sievers, E. Kieras, W. Stochaj, L. Mosyak, J. McKew and D. H. H. Tsao, *J. Med. Chem.*, 2010, **53**, 1238.
43. S. Klüter, C. Grütter, T. Naqvi, M. Rabiller, J. R. Simard, V. Pawar, M. Getlik and D. Rauh, *J. Med. Chem.*, 2010, **53**, 357.
44. P. Ranjitkar, A. M. Brock and D. J. Maly, *Chem. Biol.*, 2010, **17**, 195.
45. G. L. Simpson, J. A. Hughes, Y. Washio, and S. M. Bertrand, *Curr. Opin. Drug Discovery Dev.*, 2009, **12**, 585.
46. A. L. Gloyn, *Hum. Mutat.*, 2003, **22**, 353.
47. J. Grimsby, S. J. Berthel and R. Sarabu, *Curr. Top. Med. Chem.*, 2008, **8**, 1524.
48. M. Engel, V. Hindie, L. A. Lopez-Garcia, A. Stroba, F. Schaeffer, I. Adrian, J. Imig, L. idrissova, W. Nastainczyk, S. Zeuzem, P. M. Alzari, R. W. Hartmann, A. Piiper and R. M. Biondi, *EMBO J.*, 2006, **25**, 5469.
49. B. Cool, B. Zinker, W. Chiou, N. Cao, M. Parham, R. Dickinson, A. Adler, G. gagne, R. Iyengar, G. Zhao, K. Marsh, P. Kym, P. Jung, H. S. Camp and E. Frevert, *Cell Metab.*, 2006, **3**, 403.
50. A. Baldo, A. D. Sniderman, St, X. J. Zhang and K. Cianflone, *J. Lipid Res.*, 1995, **36**, 1415.
51. E. Zeqiraj and D. M. F. van Aalten, *Curr. Opin. Struct. Biol.*, 2010, **20**, 1.
52. T. Rajakulendran and F. Sicheri, *Sci. Signal.*, 2010, **3**, 1.
53. J. Boudeau, D. Miranda-Saavedra, G. J. Barton and D. R. Alessi, *Trends Cell Biol.*, 2006, **16**, 443.
54. J. Baselga and S. M. Swain, *Nat. Rev. Cancer*, 2009, **9**, 463.
55. E. J. Baxter, L. M. Scott, P. J. Campbell, C. East, N. Fourouclas, S. Swanton, G. S. Vassiliou, A. J. Bench, E. M. Boyd, N. Curtin, M. A. Scott, W. N. Erber and A. R. Green, *Lancet*, 2005, **2005**, 1054.
56. L. M. Scott, P. J. Campbell, E. J. Baxter, T. Todd, P. Stephens, S. Edkins, R. Wooster, M. R. Stratton and P. A. Futreal, *Blood*, 2005, **106**, 2920.
57. E. Zeqiraj, B. M. Filippi, M. Deak, D. R. Alessi and D. M. F. van Aalten, *Science*, 2009, **326**, 1707.

Subject Index